RECENT ADVANCES IN CANCER DIAGNOSTICS AND THERAPY

RECENT ADVANCES IN CANCER DIAGNOSTICS AND THERAPY
A Nano-Based Approach

Anjana Pandey and Saumya Srivastava

CRC Press
Taylor & Francis Group
Boca Raton London New York

CRC Press is an imprint of the
Taylor & Francis Group, an **informa** business

First edition published 2022
by CRC Press
6000 Broken Sound Parkway NW, Suite 300, Boca Raton, FL 33487-2742

and by CRC Press
2 Park Square, Milton Park, Abingdon, Oxon, OX14 4RN

© 2022 Taylor & Francis Group, LLC

CRC Press is an imprint of Taylor & Francis Group, LLC

Library of Congress Cataloging-in-Publication Data
Names: Pandey, Anjana, author. | Srivastava, Saumya, author.
Title: Recent advances in cancer diagnostics and therapy : a nano-based approach / by Anjana Pandey, Saumya Srivastava.
Description: First edition. | Boca Raton, FL : CRC Press, 2022. | Includes bibliographical references and index. | Summary: "The book provides information about different types and stages of cancers and their subtypes with their respective molecular mechanisms, etiology, histopathology and cellular origins of cancer"-- Provided by publisher.
Identifiers: LCCN 2021037602 (print) | LCCN 2021037603 (ebook) | ISBN 9781032062174 (hardback) | ISBN 9781032063676 (paperback) | ISBN 9781003201946 (ebook)
Subjects: LCSH: Cancer--Diagnosis. | Cancer--Treatment.
Classification: LCC RC270 .P36 2022 (print) | LCC RC270 (ebook) | DDC 616.99/406--dc23
LC record available at https://lccn.loc.gov/2021037602
LC ebook record available at https://lccn.loc.gov/2021037603

ISBN: 978–1–032–06217–4 (hbk)
ISBN: 978–1–032–06367–6 (pbk)
ISBN: 978–1–003–20194–6 (ebk)

DOI: 10.1201/9781003201946

Typeset in Times
by MPS Limited, Dehradun

Contents

Preface

Cancer is characterized by uncontrolled growth of irregular or abnormal cells in the body. It develops when the normal control mechanism of the body fails and stops working, abruptly leading to the growth of old cells without dying and the formation of tumor cells. There are also some cancers that do not form tumors, such as leukemia. In the current era, cancer has become a very common disease in men and women, assuming that half of men and one-third of women will be diagnosed with cancer in their lifetime. Living with the disease is the prime challenge for people who have encountered this disease because it brings changes in their daily routines, life, and relationships. It can weaken a person financially as well as mentally. There are different treatment methods available for cancer including chemotherapy, radiation, surgery, immune therapy, etc., but all of these treatment methods are not proven sufficient to prolong the patient's life. Most people with cancer struggle with pain either through cancer itself or from the treatment procedures like chemotherapy, radiation, and surgery. The treatment procedure changes the person's vision on how they feel and look. Some patients face conditions like depression, anxiety and distress while struggling with cancer. Early diagnosis is the only vital element that can evade the lethal effects of cancer.

Currently, nanotechnology has been developed as the most efficient discipline in the medical field for diagnosis and treatment by observing molecular levels of biological systems using different nanostructures. It has the potential to drastically change the cancer diagnosis procedure and treatment. Presently, there is much research going on for developing the nanodevices skilled in detecting cancer. It can be applied for vast purposes, from early diagnosis and detection to treatment.

The present book covers the reader's interests from the field of cancer science and nanotechnology or bio-nanotechnology. This book will give a complete description of cancer and its associated molecular markers including proteins, enzymes, DNA, RNA, circulating miRNA, etc., that are proven very effective candidates for cancer detection. This book will also highlight the different low cost, accurate, sensitive, and rapid detection methods for cancer, with a special emphasis on electrochemical biosensors.

For instance, Chapter 1 provides information about the introduction to cancer, regulatory molecules and pathways involved in cancer, cancer characteristics, cancer types and stages, and molecular pathways involved in carcinogenesis and progression alongside factors affecting carcinogenesis in humans.

Chapter 2 includes information about the incidence and mortality rates of cancer in India and other parts of the world, along with worldwide cancer patterns.

Chapter 3 provides information about cancer screening and diagnosis techniques. The available cancer diagnostic and screening techniques are discussed in detail, along with the different classes of tumor markers used in cancer diagnosis and screening. Also, newly identified techniques based on nanosensors for cancer diagnosis and screening have been discussed.

Chapter 4 includes different tools, techniques, and methodology required in development of biosensors for cancer screening and diagnosis. The chapter provides detailed information on different parts of biosensor, their classification, and their working mechanisms.

Chapter 5 includes information about recent advancements in biosesonsors in cancer screening with an emphasis on more sensitive electrochemical nanosensors, their classification, and detailed discussion about cancer-associated molecules employed as different biomarkers for electrochemical biosensors.

Chapter 6 discusses different classes of anticancer drugs employed in cancer therapy and their modes and molecular mechanisms of action.

Chapter 7 provides information about applications of different computational and bioinformatics techniques employed in cancer screening and therapeutics. This chapter discusses in detail the mathematical and computational models' biosensor development and cancer screening based on artificial neural networks and machine learning techniques along with their applications in cancer diagnosis and screening.

Overall, this book will provide readers with detailed insights into cancer, molecular mechanisms of carcinogenesis and cancer progession, conventional and emerging tools and teechniques used in cancer screening and diagnosis, cancer therapies and anticancer drugs, along with applications of computational and mathematical tools used in cancer screening and research with an intent to provide knowledge about cancer, screening, diagnosis, and therapies.

Authors

Prof. Anjana Pandey completed her M.Sc and Ph.D. in biochemistry from Banaras Hindu University, Varanasi, India. Presently, she is working as a professor in the Department of Biotechnology, MNNIT Allahabad, Prayagraj, India. She has 70 publications in national and international journals and 18 book chapters. She has research interests in the synthesis and application of nanoparticles and disease (microbial and genetic) identification, renewable energy production, biowaste to biofuel (biohydrogen, biodiesel, etc.) generation and technology development, genetically engineered microbes for the enhancement of the yield of biofuel, physiological and molecular changes of *vigna mungo* under abiotic and biotic stresses, and identification of molecular markers for screening and generation of elite varieties.

Ms. Saumya Srivastava completed her M.Sc in biochemistry from Banaras Hindu University (BHU), Varanasi, India. She is pursuing a Ph.D. in Biotechnology from MNNIT Allahabad, Prayagraj, India. She has nine publications in national and international journals and five book chapters. She has research interests in cancer research, nanotechnology, synthesis and application of nanoparticles, nanomaterial characterization, and biosensor development for disease diagnosis.

1 Introduction to Cancer

1.1 INTRODUCTION

The term *cancer* refers to a group of ailments that share equivalent traits. It is a disorder in which some body cells start dividing uncontrollably and spread into neighboring body parts. In normal conditions, old or damaged cells are replaced by new body cells while in the cancerous condition, the body starts producing new cells even when it is not needed, and form a tumor. There are approximately more than 100 cancer types and are named according to their origin organ, tissue, or location. Symptoms of cancer also vary, depending on the location and stage. The cause is multifactorial and the sickness procedure differs at distinct sites. Tobacco is one of the most important identified dangerous elements for cancer. Several other environmental exposures and specified infections play a fundamental position in carcinogenesis (Kontis et al., 2019; Nair et al., 2005). According to the WHO 2014 Report, the annual 14 million new cases of cancer could reach 22 million per year within two decades. To prevent the spread of this disease, early and robust methods of screening are required. In the past several decades, a lot of effort has been made to search and introduce reliable markers for diagnosis, screening, and treatment of cancer. These efforts resulted in the discovery of circulating blood markers for tumors. The convenience and feasibility of established blood drawing and urine sampling allowed for the progress of circulating tumor markers for the diagnosis and management of cancer sufferers (Rapisuwon et al., 2016; Wu and James T., 1999). For many cancer screeningss, sensitivity remains in the 70% to 80% range with specificity slightly lower at approximately 60% to 70% in general. No screening test can have 100% specificity or sensitivity (Schiffman et al., 2015). Bence-Jones protein was the first circulating tumor marker. By the discovery of this, people started thinking that the concentration of some serum proteins could reflect the activity and mass of the tumor and their quantification could influence the determination of treatment. It could be an enzyme, hormone, or nucleic acid (Beetham, 2000; Rapisuwon et al., 2016; Ramakrishnan and Jialal, 2019). In different studies, circulating tumor markers were found to be very effective tumor markers because at a very low concentration they can be detected in blood and, in comparison to other markers, provide greater accuracy and efficacy (Schiffman et al., 2015). In this chapter, characteristic features of cancer, its staging, types, causes, cancer etiology and molecular perspectiveare discussed.

1.2 FEATURES OF CANCER

In our bodies, the regulatory system is present to control cell proliferation and homeostasis. Cancer cells have a defect that enables them to violate the normal process. Six essential alterations are there that collectively define cancer cell genotype and are responsible for malignant growth, self-sufficiency in growth signals, insensitivity to antigrowth signals, programmed cell death evasion, unlimited replicative skills, prolonged angiogenesis, and metastasis (Figure 1.1) (Hanahan and Weinberg, 2000; Fouad and Aanei, 2017).

1.2.1 SELF-SUFFICIENT SIGNALS FOR GROWTH

Mitogenic growth signals that are required for normal cell growth are transmitted into the cell by transmembrane receptors that bind with different classes of signaling molecules, extracellular matrix components, transmissible growth factors, and cell-to-cell interaction adhesion molecules (Witsch et al., 2010; Di Domenico and Giordano, 2017).

DOI: 10.1201/9781003201946-1

FIGURE 1.1 Cancer characteristics.

Different oncogenes related to cancer act by mimicking the normal growth signaling patterns (Argyle and Khanna, 2013). Normal cell propagation depends on growth signaling in culture. Their proliferation requires appropriate transmissible mitogenic factors and a proper base for their integrins. In contrast to this, cancer cells do not show dependence on exogenous growth signaling; they generate the growth signals on their own, which disrupts the homeostatic environment of a cell within the tissue. This character of tumor cells is acquired GS (growth signal) autonomy, which plays a very important role in the cancer cell to modulate dominant oncogenes. Many cancer cells have the ability to synthesize mitogenic growth factors (GFs) and in this way, they create a positive feedback loop that leads to autocrine stimulation (Jiang et al., 2019).

The binding of GFs and integrins to extracellular matrix components results in the activation of the SOS-Ras-Raf-MAP kinase pathway (Lu and Malemud, 2019). This pathway plays the central role in the alteration of the downstream cytoplasmic cascade, the most complex mechanism of GS autonomy. It is found that Ras proteins are present in modified forms in 25% of tumor cells that are responsible for the release of mitogenic signals into cells, without stimulation by normal upstream regulators.

1.2.2 INSENSITIVITY TO ANTIGROWTH SIGNALS

There are multiple antigrowth signals which function in order to maintain quiescent cellular state and tissue homeostasis within a normal tissue (Zefferino et al., 2019). Soluble growth inhibitors and immobilized inhibitors both are included in these signals. The system that enables normal cells to respond to these signals is part of the cell cycle clock, especially those components that govern the transfer of the cell through the G_1 phase of its growth cycle. During this period, cells sense these signals and decide whether to proliferate or remain in the dormancy period. All of the antiproliferative signals are guided by the retinoblastoma protein (pRb) and its relative proteins p107 and p130 at the molecular level. In a hyperphosphorylated state, pRb alters the E2F transcription factor, which leads to blocking of the proliferation. In order to proliferate, cancer cells avoid these signals. If pRb signaling is disrupted, E2F will become free and leads to proliferation. TGFβ and some other factors govern the pRb signaling that was found to be disrupted in many types of tumor cells, either by showing mutant state or downregulation of TGFβ receptors. In this way, we can say that antiproliferative signaling is altered in cancer cells, which causes a loss of tumor suppressors (Otto and Sicinski, 2017; Ali, 2019).

1.2.3 PROGRAMMED CELL DEATH EVASION

Expansion of tumor cells is decided by both the rate of cell proliferation as well as attenuation (Meliala et al., 2020). Apoptosis or programmed cell death is majorly responsible for this attenuation. Cancer cells acquire resistance to apoptosis through different strategies. Mutation in the *p53* tumor suppressor gene is the most common loss of proapoptotic signals (Klimovich et al., 2019). In studies it was found that the majority of cancer types show the absence of the p53 protein as a result of a mutation in the *p53* gene, which leads to loss of DNA damage machinery responsible for initializing apoptotic cascade.

It can be expected that almost all the cancer cells adapt some modifications that enable them to avoid apoptosis (Hanahan and Weinberg, 2000).

1.2.4 UNLIMITED REPLICATIVE SKILLS

All types of mammalian cells undergo an intrinsic program that checks their multiplication. The ends of chromosomes, called telomeres, consist of several thousand repeats of a short 6 bp sequence. During each cell cycle, telomeres undergo a loss of 50–100 bp. By counting these base pairs, the number of replicative generations can be accounted (Muraki et al., 2012).

Malignant cells maintain the telomere activity either by telomerase enzyme upregulation or by activating ALT, which maintains telomeres through recombination-based interchromosomal exchanges of sequence information (Jafri et al., 2016). Therefore, by using different mechanisms, cancer cells maintain the telomerase enzyme value above its critical level. Both mechanisms are found to be completely absent in normal cells (Hanahan and Weinberg, 2000).

1.2.5 PROLONGED ANGIOGENESIS

To function and survive, cells depend on the nutrients and oxygen supplied by the vascular system. The growth of a new blood vessel angiogenesis is carefully controlled once the tissue is formed by positive and negative signaling (Hanahan and Weinberg, 2000; Krill and Tewari, 2015; Giampieri et al., 2016; Wu et al., 2016; Tang et al., 2019).

One class of these signals is transmitted by soluble factors and their receptors; others bind on the surface of endothelial cells, integrins, and adhesion molecules. VEGF (vascular endothelial growth factor) and FGF (fibroblast growth factor) represent the initiating signals of angiogenesis, while the thrombospondin-1 is an inhibitor of angiogenesis (Teleanu et al., 2020).

Integrin signaling also plays a role in angiogenesis regulation. Interference with the signaling of integrins can inhibit angiogenesis (Pons-Cursach and Casanovas, 2019).

Tumor cells acquire the ability to induce angiogenesis by an angiogenic switch. By changing the balanced level of activators and inhibitors of angiogenesis, tumor cells activate the angiogenic switch (Nishida et al., 2006; Xia et al., 2016; Zhao et al., 2016; Biziota et al., 2017; Gaikwad et al., 2017; Lin et al., 2017; Sakthivel and Guruvayoorappan, 2018).

Another way of sustained angiogenesis is a decreased level of thrombospondin-1 caused by the loss of p53 function, which occurs in most of the human cancerous cells, so it cannot show its inhibitory effect (Sundaram et al., 2011) and angiogenesis in tumor cells sustained.

1.2.6 Metastasis

Cancer cells have a property to invade other tissues called metastasis, which is responsible for 90% of cancer deaths (Seyfried and Huysentruyt, 2013). This ability of metastasis enables the cancer cells to colonize new sites in the body.

In the metastatic processes, different classes of proteins are altered in tethering of cells to their adjacent tissues or other sites in the body. Mainly the immunoglobulin and cadherin protein families (cell-cell adhesion molecules) are altered (Friedl and Mayor, 2017).

The major alteration was found in the level of E-cadherin in different studies. Loss of E-cadherin function was found in the majority of cancer cells by inactivation of E-cadherin genes, repression of transcription, or inactivation by proteolysis of the extracellular domain.

Another important factor involved in metastasis is extracellular proteases (Martin et al., 2013). Due to the upregulation of particular proteases, it is converted from its inactive zymogen form to active zymogen form. By binding to specific protease receptor matrix-degrading proteases can associate with the cell surface, which can be considered as a reason for cancer cells to invade the epithelial surface and blood vessel crossing (Sevenich and Joyce, 2014).

In the evaluation of malignancy of cancer cells, metastasis is defined as the ability of cancer cells to invade neighboring tissues, restrain secondary organs, and finally to occupy these organs completely. Metastasis is responsible for approximately 90% of cancer deaths due to its incurableness by surgical incision and resistance to chemotherapeutic agents. Metastatic cells can spread inside the body via different ways such as hematogenous spreading, lymphatic spreading, or through the body cavities. The lymphatic spread is the most common way observed in metastasis, while the hematogenous spread is marked as a critical mode in human tumors. Body cavity seeding is mostly observed in the case of colorectal and ovarian cancer (Ahmad, 2016; Calibasi-Kocal and Basbinar, 2018).

Cancer cells can change in a range of phenotypic states required to invade secondary sites from the primary lesion into the circulatory system and eventually for colonization to secondary sites. Metastatic lesions can tolerate and avoid different biophysical and molecular hurdles (Aceto et al., 2014; Lambert et al., 2017; Wang et al., 2017).

Many of these properties may depend on cell behaviors, which are considered incompatible with cells required to establish the primary tumor. For example, it has been observed that despite the proliferative nature of initial tumor cell, invasive cells undergoing EMT (epithelial-mesenchymal transition) suspend the proliferation (Vega et al., 2004; Zheng et al., 2015; Rojas-Puentes et al., 2016; Chakraborty et al., 2020; Ribatti et al., 2020; Wu et al., 2020).

Metastatic tumor cells show an anti-correlation among the phenotypic states of "grow" and "go," though, after detaining in the metastatic site, it could recommence proliferation, despite remaining dormant for long periods (Giese et al., 1996; Hatzikirou et al., 2012; Nakano et al., 2020; Ghosh et al., 2021). In a study, it was found that proliferation is associated with metabolism and stress in metastatic cancers. However, increased inflammation was found in cancers with more EMT phenotypic characteristics (Robinson et al., 2017; Ahmed, 2020).

1.3 INITIATION OF METASTASIS

The idea that metastatic abrasions are produced from tumor cells dispersed from primary tumors has been significantly revised after the study of disseminated tumor cells in patients of early-stage bone marrow cancer (Klein et al., 1999; Schardt et al., 2005; Klein, 2020).

This phenomenon was also observed in the breast cancer animal models that were impulsively metastasizing. In this case, tumor cells lived in distal organs after shedding from premalignant lesions and later form micro-metastases (Harper et al., 2016; Bakir et al., 2020; Phan and Croucher, 2020).

Cancer cells have the adaptation capability to survive in different constraints of the tumor microenvironment, such as distortion of the nucleus leading to chromosomal instability, changed expressions of genes, and metastasis (Bakhoum 2018; Vishwakarma and McManus, 2020; Deville et al., 2021; Ma et al., 2021). Such chromosomal instability induced by invasion recommends a possible mechanism for improved genomic heterogeneity at metastasis sites. The hypothesis that epithelial cells must be of a mesenchymal phenotype for metastasis has been disputed. In a study of breast cancer collective epithelial cell migration, the principal cell was positive for cytokeratin 14, a marker for basal epithelial cell (Cheung et al., 2013; Amintas et al., 2020; Wrenn et al., 2020). In another study, imaging of cancer cell clusters migrating to the tumor has shown E-cadherin (Friedl and Gilmour, 2009; Aceto, 2020; Mizukoshi et al., 2020). Studies conducted on pancreatic and breast cancer animal models have suggested that repression of EMT via genetic alteration of transcription factors has no result on metastasis rate (Zheng et al., 2015; Dudás et al., 2020; Hapke and Haake, 2020). On the contrary, a recent study has shown the requirement of Zeb1, an EMT transcription factor (Krebs et al., 2017; Drápela et al., 2020; Bakir et al., 2020; Sheng et al., 2020) and mesenchymal cells collaborate with epithelial cells after EMT migration to become metastatic through paracrine signaling (Neelakantan et al., 2017; Shen and Reedijk, 2021). Metastasis initiation is not considered only a cell-autonomous process; the tissue microenvironment also affects this event. It has already been identified that connections among the cancer cells, endothelial cells, fibroblasts, immune cells, and the extracellular matrix architecture greatly influence tumor progression. Due to oxygen tension fluctuation caused via the hypoxia-inducible factor (HIF), tumor cells can switch to an amoeboid from collective migration. This state can stimulate the reciprocal signaling between the tumor and stem cells, leading to the metastatic phenotype (Chaturvedi et al., 2012; Lehmann et al., 2017). Cancer cells secrete lactate, which enables the tumor-associated macrophages (TAMs) to encourage angiogenesis (Colegio et al., 2014; Zhang and Li, 2020; Zhang et al., 2021), a necessity for distant metastasis (Linde et al., 2018; Thomas et al., 2020; Shen et al., 2021). Cancer cell migration and invasion are dependent on the autonomous nature of the tumor microenvironment and tissue settings (Spill et al., 2016; Zanotelli et al., 2021). The complex association of tumor cells and cancer microenvironment leading to metastasis requires integrating from an invitro and invivo molecular characterization experimental data model. For example, data collected from protein phosphorylation modeling in the case of pancreatic cancer helps to understand mutual molecular interactions between the stroma and tumor cells, allowing the description of the heterotypic cell-cell interactions in cancer growth (Tape, 2016; Tape et al., 2016).

The process of cancer cells invasion to distant body parts, known as metastasis, includes a sequential and interconnected step: (a) spread to local tissues, (b) invasion into stroma through blood vessels, (c) survival in vascular system, (d) extravasation to distant tissues, and (e) persistence and colonization in the new microenvironment and leads to micro and macro metastasis (Zubair and Ahmad, 2017; Calibasi-Kocal and Basbinar, 2018).

1.3.1 DISSEMINATION

This process involves the initial phase of the invasion metastatic cascade process. During the dissemination process, cancer cells start invading near neighboring tissues from the starting point and transport them to secondary tumor sites (Goyal et al., 2018; Subbotin, 2018; Stephens et al., 2019). This invasion capacity of cancer cells is employed to discriminate malignant tumors from benign ones. In the metastasis process and progression of tumor cells, the invasion signaling pathway plays a key role. It involves the entry of cancer cells to the surrounding tumor microenvironment and tissue parenchyma. Cancer cells migrate to adjacent cells by losing their adhesion ability and leave the original (primary) tumor site. Hence, the metastasis starts with tumor cell migration (Chaffer and Weinberg, 2011; Asano et al., 2017; Chen et al., 2019; Subbotin, 2019; Triki et al., 2019).

At this stage, epithelial-mesenchymal transition (EMT), a process involving detachment from epithelium stratum and acquiring motility, plays a crucial role in cancer progression. Dissemination and gain of EMT character occur at the early stage of metastasis. Several transcription factors such as Snail, Twist, Zeb1, Zeb2, etc., are responsible for the EMT process (Harper et al., 2016).

Cancer cells change the gene expression patterns by altering the EMT transcription factors. Migratory tumor cells lose their apical-basal polarity after losing adhesion molecules and intercellular connections. Due to remodeling, the cytoskeleton starts protruding to form invasive growth that penetrates the adjacent stromal matrix. Tumor cells with migratory features show more resistance towards anticancer agents than other tumor cells (Hosseini et al., 2016; Wang et al., 2017; Punzi et al., 2018; Wong et al. 2018; Bornes et al., 2019; Multinu et al., 2019; Vega-Arroyo et al., 2019; Yamamoto et al., 2019).

1.3.2 Intravasation

Motile tumor cells enter into the blood vessels to travel via circulation and access other metastatic locations to form micro and macro metastasis. During this process, cancer cells transport to endothelial vessels through the tissues (Xue et al., 2006; Calibasi-Kocal and Basbinar, 2018).

The intravasation process is promoted by specific signaling molecules (Arwert et al., 2018; Kai, Drain and Weaver, 2019; Borriello et al., 2020; Deryugina et al., 2020; Smeda et al., 2020), biochemical and biophysical conditions of the vascular system and the microenvironment. The intravasation route is controlled by the structural alterations between lymphatic and blood vessels. The blood vessels' invasion is limited because they possess tighter and more tight junctions than the lymph vessels (Chiang et al., 2016).

1.3.3 Survival in the Circulatory System

In intravasation, cancer cells become circulating tumor cells (CTCs) after coming into the blood vessels. Survival of the cancer cells into the circulation is a significant step for metastasis. Millions of cancer cells circulate through the vascular system inside the body. CTCs can travel either as clusters or single cells. However, immune cells such as macrophages, neutrophils, natural killer cells, etc., show their effectiveness by clearing these CTCs from the circulation. Hence, the metastasis success rate is lower due to infrequent CTCs (Steeg, 2007; Calibasi-Kocal and Basbinar, 2018; Font-Clos et al., 2020).

Cancer cells have advanced themselves by incorporating numerous strategies for escaping the immune system. This is done by losing the immunostimulatory molecules and gaining the immunoinhibitory components and higher expression of molecules or factors responsible for apoptosis (Steeg, 2007; Calibasi-Kocal and Basbinar, 2018; Font-Clos et al., 2020). Blood cells such as platelets help CTCs survive by fighting against the natural killer cells. Coagulating molecules like tissue factor and thrombin activate the platelets and, due to this, platelet-cancer cell aggregates are formed (Steeg, 2007; Calibasi-Kocal and Basbinar, 2018; Font-Clos et al., 2020). In recent times, CTCs are considered very promising biomarkers for cancer diagnosis and prognosis.

1.3.4 Extravasation

Tumor cells are detained at the secondary metastatic sites after surviving in the harsh environment of circulation and extravasate into parenchyma tissue. This process requires transendothelial migration (Zhao et al., 2017; Cui et al., 2018; de Oliveira et al., 2018; Liu et al., 2018; Ward et al., 2018) of cancer cells to the endothelial wall. It is carried out by adhering the cancerous cells to endothelial cells for migration through the endothelial barrier. Endothelial cells can control the adhesion of cancer cells; therefore, they have an essential role in metastatic body formation (Foss et al., 2020).

For the transmigration, vascular permeabilization is done by ATP produced by the platelets and angiopoietin-like 4 (ANGPTL4). Other molecules such as VEGF, CCL2, MMPs, etc., also help transendothelial migration by disrupting the vascular integrity (Cheema et al., 2020; Rahmani et al., 2020).

1.4 CLASSIFICATION AND TYPES OF CANCER

Cancer classification is highly complicated, based on their primary site of origin and tissue types. There are mainly six different cancer types (Kunnumakkara et al., 2018).

1.4.1 CARCINOMA

Cancer of epithelial tissues, responsible for 80–90% of all cancer incidence, is called carcinoma. It commonly affects the secretory glands and organs like the breast, colon, prostate, etc. Further, it can classify into adeno or squamous cell carcinoma (Padmavathi et al., 2017).

1.4.2 SARCOMA

Cancer that arises in connective tissues such as muscle, cartilage, chondrosarcoma, etc., is called a sarcoma (Le Cesne et al., 2015; Taieb et al., 2015; Ballinger et al., 2016; Benson et al., 2016; Ramlawi et al., 2016; Smith et al., 2016; Dancsok et al., 2017).

1.4.3 MYELOMA

This type of cancer arises in the plasma cells of bone marrow. It is associated with the over-production of adolescent white blood cells such as granulocytic leukemia or lymphoblastic leukemia, etc. (Rajkumar and Kumar, 2016; Campbell et al., 2017; Terpos, 2018).

1.4.4 LYMPHOMA

Lymphoma is also known as "solid cancer." It develops in the lymphatic nodes or glands, involved in body fluids purification and lymphocyte production (Mugnaini and Ghosh, 2016).

There are more than 100 types of cancer; some specific cancer types related to their sites and carcinogenic agents responsible for it are mentioned in Table 1.1.

1.5 STAGING OF CANCER

To know the size of the tumor and to check whether the tumor cells have spread into the body, some tests are done called the staging of cancer. It provides information about cancer whether it has started spreading from its starting point or not. Staging is denoted by roman numerals. Mainly it is labeled with 0 and I-IV. Small localized cancers come under the category of 0. Most advanced or metastatic cancers are of stage IV, while the II and III are locally advanced cancer (http://cancerguide.org/basic.html).

1.5.1 TNM STAGING

Stage 0–IV is actually described in terms of "TNM" system: T stands for tumor, N stands for node, and M stands for metastasis (Bobdey et al., 2018; Bron et al., 2019; Lin et al., 2019; Piñeros et al., 2019; Proto et al., 2019; J. Y. Yoon et al., 2019).

TABLE 1.1

Different Types of Cancer

Cancer Site	Possible Carcinogenic Agents	References
Oral site		
Oral cavity	Alcoholic beverages, Tobacco	Ustrell-Borràs et al., 2020
Pharynx	Alcoholic beverages, Tobacco	Ustrell-Borràs et al., 2020
Digestive organs		
Oesophagus	Acetaldehyde, Tobacco	Prabhu et al., 2014
Stomach	Tobacco, Radiation	Tey et al., 2019
Colon	Alcoholic Beverages, Processed meat	Hur et al., 2019
Liver and bile duct	Alcohol, Hepatitis B virus Hepatitis C virus	Sarin et al., 2019
Gall bladder	Thorium-232	Keith et al., 2019
Pancreas	Tobacco	Stuart et al., 2020
Respiratory organs		
Nasal cavity	Radium-226, Wood dust, Smoking	Krewski et al., 2019
Lung	Tobacco smoke, X-radiation, Aluminium Production, Arsenic, Cadmium compounds, Coal tar pitch, etc.	Krewski et al., 2019
Bone & Skin		
Bone	Plutonium Radium-224 Radium-226 Radium-228	Priest, 2019
Skin (melanoma)	Solar radiation Ultraviolet-emitting tanning devices	Miligi, 2020
Breast & female genital organs		
Breast	Alcoholic beverages, Estrogen-progesterone contraceptives, X-radiation, gamma-radiation	Labrèche et al., 2020
Vulva	Human papillomavirus type 16	Wang et al., 2020
Vagina	Human papillomavirus type 16	Wang et al., 2020
Uterine cervix	Estrogen-progesterone contraceptives Human papillomavirus types 16, 18, 52, 56, 58, Tobacco smoking	Lyu et al., 2019
Ovary	Asbestos (all forms) Estrogen menopausal therapy Tobacco smoking	Krewski et al., 2019
Male genital organs		
Penis	Human immunodeficiency virus type 1 Human papillomavirus type 18	Van Bilsen et al., 2019
Prostate	Androgenic (anabolic) steroids, Arsenic and inorganic arsenic compounds, Cadmium and cadmium compounds, X-radiation, gamma-radiation, Red meat	Hall and Greco, 2020
Urinary tract		
Kidney	Arsenic and inorganic arsenic compounds, Cadmium and cadmium compounds Perfluorooctanoic acid etc.	Michalek et al., 2019
Urinary bladder	Aluminum production 4-Aminobiphenyl, Arsenic and inorganic arsenic compounds, Auramine production, Benzidine, Chlornaphazine, Cyclophosphamide, etc.	Birkett et al., 2019
Eye, brain, & central nervous system		
Eye	Human immunodeficiency virus type1, Ultraviolet-emitting tanning devices Welding	Miligi, 2020
Brain & spinal cord	X-radiation, gamma-radiation	https://www.drugs.com/prednisone.html
Lymphoid organs		
Leukaemia and/or lymphoma	Cyclophosphamide, Epstein-Barr virus, etc.	Yamaguchi et al., 2019

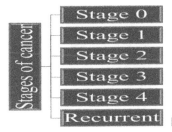

FIGURE 1.2 Different stages of cancer.

- **T: Tumors:** It gives information about the extent of the primary tumor. T0 stage is about the lack of invasion of tumor cells to adjacent tissues while the T4 stage defines the invasion of tumor cells to other sites in the body.
- **N: Node:** It describes the involvement of the regional lymph node. N0 means no involvement of the lymph node and N4 means huge involvement.
- **M: Metastasis:** M0 means no metastasis while M1–4 shows metastasis.

1.5.1.1 Stages of Cancer

There are different stages of cancer (Fang et al., 2021; Herbert et al., 2021; Li et al., 2021; Oki et al., 2021; Zhang et al., 2021):

a. **Stage 0:** Carcinoma is in the early stage. The abnormal cells are present in only the starting layer of the primary site, not in the deeper tissues (Figure 1.2).
b. **Stage I:** This stage of cancer involves only the primary site and is curable.
c. **Stage II:** It may involve lymph nodes and is larger in size than stage I cancer.
d. **Stage III:** It involves a primary site and nearby lymph nodes.
e. **Stage IV:** The spread of cancer to other parts or sites of the body.
f. **Recurrent:** Sometimes cancer can come back to the same site or another site again after it has been treated.

1.6 CANCER FROM THE MOLECULAR PERSPECTIVE

Oncogene generation and associated genetic disorders originate inside the body due to genetic alterations that include chromosomal translocation, deletion, point mutation, and insertion activation. For example, chronic blood cancer is the result of genetic material exchange between chromosome numbers 9 and 22. This condition produces a biomarker i.e., ph1, found in 95% of elderly patients and helps in correct diagnosis. Bcr gene to Abl oncogene connection leads to the generation of a new combination of the gene, which ultimately translated to a protein having kinase activity (Joensuu and Dimitrijevic, 2001; Thomas et al., 2007). Likewise, the *p53* gene mutation causes an unusual protein formation and disturbs the molecular processes associated with p53. These abnormal molecular and biological processes lead to cancer cell formation and p53 abnormality has been reported in about 60% of cancer cases. The *p53* gene is involved in cell division, apoptosis, angiogenesis, and DNA replication in normal conditions. The association of p53 with CDK1-P2 and CDC2 retains cancer cells in the G_1 and G_2 stages of the cell cycle (Taylor and Stark, 2001; Chae et al., 2011). In fact, it can be considered as either an inhibitor or promoter of tumor cells. The *p53* gene exhibits its anticancer activity by choosing three pathways: (a) during repairing of DNA proteins, (b) in apoptosis induction, and (c) cell cycle arrest in the G_1/S phase (Matlashewski et al., 1984; May and May, 1999; Selivanova, 2004). Generally, hypomethylation is the property of repeating sequences, which ultimately leads to addiction, and deletion of genes and chromosomal instability (Goelz et al., 1985; Kanai and Hirohashi, 2007). Hypomethylation due to member of the LINE family, L1, is an important example that has been witnessed in different types of cancer including breast and lung cancer (Wilson et al., 2007). In the specific promoters, ectopic expression of oncogenes can be activated by hypomethylation. This condition can be seen for maspin in breast cancer where it acts as a tumor suppressor gene. Other examples are S100P and MAGE in pancreatic cancer and melanoma, respectively (Futscher et al., 2004).

Unlike the hypomethylation, hypermethylation is a property shown by the specific CpG region only (Bastian et al., 2004; Ellinger et al., 2008; Kvasha et al., 2008; Liu et al., 2010; Xi et al., 2013; Fujii et al., 2015; Skvortsova et al., 2019). Due to that transcriptional inactivation of promoter genes involved in cell repair, cell cycle mechanism, and apoptosis process occurs. This leads to cancer induction (Esteller, 2007), making the hypermethylated promoters as a cancer biomarker for prognostic and diagnostic purposes due to the involvement of CpG regions of mostly promoter genes in hypermethylation. DNA methylation patterns can be disturbed by the impaired functioning of DNMT (DNA methyltransferase) due to getting evidence of higher expressions of DNMT1 and DNMT3b in different tumors (Miremadi et al., 2007).

Deacetylation is another condition, which is found in different tumors, catalyzed by histone deacetylases (HDAC). Regulation of HDACs expression was found to be regulated by microRNAs: for example, in prostate cancer, cell growth and survival are regulated by miR-449a via inhibition of HDAC1 expression. Besides the changes in HDAC expression in different types of cancer like colon, lung, and leukemia, disruption in histone acetylation occurs also via deletion and ectopic mutations in HAT (Histone acetyltransferase) and associated genes. These all conditions lead to tumor formation (Noonan et al., 2009; Yen et al., 2016; Sato et al., 2017; Spencer et al., 2017; Quintela et al., 2019; Dai et al., 2021; Hogg et al., 2021). Histone methylation alterations occur mainly due to histone methyltransferases and demethylase expressions. In the case of leukemia, H3K4 (Histone3 lysine4) and H3K29 abnormal methylation patterns are observed due to MLL oncoprotein. This leads to altered expressions of MLL target genes (Wang et al., 2009). ATPase subunits (BRG1 and BRM), associated with SWI/SNF complex, are known tumor suppressors and found to be responsible for 15–20% lung cancer cases. SWI/SNF complex is found to be involved in the generation of different types of cancer by interacting with the retinoblastoma protein gene, p53, MYC, and BRCA1. Hence, the dysfunctional SWI/SNF complex leads to a disturbance in cell growth (Mohd-Sarip et al., 2017; Pierre and Kadoch, 2017; Szymanski et al., 2017; Yoshimoto, Matsubara and Niki, 2017; Agaimy et al., 2018; Alpsoy and Dykhuizen, 2018; Moreno et al., 2018; Savas and Skardasi, 2018; Sinha et al., 2018).

Additionally, modifications in nucleosome positions result in transcription suppression due to promoter hypermethylation. During hypermethylation, genes encoding changing complexes subunits of chromatin are the principal target of CpG hypermethylation for causing cancer. Along with the nucleosome positions, histone modifications are also responsible for cancer, for example, expression of MacroH2A during the process of senescence in the case of lung cancer (Cantariño, Douet and Buschbeck, 2013; Huidobro et al., 2021; Lee and Kim, 2021; Hassanpour and Dehghani, 2017). Apparently, during the last three decades, researchers have investigated and reported extensive information about the roles of genes and proteins in the production of tumor cells. Among them, the role of mutated genes is the most potent discovery for cancer research. By using different molecular techniques, gene expression analysis and novel biomarker detection can be done.

1.7 CANCER ETIOLOGY

It is already proven that the two factors that have the dominant role for cancer development are genes and the environment. Across the populations, despite unequal distribution, genes do not explain the cancer incidence differences. For example, stomach cancer incidence is six times higher in Japanese than in Americans. However, Japanese who have settled in America have shown a comparable rate to Americans, supporting the environmental facts for cancer development (Figure 1.3).

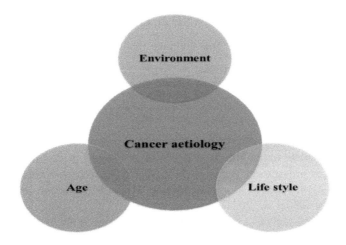

FIGURE 1.3   Cancer etiology.

1.7.1   ENVIRONMENT AND LIFESTYLE

English physician John Hill, who observed the relation between nasal cancer and tobacco, proved the first evidence of involvement of lifestyle in cancer. In 1950, heavy chain-smokers had shown a higher risk of cancer development than non-smokers. Since then, alcohol consumption and smoking have been related to huge numbers of deaths due to throat and mouth cancer (Ishiguro et al., 2009; Johansen et al., 2009; Talamini et al., 2010; Kabat et al., 2012; Colussi et al., 2018; Jarosz and Rychlik, 2019; Lee et al., 2019; Hejmadi, 2004; Huang et al., 2021). It is analyzed that around more than half of a million deaths are due to the bad lifestyle like obesity, lack of physical activity, diet problems (low vitamin, high salt, low fibers intake that are directly linked to breast, esophageal, breast, and prostate cancer). Viruses like the hepatitis B virus (HBV) and human papillomavirus (HPV) have also increased cancer risks including nasopharyngeal, cervical carcinomas, etc. (Loechler, 2001). In Australia, skin cancer is spreading its way continuously because of UV rays, which can be get reduced by using different types of vaccines and changing lifestyles (J. H. Yoon et al., 2019; Beecher et al., 2020; Duarte et al., 2020; Chen et al., 2021; Ke and Wang, 2021; Saji et al., 2021; Shahid et al., 2021).

Additionally, there are different cancer-causing agents called carcinogens and mutagens, present in food, water, air, and in the chemical forms. As we know, epithelial cells cover the entire skin, respiratory canals, and internal organs. It metabolizes the ingested carcinogens; therefore, more than 90% of cancer arises from epithelial cells known as carcinomas.

1.7.2   AGE

Although the cancerous condition can occur at any age in the life span of a person, it is reported to be most common in elderly persons. Nearly two-thirds of the total deaths due to cancer occur in populations of more than the age of 60 years. This is due to the altered immune system with age and accumulation of genetic mutations of a lifetime of exposure to different types of carcinogens.

Multiple genetic changes lead to the progression of cancer development. The exponential cancer rise with age correlates with an increased vulnerability to the late stage carcinogenesis with environmental exposures. Breast or uterine cancer can arise from lifetime exposure to estrogen while lifetime testosterone exposure results in prostate cancer (Hejmadi, 2004). Impaired cellular immunity is also responsible for certain types of immunogenic cancer such as lymphomas, melanomas, etc.

1.8 CANCER HISTOPATHOLOGY AND IDENTIFICATION

There are different histopathological and staining methods for the identification of cancer (Hejmadi, 2004):

  a. **Diagnosis:** The nature of the tumor cell whether it is benign or malignant can be determined by microscopic studies. The classification of cancerous cells is based on cellular morphology and tissue-specific markers.
  b. **Therapy:** Pathology is also used for confirmation of prognosis. For example, we can check whether the complete tumor has been removed or not and prediction of cancer rate progression. Cancer progression can also be predicted by histotype (Munzone et al., 2015; Grandi et al., 2019; Jones et al., 2020; Marano et al., 2020; Tappero et al., 2020; Nameki et al., 2021; Yamada et al., 2021).
  c. **Cellular origin (histogenesis):** Histopathological classifications of cancer tissues can help in determining the origin of cancer (Saxena and Gyanchandani, 2020; Beck et al., 2021; Chan et al., 2021; Hernes et al., 2021; Sandgren et al., 2021; Seierstad et al., 2021; Zhou et al., 2021). It can be a primary or secondary tumor. For example, tumors present in any organ can be metastasized from elsewhere. Also, the source can also be determined for example: in the case of lung cancer is epithelial due to smoking but it becomes mesothelial because of asbestos exposure.

1.9 CANCER EPIDEMIOLOGY

Each year, approximately 10 million people are diagnosed with cancer and over 5 million cancer deaths worldwide. Cancer comes at number two after heart disease, causing over 10% of all deaths in developed countries. However, cancer is considered an ailment of the developed world. More than half of cancer cases arise in developing countries, including three-quarters of the population. In developed countries, every third person will develop this disease during their lifetime and it will switch to one in two persons with time (DeVita et al., 1997; DeVita et al., 2012; Tarver, 2012; In, 2019; Viale, 2020).

Cancer covers around 200 different entities, divergent in their genetic characteristics, clinical characteristics, etiology, and progression patterns. At the broader level, cancer can be classified according to germ layers from which it arises, i.e., sarcoma and carcinoma. Carcinomas originate within the tissues themselves, derived from the ecto- or endoderm of the fetus and consist of most common adult cancers, whereas the sarcomas arise from the fetal mesoderm such as bone, connective tissues, muscle, etc. and are more common in children. About 50% of cancer cases in developed countries are lung, breast, prostate, and colon carcinomas though hematological cancers such as leukemia and lymphomas are responsible for 8–10% of total cancer cases (Tarver, 2012; In, 2019; Viale, 2020). The genetic basis of cancer is well recognized with the carcinogenetic models, somatic cell genetics, and tumor viruses' studies. Cumulative mutation in the oncogenes, DNA repair, and tumor suppressor genes of the somatic cells result in cancer in the beginning (Vogelstein et al., 1988; Hussain and Harris, 1998; Hanahan and Weinberg, 2000; Ponder, 2001; Braga et al., 2002; Goldgar, 2002; Potter, 2003; Vähäkangas, 2003; Bocchetta and Carbone, 2004; Hemminki et al., 2004; Vogelstein and Kinzler, 2004; Rohan et al., 2006; Fischer, 2014; Cross et al., 2018). However, some cancers arise from germline mutations and are responsible for genetic predisposition. Cancer follows typical natural antiquity (DeVita et al., 1997; DeVita et al., 2012; Tarver, 2012; In, 2019; Viale, 2020).

Normal cells start becoming dysplastic with morphological abnormalities that show the beginning of the transformation. These abnormalities after proliferation lead to carcinoma. In this early phase, the tumor does not invade the original tissue and is curable significantly. Stage I of cancer remains localized where the disrupted tissues form a primary lesion. Then in stage II, tumor cells start invading distant regions such as lymph nodes as secondary cancers. Finally, cancer cells enter into the bloodstream initiating the metastasis, a characteristic of cancer stage IV (Hussain and Harris, 1998; Potter, 2003; Vähäkangas, 2003; Bocchetta and Carbone, 2004; Hemminki et al., 2004). There is a strong correlation between survival and stage of cancer and it also helps in the prognosis of the disease.

REFERENCES

Aceto, N., 2020. Bring along your friends: Homotypic and heterotypic circulating tumor cell clustering to accelerate metastasis. *Biomedical Journal*, *43*(1), pp. 18–23.

Aceto, N., Bardia, A., Miyamoto, D.T., Donaldson, M.C., Wittner, B.S., Spencer, J.A., Yu, M., Pely, A., Engstrom, A., Zhu, H., and Brannigan, B.W., 2014. Circulating tumor cell clusters are oligoclonal precursors of breast cancer metastasis. *Cell*, *158*(5), pp. 1110–1122.

Agaimy, A., Amin, M.B., Gill, A.J., Popp, B., Reis, A., Berney, D.M., Magi-Galluzzi, C., Sibony, M., Smith, S.C., Suster, S., Trpkov, K., Hes, O., and Hartmann, A., 2018. SWI/SNF protein expression status in fumarate hydratase–deficient renal cell carcinoma: immunohistochemical analysis of 32 tumors from 28 patients. *Human Pathology*, *77*, pp. 139–146.

Ahmad, A. ed., 2016. *Introduction to Cancer Metastasis*. Academic Press.

Ahmed, Z., 2020. Practicing precision medicine with intelligently integrative clinical and multi-omics data analysis. *Human Genomics*, *14*(1), pp. 1–5.

Ali, M., 2019. Cell cycle regulation in cancer: A noncoding perspective. doi:10.13140/RG.2.2.16083. 40482.

Alpsoy, A., and Dykhuizen, E.C., 2018. Glioma tumor suppressor candidate region gene 1 (GLTSCR1) and its paralog GLTSCR1-like form SWI/SNF chromatin remodeling subcomplexes. *Journal of Biological Chemistry*, *293*(11), pp. 3892–3903.

Amintas, S., Bedel, A., Moreau-Gaudry, F., Boutin, J., Buscail, L., Merlio, J.P., Vendrely, V., Dabernat, S., and Buscail, E., 2020. Circulating tumor cell clusters: United we stand divided we fall. *International Journal of Molecular Sciences*, *21*(7), p. 2653.

Argyle, D.J., and Khanna, C., 2013. Tumor biology and metastasis. *Small Clinical Veterinary Oncology, fifth ed. Saunders Elsevier, St Louis, MO, USA*, pp. 30–50.

Arwert, E.N., Harney, A.S., Entenberg, D., Wang, Y., Sahai, E., Pollard, J.W., and Condeelis, J.S., 2018. A unidirectional transition from migratory to perivascular macrophage is required for tumor cell intravasation. *Cell Reports*, *23*(5), pp. 1239–1248.

Asano, Y., Odagiri, T., Oikiri, H., Matsusaki, M., Akashi, M., and Shimoda, H., 2017. Construction of artificial human peritoneal tissue by cell-accumulation technique and its application for visualizing morphological dynamics of cancer peritoneal metastasis. *Biochemical and Biophysical Research Communications*, *494*(1–2), pp. 213–219.

Bakhoum, S.F., Ngo, B., Laughney, A.M., Cavallo, J.A., Murphy, C.J., Ly, P., Shah, P., Sriram, R.K., Watkins, T.B., Taunk, N.K., and Duran, M., 2018. Chromosomal instability drives metastasis through a cytosolic DNA response. *Nature*, *553*(7689), pp. 467–472.

Bakir, B., Chiarella, A.M., Pitarresi, J.R., and Rustgi, A.K., 2020. EMT, MET, plasticity, and tumor metastasis. *Trends in Cell Biology*. doi: 10.1016/j.tcb.2020.07.003.

Ballinger, M.L., Goode, D.L., Ray-Coquard, I., James, P.A., Mitchell, G., Niedermayr, E., Puri, A., Schiffman, J.D., Dite, G.S., Cipponi, A., Maki, R.G., Brohl, A.S., Myklebost, O., Stratford, E.W., Lorenz, S., Ahn, S.M., Ahn, J.H., Kim, J.E., Shanley, S., Beshay, V., Randall, R.L., Judson, I., Seddon, B., Campbell, I.G., Young, M.A., Sarin, R., Blay, J.Y., O'Donoghue, S.I., and Thomas, D.M., 2016. Monogenic and polygenic determinants of sarcoma risk: an international genetic study. *The Lancet Oncology*, *17*(9), pp. 1261–1271.

Bastian, P.J., Yegnasubramanian, S., Palapattu, G.S., Rogers, C.G., Lin, X., De Marzo, A.M., and Nelson, W.G., 2004. Molecular biomarker in prostate cancer: The role of CpG island hypermethylation. *European Urology*, *46*(6) pp. 698–708.

Beck, T., Zhang, N., Shah, A., Khoncarly, S., McHenry, C., and Jin, J., 2021. Thyroid cancer identified after positron emission tomography (PET) shows aggressive histopathology. *Journal of Surgical Research*, *260*, pp. 245–250.

Beecher, M., Kumar, N., Jang, S., Rapić-Otrin, V., and Van Houten, B., 2020. Expanding molecular roles of UV-DDB: Shining light on genome stability and cancer. *DNA Repair*, *94*. doi: 10.1016/j.dnarep. 2020.102860.

Beetham, R., 2000. Detection of Bence-Jones protein in practice. *Annals of Clinical Biochemistry, 37*(5), pp. 563–570.

Benson, C., Ray-Coquard, I., Sleijfer, S., Litière, S., Blay, J.Y., Le Cesne, A., Papai, Z., Judson, I., Schöffski, P., Chawla, S., Gil, T., Piperno-Neumann, S., Marréaud, S., Dewji, M.R., and Van Der Graaf, W.T.A., 2016. Outcome of uterine sarcoma patients treated with pazopanib: A retrospective analysis based on two European Organisation for Research and Truatment of Cancer (EORTC) Soft Tissue and Bone Sarcoma Group (STBSG) clinical trials 62043 and 62072. *Gynecologic Oncology, 142*(1), pp. 89–94.

Bettegowda, Chetan, et al., 2014. Detection of circulating tumor DNA in early-and late-stage human malignancies. *Science Translational Medicine, 6*(224), pp. 224ra24–224ra24.

Birkett, N., Al-Zoughool, M., Bird, M., Baan, R.A., Zielinski, J., and Krewski, D., 2019. Overview of biological mechanisms of human carcinogens. *Journal of Toxicology and Environmental Health, Part B, 22*(7-8), pp. 288–359.

Biziota, E., Mavroeidis, L., Hatzimichael, E., and Pappas, P., 2017. Metronomic chemotherapy: A potent macerator of cancer by inducing angiogenesis suppression and antitumor immune activation. *Cancer Letters, 400*, pp. 243–251.

Bobdey, S., Mair, M., Nair, S., Nair, D., Balasubramaniam, G., and Chaturvedi, P., 2018. A Nomogram based prognostic score that is superior to conventional TNM staging in predicting outcome of surgically treated T4 buccal mucosa cancer: Time to think beyond TNM. *Oral Oncology, 81*, pp. 10–15.

Bocchetta, M., and Carbone, M., 2004. Epidemiology and molecular pathology at crossroads to establish causation: molecular mechanisms of malignant transformation. *Oncogene, 23*(38), pp. 6484–6491.

Bornes, L., van Scheppingen, R.H., Beerling, E., Schelfhorst, T., Ellenbroek, S.I.J., Seinstra, D., and van Rheenen, J., 2019. Fsp1-mediated lineage tracing fails to detect the majority of disseminating cells undergoing EMT. *Cell Reports, 29*(9), pp. 2565–2569.e3.

Borriello, L., Karagiannis, G.S., Duran, C.L., Coste, A., Oktay, M.H., Entenberg, D., and Condeelis, J.S., 2020. The role of the tumor microenvironment in tumor cell intravasation and dissemination. *European Journal of Cell Biology, 99*(6). pp. 151098.

Braga, P.E., Latorre, M.D.R.D.D.O., and Curado, M.P., 2002. Childhood cancer: a comparative analysis of incidence, mortality, and survival in Goiania (Brazil) and other countries. *Cadernos de saude publica, 18*(1), pp. 33–44.

Bron, G., Scemama, U., Villes, V., Fakhry, N., Salas, S., Chagnaud, C., Bendahan, D., and Varoquaux, A., 2019. A new CT dynamic maneuver "Mouth Opened with Tongue Extended" can improve the clinical TNM staging of oral cavity and oropharynx squamous cell carcinomas. *Oral Oncology, 94*, pp. 41–46.

Calibasi-Kocal, G., and Basbinar, Y., 2018. Introductory Chapter: Cancer Metastasis. *In: Cancer Metastasis.* IntechOpen.

Campbell, J.P., Heaney, J.L., Pandya, S., Afzal, Z., Kaiser, M., Owen, R., Child, J.A., Cairns, D.A., Gregory, W., Morgan, G.J., & Jackson, G.H., 2017. Response comparison of multiple myeloma and monoclonal gammopathy of undetermined significance to the same anti-myeloma therapy: a retrospective cohort study. *The Lancet Haematology, 4*(12), pp. e584–e594.

Cantariño, N., Douet, J., and Buschbeck, M., 2013. MacroH2A – an epigenetic regulator of cancer. *Cancer Letters, 336*(2), pp. 247–252.

Cardona, A.F., Carranza, H., Vargas, C., Zea, D., Cetina, L., Wills, B., Ruiz García, E., Arrieta, O., Jaramillo García, L.F., and Rojas Puentes, L.L., 2016. Epithelial–mesenchymal transition, proliferation, and angiogenesis in locally advanced cervical cancer treated with chemoradiotherapy. *Cancer Medicine 5*, pp. 1989–1999.

Chae, S.W., Sohn, J.H., Kim, D.H., Choi, Y.J., Park, Y.L., Kim, K., Cho, Y.H., Pyo, J.S., and Kim, J.H., 2011. Overexpressions of Cyclin B1, cdc2, p16 and p53 in human breast cancer: the clinicopathologic correlations and prognostic implications. *Yonsei Medical Journal, 52*(3), pp. 445–453.

Chaffer, C.L., and Weinberg, R.A., 2011. A perspective on cancer cell metastasis. *Science, 331*(6024), pp. 1559–1564.

Chakraborty, S., Mir, K.B., Seligson, N.D., Nayak, D., Kumar, R., and Goswami, A., 2020. Integration of EMT and cellular survival instincts in reprogramming of programmed cell death to anastasis. *Cancer and Metastasis Reviews, 39*(2), pp. 553–566.

Chan, R.C., Chandra, A., and Varma, M., 2021. Clinical utility of histopathology data: cancers of the testis and urinary bladder. *Diagnostic Histopathology, 27*(7), pp. 290–296.

Chaturvedi, P., Gilkes, D.M., Wong, C.C.L., Luo, W., Zhang, H., Wei, H., Takano, N., Schito, L., Levchenko, A., and Semenza, G.L., 2012. Hypoxia-inducible factor–dependent breast cancer–mesenchymal stem cell bidirectional signaling promotes metastasis. *The Journal of Clinical Investigation, 123*(1), pp. 189–205.

Cheema, P.S., Kumar, G., Mittal, S., Parashar, D., Geethadevi, A., Jadhav, K., and Tuli, H.S., 2020. Metastasis: A Major Driver of Cancer Pathogenesis. *In Drug Targets in Cellular Processes of Cancer: From Nonclinical to Preclinical Models* (pp. 185–211). Springer, Singapore.

Chen, Q., Li, X., Xie, Y., Hu, W., Cheng, Z., Zhong, H., and Zhu, H., 2021. Azo modified hyaluronic acid based nanocapsules: CD44 targeted, UV-responsive decomposition and drug release in liver cancer cells. *Carbohydrate Polymers*, *267*, p. 118152.

Chen, Y., Sumardika, I.W., Tomonobu, N., Winarsa Ruma, I.M., Kinoshita, R., Kondo, E., Inoue, Y., Sato, H., Yamauchi, A., Murata, H., Yamamoto, K. ichi, Tomida, S., Shien, K., Yamamoto, H., Soh, J., Liu, M., Futami, J., Sasai, K., Katayama, H., Kubo, M., Putranto, E.W., Hibino, T., Sun, B., Nishibori, M., Toyooka, S., and Sakaguchi, M., 2019. Melanoma cell adhesion molecule is the driving force behind the dissemination of melanoma upon S100A8/A9 binding in the original skin lesion. *Cancer Letters*, *452*, pp. 178–190.

Cheung, K.J., and Ewald, A.J., 2016. A collective route to metastasis: Seeding by tumor cell clusters. *Science*, *352*(6282), pp. 167–169.

Cheung, K.J., Gabrielson, E., Werb, Z., and Ewald, A.J. (2013). Collective invasion in breast cancer requires a conserved basal epithelial program. *Cell*, 155(7), pp. 1639–1651.

Chiang, S.P., Cabrera, R.M., and Segall, J.E., 2016. Tumor cell intravasation. *American Journal of Physiology-Cell Physiology*, *311*(1), pp. C1–C14.

Colegio, O.R., Chu, N.Q., Szabo, A.L., Chu, T., Rhebergen, A.M., Jairam, V., Cyrus, N., Brokowski, C.E., Eisenbarth, S.C., Phillips, G.M., and Cline, G.W., 2014. Functional polarization of tumour-associated macrophages by tumour-derived lactic acid. *Nature*, *513*(7519), pp. 559–563.

Colussi, D., Fabbri, M., Zagari, R.M., Montale, A., Bazzoli, F., and Ricciardiello, L., 2018. P.05.14 Obesity, smoking, heavy alcohol consumption, diabetes mellitus and risk for colorectal polyps and cancer at index colonoscopy in a fit-positive screening population. *Digestive and Liver Disease*, *50*(2), p. e171.

Cross, W., Kovac, M., Mustonen, V., Temko, D., Davis, H., Baker, A.M., Biswas, S., Arnold, R., Chegwidden, L., Gatenbee, C., and Anderson, A.R., 2018. The evolutionary landscape of colorectal tumorigenesis. *Nature Ecology & Evolution*, *2*(10), pp. 1661–1672.

Cui, X., Tjønnfjord, G.E., Kanse, S., Iversen, N., Dahm, A.E., Myklebust, C.F., Sun, L., Jiang, Z.X., Ho, M., and Sandset, P.M., 2018. Tissue factor pathway inhibitor enhances transendothelial migration of chronic lymphocytic leukemia cells through binding to Glypican-3. *Blood*, *132*(Supplement 1), pp.2452–2452.

Dai, M., Yang, B., Chen, J., Liu, F., Zhou, Y., Zhou, Y., Xu, Q., Jiang, S., Zhao, S., Li, X., Zhou, X., Yang, Q., Li, J., Wang, Y., Zhang, Z., and Teng, Y., 2021. Nuclear-translocation of ACLY induced by obesity-related factors enhances pyrimidine metabolism through regulating histone acetylation in endometrial cancer. *Cancer Letters*, *513*, pp. 36–49.

Dancsok, A.R., Asleh-Aburaya, K., and Nielsen, T.O., 2017. Advances in sarcoma diagnostics and treatment. *Oncotarget*, *8*(4), p. 7068.

de Oliveira, A.S., de Almeida, V.H., Gomes, F.G., Rezaie, A.R., and Monteiro, R.Q., 2018. TR47, a PAR1-based peptide, inhibits melanoma cell migration in vitro and metastasis in vivo. *Biochemical and Biophysical Research Communications*, *495*(1), pp. 1300–1304.

Deryugina, E., Carré, A., Ardi, V., Muramatsu, T., Schmidt, J., Pham, C., and Quigley, J.P., 2020. Neutrophil elastase facilitates tumor cell intravasation and early metastatic events. *iScience*, *23*(12), p.101799.

Deville, S., Berckmans, P., Van Hoof, R., Lambrichts, I., Salvati, A., and Nelissen, I. 2021. Comparison of extracellular vesicle isolation and storage methods using high-sensitivity flow cytometry. *PloS One*, 16(2), p. e0245835.

DeVita, N.T., Hellman, S., and Rosenberg, S.A., 1997. *Cancer: Principles and Practice of Oncology.* Lippencott-Raven, Philadelphia, PA.

DeVita, V.T., Lawrence, T.S., and Rosenberg, S.A., 2012. *Cancer: Principles & Practice of Oncology: Primer of the Molecular Biology of Cancer.* Lippincott Williams & Wilkins, Philadelphia, PA.

Deyell, M., Garris, C.S., and Laughney, A.M., 2021. Cancer metastasis as a non-healing wound. *British Journal of Cancer*, *124*, pp. 1–12.

Di Domenico, M., and Giordano, A., 2017. Signal transduction growth factors: The effective governance of transcription and cellular adhesion in cancer invasion. *Oncotarget*, *8*(22), p. 36869.

Drápela, S., Bouchal, J., Jolly, M.K., Culig, Z., and Souček, K., 2020. ZEB1: A critical regulator of cell plasticity, DNA damage response, and therapy resistance. *Frontiers in Molecular Biosciences*, *7*, p.36.

Drugs.com. Prednisone Information from Drugs.com; c1996–2018 [Updated: 13 February 2018, Cited: 19 June 2018]. Available from: https://www.drugs.com/prednisone.html.

Duarte, A.F., Mota, I., Campo, M., and Correia, O., 2020. Skin cancer and UV literacy - Outdoor workers study. *Actas Dermo-Sifiliográficas (English Edition)*, *111*(6), pp. 531–533.

Dudás, J., Ladányi, A., Ingruber, J., Steinbichler, T.B., and Riechelmann, H., 2020. Epithelial to mesenchymal transition: A mechanism that fuels cancer radio/chemoresistance. *Cells*, *9*(2), p. 428.

Ellinger, J., Bastian, P.J., Jurgan, T., Biermann, K., Kahl, P., Heukamp, L.C., Wernert, N., Müller, S.C., and von Ruecker, A., 2008. CpG island hypermethylation at multiple gene sites in diagnosis and prognosis of prostate cancer. *Urology*, *71*(1), pp. 161–167.

Esteller, M., 2007. Epigenetic gene silencing in cancer: The DNA hypermethylome. *Human Molecular Genetics*, *16*(R1), pp. R50–R59.

Fang, C., Zhang, P., Yu, A., Yang, Y., and Zhang, J., 2021. Different prognosis of stage IIIB cervical cancer patients with lower third of vaginal invasion and those without. *Gynecologic Oncology*, *162*(1), pp. 50–55.

Fischer, D.J., 2014. *Analysis of metabolic alterations during colorectal cancer development* (Doctoral dissertation, University of Zurich).

Font-Clos, F., Zapperi, S., and La Porta, C.A., 2020. Blood flow contributions to cancer metastasis. *Iscience*, *23*(5), p. 101073.

Foss, A., Muñoz-Sagredo, L., Sleeman, J., and Thiele, W., 2020. The contribution of platelets to intravascular arrest, extravasation, and outgrowth of disseminated tumor cells. *Clinical & experimental metastasis*, *37*(1), pp. 47–67.

Fouad, Y.A., and Aanei, C., 2017. Revisiting the hallmarks of cancer. *American Journal of Cancer Research*, *7*(5), p. 1016.

Friedl, P., and Gilmour, D., 2009. Collective cell migration in morphogenesis, regeneration and cancer. *Nature Reviews Molecular Cell Biology*, *10*(7), pp. 445–457.

Friedl, P., and Mayor, R., 2017. Tuning collective cell migration by cell-cell junction regulation. *Cold Spring Harbor Perspectives in Biology*, *9*(4), p. a029199.

Fujii, A., Harada, T., Iwama, E., Ota, K., Furuyama, K., Ijichi, K., Okamoto, T., Okamoto, I., Takayama, K., and Nakanishi, Y., 2015. Hypermethylation of the CpG dinucleotide in epidermal growth factor receptor codon 790: Implications for a mutational hotspot leading to the T790M mutation in non-small-cell lung cancer. *Cancer Genetics*, *208*(5), pp. 271–278.

Futscher, B.W., O'Meara, M.M., Kim, C.J., Rennels, M.A., Lu, D., Gruman, L.M., Seftor, R.E., Hendrix, M.J., and Domann, F.E., 2004. Aberrant methylation of the maspin promoter is an early event in human breast cancer. *Neoplasia (New York, NY)*, *6*(4), p. 380.

Gaikwad, D., Shewale, R., Patil, V., Mali, D., Gaikwad, U., and Jadhav, N., 2017. Enhancement in in vitro anti-angiogenesis activity and cytotoxicity in lung cancer cell by pectin-PVP based curcumin particulates. *International Journal of Biological Macromolecules*, *104*, pp. 656–664.

Ghosh, K., Ghosh, S., Chatterjee, U., Bhattacharjee, P., and Ghosh, A., 2021. Dichotomy in growth and invasion from low-to high-grade glioma cellular variants. *Cellular and Molecular Neurobiology*, pp. 1–16. doi: 10.1007/s10571-021-01096-1.

Giampieri, R., Caporale, M., Pietrantonio, F., De Braud, F., Negri, F. V., Giuliani, F., Pusceddu, V., Demurtas, L., Restivo, A., Fontanella, C., Aprile, G., Cascinu, S., and Scartozzi, M., 2016. Second-line angiogenesis inhibition in metastatic colorectal cancer patients: Straightforward or overcrowded? *Critical Reviews in Oncology/Hematology*, *100*, pp. 101–106.

Giese, A., Loo, M.A., Tran, N., Haskett, D., Coons, S.W., and Berens, M.E., 1996. Dichotomy of astrocytoma migration and proliferation. *International Journal of Cancer*, *67*(2), pp. 275–282.

Goelz, S.E., Vogelstein, B., and Feinberg, A.P., 1985. Hypomethylation of DNA from benign and malignant human colon neoplasms. *Science*, *228*(4696), pp. 187–190.

Goldgar, D.E., 2002. Population aspects of cancer genetics. *Biochimie*, *84*(1), pp. 19–25.

Goyal, A., Cajigas, I., Ibrahim, G.M., Brathwaite, C.D., Khatib, Z., Niazi, T., Bhatia, S., and Ragheb, J., 2018. Surgical treatment of intramedullary spinal metastasis in medulloblastoma: Case report and review of the literature. *World Neurosurgery*, *118*, pp. 42–46.

Grandi, G., Perrone, A.M., Chiossi, G., Friso, S., Toss, A., Sammarini, M., Facchinetti, F., Botticelli, L., Palma, F., and De Iaco, P., 2019. Increasing BMI is associated with both endometrioid and serous histotypes among endometrial rather than ovarian cancers: a case-to-case study. *Gynecologic Oncology*, *154*(1), pp. 163–168.

Gray, Elin S., Rizos, H., Reid, A.L., Boyd, S.C., Pereira, M.R., Lo, J., Tembe, V., Freeman, J., Lee, J.H., Scolyer, R.A., Siew, K., Lomma, C., Cooper, A., Khattak, M.A., Meniawy, T.M., Long, G.V., Carlino, M.S., Millward, M., and Ziman, M., 2015. Circulating tumor DNA to monitor treatment response and detect acquired resistance in patients with metastatic melanoma. *Oncotarget* *6*(39), pp. 42008.

Hall, J.M., and Greco, C.W., 2020. Perturbation of nuclear hormone receptors by endocrine disrupting chemicals: Mechanisms and pathological consequences of exposure. *Cells*, *9*(1), p. 13.

Hanahan, D., and Weinberg, R.A. 2000. The hallmarks of cancer. *Cell*, 100(1), pp. 57–70.

Hapke, R.Y., and Haake, S.M., 2020. Hypoxia-induced epithelial to mesenchymal transition in cancer. *Cancer Letters*, *487*, pp. 10–20.

Harper, K.L., Sosa, M.S., Entenberg, D., Hosseini, H., Cheung, J.F., Nobre, R., Avivar-Valderas, A., Nagi, C., Girnius, N., Davis, R.J., and Farias, E.F., 2016. Mechanism of early dissemination and metastasis in Her2+ mammary cancer. *Nature*, *540*(7634), pp. 588–592.

Harris, C.C., 1996. p53 tumor suppressor gene: from the basic research laboratory to the clinic – an abridged historical perspective. *Carcinogenesis 17*, pp. 1187–1198.

Hassanpour, S.H., and Dehghani, M., 2017. Review of cancer from perspective of molecular. *Journal of Cancer Research and Practice*, *4*(4), pp. 127–129.

Hatzikirou, H., Basanta, D., Simon, M., Schaller, K., and Deutsch, A., 2012. 'Go or grow': the key to the emergence of invasion in tumour progression?. *Mathematical Medicine and Biology: A Journal of the IMA*, *29*(1), pp. 49–65.

Hejmadi, M., 2014. *Introduction to Cancer Biology*. Bookboon.

Hemminki, K., Rawal, R., Chen, B., and Bermejo, J.L., 2004. Genetic epidemiology of cancer: From families to heritable genes. *International Journal of Cancer*, *111*(6), pp. 944–950.

Herbert, A., Barclay, M.E., Koo, M.M., Rous, B., Greenberg, D.C., Abel, G., and Lyratzopoulos, G., 2021. Stage–specific incidence trends of renal cancers in the East of England, 1999–2016. *Cancer Epidemiology*, *71*. doi: 10.1016/j.canep.2020.101883.

Hernes, E., Revheim, M.E., Hole, K.H., Tulipan, A.J., Strømme, H., Lilleby, W., and Seierstad, T., 2021. Prostate-specific membrane antigen PET for assessment of primary and recurrent prostate cancer with histopathology as reference standard: A systematic review and meta-analysis. *PET Clinics*, *16*(2), pp.147–165.

Hogg, S.J., Motorna, O., Cluse, L.A., Johanson, T.M., Coughlan, H.D., Raviram, R., Myers, R.M., Costacurta, M., Todorovski, I., Pijpers, L., Bjelosevic, S., Williams, T., Huskins, S.N., Kearney, C.J., Devlin, J.R., Fan, Z., Jabbari, J.S., Martin, B.P., Fareh, M., Kelly, M.J., Dupéré-Richer, D., Sandow, J.J., Feran, B., Knight, D., Khong, T., Spencer, A., Harrison, S.J., Gregory, G., Wickramasinghe, V.O., Webb, A.I., Taberlay, P.C., Bromberg, K.D., Lai, A., Papenfuss, A.T., Smyth, G.K., Allan, R.S., Licht, J.D., Landau, D.A., Abdel-Wahab, O., Shortt, J., Vervoort, S.J., and Johnstone, R.W., 2021. Targeting histone acetylation dynamics and oncogenic transcription by catalytic P300/CBP inhibition. *Molecular Cell*, *81*(10), pp. 2183–2200.e13.

Hosseini, H., Obradović, M.M., Hoffmann, M., Harper, K.L., Sosa, M.S., Werner-Klein, M., Nanduri, L.K., Werno, C., Ehrl, C., Maneck, M., and Patwary, N., 2016. Early dissemination seeds metastasis in breast cancer. *Nature*, *540*(7634), pp. 552–558.

Huang, J., Leung, D.K.W., Chan, E.O.T., Lok, V., Leung, S., Wong, I., Lao, X.Q., Zheng, Z.J., Chiu, P.K.F., Ng, C.F., Wong, J.H.M., Volpe, A., Merseburger, A.S., Powles, T., Teoh, J.Y.C., and Wong, M.C.S., 2021. A global trend analysis of kidney cancer incidence and mortality and their associations with smoking, alcohol consumption, and metabolic syndrome. *European Urology Focus*, *S2405-4569*(21), pp.00001-8.

Huidobro, C., Martín-Vicente, P., López-Martínez, C., Alonso-López, I., Amado-Rodríguez, L., Crespo, I., and M Albaiceta, G., 2021. Cellular and molecular features of senescence in acute lung injury. *Mechanisms of Ageing and Development*, *193*, pp.111410.

Hur, S.J., Yoon, Y., Jo, C., Jeong, J.Y., and Lee, K.T., 2019. Effect of dietary red meat on colorectal cancer risk – a review. *Comprehensive Reviews in Food Science and Food Safety*, *18*(6), pp. 1812–1824.

Hussain, S.P., and Harris, C.C., 1998. Molecular epidemiology of human cancer. *Genes and Environment in Cancer*, *154*, pp. 22–36.

In, T., 2019. Facts & figures 2019: US cancer death rate has dropped 27% in 25 years. American Cancer Society. https://www.cancer.org/research/cancer-facts-statistics.html

Ishiguro, S., Sasazuki, S., Inoue, M., Kurahashi, N., Iwasaki, M., and Tsugane, S., 2009. Effect of alcohol consumption, cigarette smoking and flushing response on esophageal cancer risk: A population-based cohort study (JPHC study). *Cancer Letters*, *275*(2), pp. 240–246.

Jafri, M.A., Ansari, S.A., Alqahtani, M.H., and Shay, J.W., 2016. Roles of telomeres and telomerase in cancer, and advances in telomerase-targeted therapies. *Genome Medicine*, *8*(1), p. 69.

Jarosz, M., and Rychlik, E., 2019. Alcohol consumption and tobacco smoking and selected gastrointestinal cancers morbidity rates in Poland. *Annals of Oncology*, *30*, pp. iv37.

Jiang, X., Xie, H., Dou, Y., Yuan, J., Zeng, D., and Xiao, S., 2019. Expression and function of FRA1 protein in tumors. *Molecular Biology Reports*, *47*(1), pp. 1–16.

Joensuu, H., and Dimitrijevic, S., 2001. Tyrosine kinase inhibitor imatinib (STIS71) as an anticancer agent for solid tumors. *Annals of Medicine*, *33*(7), pp. 451–455.

Johansen, D., Borgström, A., Lindkvist, B., and Manjer, J., 2009. Different markers of alcohol consumption, smoking and body mass index in relation to risk of pancreatic Cancer: A prospective cohort study within the Malmö preventive project. *Pancreatology*, *9*(5), pp. 677–686.

Jolly, M.K., Mani, S.A., and Levine, H., 2018. Hybrid epithelial/mesenchymal phenotype(s): The 'fittest' for metastasis? *Biochimica et Biophysica Acta – Reviews on Cancer*, *1870*(2), pp.151–157.

Jones, M.R., Peng, P.C., Coetzee, S.G., Tyrer, J., Reyes, A.L.P., Corona, R.I., Davis, B., Chen, S., Dezem, F., Seo, J.H., Kar, S., Dareng, E., Berman, B.P., Freedman, M.L., Plummer, J.T., Lawrenson, K., Pharoah, P., Hazelett, D.J., and Gayther, S.A., 2020. Ovarian Cancer Risk Variants Are Enriched in Histotype-Specific Enhancers and Disrupt Transcription Factor Binding Sites. *American Journal of Human Genetics*, *107*(4), pp. 622–635.

Kabat, G.C., Kim, M.Y., Wactawski-Wende, J., and Rohan, T.E., 2012. Smoking and alcohol consumption in relation to risk of thyroid cancer in postmenopausal women. *Cancer Epidemiology*, *36*(4), pp. 335–340.

Kai, F.B., Drain, A.P., and Weaver, V.M., 2019. The extracellular matrix modulates the metastatic journey. *Developmental Cell*, *49*(3), pp.332–346.

Kanai, Y., and Hirohashi, S., 2007. Alterations of DNA methylation associated with abnormalities of DNA methyltransferases in human cancers during transition from a precancerous to a malignant state. *Carcinogenesis*, *28*(12), pp. 2434–2442.

Katt, M.E., Wong, A.D., and Searson, P.C., 2018. Dissemination from a solid tumor: Examining the multiple parallel pathways. *Trends in Cancer*, *4*(1), pp.20–37.

Ke, Y., and Wang, X.J., 2021. TGFβ signaling in photoaging and UV-induced skin cancer. *Journal of Investigative Dermatology*, *141*(4 S) pp.1104–1110.

Keith, S., Wohlers, D., and Ingerman, L., 2019. Toxicological profile for thorium. Agency for Toxic Substances and Disease Registry, U.S. Department of Health and Human Services.

Klein, C.A., 2020. Cancer progression and the invisible phase of metastatic colonization. *Nature Reviews Cancer*, *20*(11), pp. 681–694.

Klein, C.A., Schmidt-Kittler, O., Schardt, J.A., Pantel, K., Speicher, M.R., and Riethmüller, G., 1999. Comparative genomic hybridization, loss of heterozygosity, and DNA sequence analysis of single cells. *Proceedings of the National Academy of Sciences*, *96*(8), pp. 4494–4499.

Klimovich, B., Mutlu, S., Schneikert, J., Elmshäuser, S., Klimovich, M., Nist, A., Mernberger, M., Timofeev, O., and Stiewe, T., 2019. Loss of p53 function at late stages of tumorigenesis confers ARF-dependent vulnerability to p53 reactivation therapy. *Proceedings of the National Academy of Sciences*, *116*(44), pp. 22288–22293.

Kontis, V., Cobb, L.K., Mathers, C.D., Frieden, T.R., Ezzati, M., and Danaei, G., 2019. Three public health interventions could save 94 million lives in 25 years: global impact assessment analysis. *Circulation*, *140*(9), pp. 715–725.

Krebs, A.M., Mitschke, J., Losada, M.L., Schmalhofer, O., Boerries, M., Busch, H., Boettcher, M., Mougiakakos, D., Reichardt, W., Bronsert, P., and Brunton, V.G., 2017. The EMT-activator Zeb1 is a key factor for cell plasticity and promotes metastasis in pancreatic cancer. *Nature cell biology*, *19*(5), pp. 518–529.

Krewski, D., Rice, J.M., Bird, M., Milton, B., Collins, B., Lajoie, P., Billard, M., Grosse, Y., Cogliano, V.J., Caldwell, J.C., and Rusyn, I.I., 2019. Concordance between sites of tumor development in humans and in experimental animals for 111 agents that are carcinogenic to humans. *Journal of Toxicology and Environmental Health, Part B*, *22*(7-8), pp. 203–236.

Krill, L.S., and Tewari, K.S., 2015. Exploring the therapeutic rationale for angiogenesis blockade in cervical cancer. *Clinical Therapeutics*, *37*(1), pp.9–19.

Kunnumakkara, A.B., Bordoloi, D., and Monisha, J. eds., 2018. *Cancer Cell Chemoresistance and Chemosensitization*. World Scientific.

Kvasha, S., Gordiyuk, V., Kondratov, A., Ugryn, D., Zgonnyk, Y.M., Rynditch, A. V., and Vozianov, A.F., 2008. Hypermethylation of the 5′CpG island of the FHIT gene in clear cell renal carcinomas. *Cancer Letters*, *265*(2), pp. 250–257.

Labrèche, F., Goldberg, M.S., Hashim, D., and Weiderpass, E., 2020. Female breast cancer. *Occupational Cancers*, p. 417.

Lambert, A.W., Pattabiraman, D.R., and Weinberg, R.A., 2017. Emerging biological principles of metastasis. *Cell*, *168*(4), pp. 670–691.

Le Cesne, A., Ray-Coquard, I., Duffaud, F., Chevreau, C., Penel, N., Bui Nguyen, B., Piperno-Neumann, S., Delcambre, C., Rios, M., Chaigneau, L., Le Maignan, C., Guillemet, C., Bertucci, F., Bompas, E., Linassier, C., Olivier, T., Kurtz, J.E., Even, C., Cousin, P., and Yves Blay, J., 2015. Trabectedin in patients with advanced soft tissue sarcoma: A retrospective national analysis of the French Sarcoma Group. *European Journal of Cancer*, *51*(6), pp. 742–750.

Lee, J.E., and Kim, M.Y., 2021. Cancer epigenetics: Past, present and future. *Seminars in Cancer Biology*. doi: 10.1016/j.semcancer.2021.03.025.

Lee, S., Woo, H., Lee, J., Oh, J.H., Kim, J., and Shin, A., 2019. Cigarette smoking, alcohol consumption, and risk of colorectal cancer in South Korea: A case-control study. *Alcohol*, *76*, pp. 15–21.

Lehmann, S., Te Boekhorst, V., Odenthal, J., Bianchi, R., van Helvert, S., Ikenberg, K., Ilina, O., Stoma, S., Xandry, J., Jiang, L., and Grenman, R., 2017. Hypoxia induces a HIF-1-dependent transition from collective-to-amoeboid dissemination in epithelial cancer cells. *Current Biology*, *27*(3), pp. 392–400.

Li, J., Ouyang, X., Gong, X., Li, P., Xiao, L., Chang, X., and Tang, J., 2021. Survival outcomes of minimally invasive surgery for early-stage cervical cancer: A single-center, one surgeon, retrospective study. *Asian Journal of Surgery*. doi: 10.1016/j.asjsur.2021.05.037.

Lin, H., Fang, Z., Su, Y., Li, P., Wang, J., Liao, H., Hu, Q., Ye, C., Fang, Y., Luo, Q., Lin, Z., Pan, C., Wang, F., and Zhang, Z.Y., 2017. DHX32 promotes angiogenesis in colorectal cancer through augmenting β-catenin signaling to induce expression of VEGFA. *EBioMedicine*, *18*, pp. 62–72.

Lin, J.X., Lin, J.P., Xie, J.W., Wang, J. Bin, Lu, J., Chen, Q.Y., Cao, L.L., Lin, M., Tu, R.H., Zheng, C.H., Huang, C.M., and Li, P., 2019. Is the AJCC TNM staging system still appropriate for gastric cancer patients survival after 5 years? *European Journal of Surgical Oncology*, *45*(6), pp. 1115–1120.

Linde, N., Casanovaacebes, M., Sosa, M.S., Mortha, A., Rahman, A., Farias, E.F., Harper, K., Tardio, E., Torres, I., and Jones, J.G., 2018. Macrophages orchestrate breast cancer early dissemination and metastasis. Nat Commun 9, pp. 21.

Liu, W. Bin Ao, L., Zhou, Z.Y., Cui, Z.H., Zhou, Y.H., Yuan, X.Y., Xiang, Y.L., Cao, J., and Liu, J.Y., 2010. CpG island hypermethylation of multiple tumor suppressor genes associated with loss of their protein expression during rat lung carcinogenesis induced by 3-methylcholanthrene and diethylnitrosamine. *Biochemical and Biophysical Research Communications*, *402*(3), pp. 507–514.

Liu, Y., Geng, Y.H., Yang, H., Yang, H., Zhou, Y.T., Zhang, H.Q., Tian, X.X., and Fang, W.G., 2018. Extracellular ATP drives breast cancer cell migration and metastasis via S100A4 production by cancer cells and fibroblasts. *Cancer Letters*, *430*, pp. 1–10.

Loechler, E.L., 2001. Environmental carcinogens and mutagens. In *Encyclopedia of Life Sciences*. American Cancer Society.

Lu, N., and Malemud, C.J., 2019. Extracellular signal-regulated kinase: a regulator of cell growth, inflammation, chondrocyte and bone cell receptor-mediated gene expression. *International journal of molecular sciences*, *20*(15), p. 3792.

Lyu, Y., Ding, L., Gao, T., Li, Y., Li, L., Wang, M., Han, Y., and Wang, J., 2019. Influencing factors of high-risk human papillomavirus infection and DNA load according to the severity of cervical lesions in female coal mine workers of China. *Journal of Cancer*, *10*(23), p. 5764.

Ma, Q., Wang, J., Qi, J., Peng, D., Guan, B., Zhang, J., Li, Z., Zhang, H., Li, T., Shi, Y., and Li, X., 2021. Increased chromosomal instability characterizes metastatic renal cell carcinoma. *Translational oncology*, *14*(1), p. 100929.

Marano, L., Polom, K., D''Ignazio, A., Marrelli, D., and Roviello, F., 2020. Incidence and prognostic impact of lymph node metastasis in gastric cancer: Deep inside of signet ring cell histotype. *European Journal of Surgical Oncology*, *46*(12), p. e10.

Martin, T.A., Ye, L., Sanders, A.J., Lane, J., and Jiang, W.G., 2013. Cancer invasion and metastasis: molecular and cellular perspective. In *Madame Curie Bioscience Database [Internet]*. Landes Bioscience.

Matlashewski, G., Lamb, P., Pim, D., Peacock, J., Crawford, L., and Benchimol, S., 1984. Isolation and characterization of a human p53 cDNA clone: Expression of the human p53 gene. *The EMBO Journal*, *3*(13), pp. 3257–3262.

May, P., and May, E., 1999. Twenty years of p53 research: Structural and functional aspects of the p53 protein. *Oncogene*, *18*(53), pp. 7621–7636.

Meliala, I.T.S., Hosea, R., Kasim, V., and Wu, S., 2020. The biological implications of Yin Yang 1 in the hallmarks of cancer. *Theranostics*, *10*(9), p. 4183.

Michalek, I.M., Martinsen, J.I., Weiderpass, E., Kjaerheim, K., Lynge, E., Sparen, P., Tryggvadottir, L., & Pukkala, E. 2019. Occupation and risk of cancer of the renal pelvis in Nordic countries. *BJU International*, 123(2), pp. 233–238.

Miligi, L., 2020. Ultraviolet radiation exposure: Some observations and considerations, focusing on some italian experiences, on cancer risk, and primary prevention. *Environments*, 7(2), p. 10.

Miremadi, A., Oestergaard, M.Z., Pharoah, P.D., and Caldas, C., 2007. Cancer genetics of epigenetic genes. *Human molecular genetics*, 16(R1), pp. R28–R49.

Mizukoshi, K., Okazawa, Y., Haeno, H., Koyama, Y., Sulidan, K., Komiyama, H., Saeki, H., Ohtsuji, N., Ito, Y., Kojima, Y., and Goto, M., 2020. Metastatic seeding of human colon cancer cell clusters expressing the hybrid epithelial/mesenchymal state. *International Journal of Cancer*, 146(9), pp. 2547–2562.

Mohd-Sarip, A., Teeuwssen, M., Bot, A.G., De Herdt, M.J., Willems, S.M., Baatenburg de Jong, R.J., Looijenga, L.H.J., Zatreanu, D., Bezstarosti, K., van Riet, J., Oole, E., van Ijcken, W.F.J., van de Werken, H.J.G., Demmers, J.A., Fodde, R., and Verrijzer, C.P., 2017. DOC1-dependent recruitment of NURD reveals antagonism with SWI/SNF during epithelial-mesenchymal transition in oral cancer cells. *Cell Reports*, 20(1), pp. 61–75.

Moreno, T., Gonzalez-Silva, L., Revilla, C., Monterde, B., Agraz-Doblas, A., Betancor, I., Freire, J., Gomez-Roman, J., Salido, E., and Varela, I., 2018. PO-376 SWI/SNF alterations as markers for prognosis and specific treatments in human cancer. *ESMO Open*, 3, pp. A169.

Mugnaini, E.N., and Ghosh, N., 2016. Lymphoma. *Primary Care: Clinics in Office Practice*, 43(4), pp. 661–675.

Multinu, F., Casarin, J., Cappuccio, S., Keeney, G.L., Glaser, G.E., Cliby, W.A., Weaver, A.L., McGree, M.E., Angioni, S., Faa, G., Leitao, M.M., Abu-Rustum, N.R., and Mariani, A., 2019. Ultrastaging of negative pelvic lymph nodes to decrease the true prevalence of isolated paraaortic dissemination in endometrial cancer. *Gynecologic Oncology*, 154(1), pp. 60–64.

Munzone, E., Giobbie-Hurder, A., Gusterson, B.A., Mallon, E., Viale, G., Thürlimann, B., Ejlertsen, B., MacGrogan, G., Bibeau, F., Lelkaitis, G., Price, N., Gelber, R.D., Coates, A.S., Goldhirsch, A., and Colleoni, M., 2015. Outcomes of special histotypes of breast cancer after adjuvant endocrine therapy with letrozole or tamoxifen in the monotherapy cohort of the BIG 1-98 trial. *Annals of Oncology*, 26(12), pp. 2442–2449.

Muraki, K., Nyhan, K., Han, L., and Murnane, J.P., 2012. Mechanisms of telomere loss and their consequences for chromosome instability. *Frontiers in Oncology*, 2, p. 135.

Nair, M. Krishnan, Cherian Varghese, and R. Swaminathan, 2005. Cancer: Current scenario, intervention strategies and projections for 2015. *NCHM Background papers-Burden of Disease in India* 219–225. doi: 10.12691/ajnr-6-2-4

Nakano, T., Okaie, Y., Dietis, N., and Odysseos, A.D., 2020, December. Growing Bio-nanomachine Networks: Application to Malignant Tumor Evolution and Progression. In *GLOBECOM 2020-2020 IEEE Global Communications Conference* (pp. 1–6). IEEE.

Nameki, R., Chang, H., Reddy, J., Corona, R.I., and Lawrenson, K., 2021. Transcription factors in epithelial ovarian cancer: histotype-specific drivers and novel therapeutic targets. *Pharmacology and Therapeutics*. doi: 10.1016/j.pharmthera.2020.107722.

Neelakantan, D., Zhou, H., Oliphant, M.U., Zhang, X., Simon, L.M., Henke, D.M., Shaw, C.A., Wu, M.F., Hilsenbeck, S.G., White, L.D., and Lewis, M.T., 2017. EMT cells increase breast cancer metastasis via paracrine GLI activation in neighbouring tumour cells. *Nature communications*, 8(1), pp. 1–14.

Nishida, N., Yano, H., Nishida, T., Kamura, T., and Kojiro, M., 2006. Angiogenesis in cancer. *Vascular health and risk management*, 2(3), p. 213.

Noonan, E., Place, R.F., Pookot, D., Basak, S., Whitson, J.M., Hirata, H., Giardina, C., and Dahiya, R., 2009. miR-449a targets HDAC-1 and induces growth arrest in prostate cancer. *Oncogene*, 28(14), pp. 1714–1724.

Oki, E., Watanabe, J., Sato, T., Kagawa, Y., Kuboki, Y., Ikeda, M., Ueno, H., Kato, T., Kusumoto, T., Masuishi, T., Yamaguchi, K., Kanazawa, A., Nishina, T., Uetake, H., Yamanaka, T., and Yoshino, T., 2021. Impact of the 12-gene recurrence score assay on deciding adjuvant chemotherapy for stage II and IIIA/B colon cancer: the SUNRISE-DI study. *ESMO Open*, 6(3), p. 100146.

Otto, T., and Sicinski, P., 2017. Cell cycle proteins as promising targets in cancer therapy. *Nature Reviews Cancer*, 17(2), p. 93.

Padmavathi, G., Harsha, C., Bordoloi, D., Banik, K., and Kunnumakkara, A.B., 2017. Mucoepidermoid Carcinoma (MEC) and Associated MAML2 Fusion Genes. *In: Fusion Genes and Cancer* (pp. 221–230). World Scientific Publishing Co Pte Ltd.

Phan, T.G., and Croucher, P.I., 2020. The dormant cancer cell life cycle. *Nature Reviews Cancer*, 20(7), pp. 398–411.

Pierre, R.S., and Kadoch, C., 2017. Mammalian SWI/SNF complexes in cancer: Emerging therapeutic opportunities. *Current Opinion in Genetics and Development*, 42, pp.56–67.

Piñeros, M., Parkin, D.M., Ward, K., Chokunonga, E., Ervik, M., Farrugia, H., Gospodarowicz, M., O'Sullivan, B., Soerjomataram, I., Swaminathan, R., Znaor, A., Bray, F., and Brierley, J., 2019. Essential TNM: A registry tool to reduce gaps in cancer staging information. *The Lancet Oncology*, *20*(2), pp.e103–e111.

Ponder, B.A., 2001. Cancer genetics. *Nature*, *411*(6835), pp. 336–341.

Pons-Cursach, R., and Casanovas, O., 2019. Mechanisms of anti-angiogenic therapy. *Tumor Angiogenesis: A Key Target for Cancer Therapy*, pp. 183–208. doi: 10.1007/978-3-319-31215-6_2-2.

Potter, J.D., 2003. Epidemiology, cancer genetics and microarrays: making correct inferences, using appropriate designs. *TRENDS in Genetics*, *19*(12), pp. 690–695.

Prabhu, A., Obi, K.O., and Rubenstein, J.H., 2014. The synergistic effects of alcohol and tobacco consumption on the risk of esophageal squamous cell carcinoma: a meta-analysis. *American Journal of Gastroenterology*, *109*(6), pp. 822–827.

Priest, N.D., 2019. A Nontarget Mechanism to Explain Carcinogenesis Following α-Irradiation. *Dose-Response*, *17*(4). doi: 1559325819893195.

Proto, C., Signorelli, D., Mallone, S., Prelaj, A., Lo Russo, G., Imbimbo, M., Galli, G., Ferrara, R., Ganzinelli, M., Leuzzi, G., Greco, F.G., Calareso, G., Botta, L., Gatta, G., Garassino, M., and Trama, A., 2019. The prognostic role of TNM staging compared with tumor volume and number of pleural sites in malignant pleural mesothelioma. *Clinical Lung Cancer*, *20*(6), pp. e652–e660.

Punzi, S., Balestrieri, C., D'Alesio, C., Bossi, D., Dellino, G.I., Gatti, E., Natoli, G., Pelicci, P.G., and Lanfrancone, L., 2018. PO-178 WDR5 promotes metastasis dissemination in breast cancer. *ESMO Open*, *3*, pp. A90.

Quintela, M., Sieglaff, D.H., Gazze, A.S., Zhang, A., Gonzalez, D., Francis, L., Webb, P., and Conlan, R.S., 2019. HBO1 directs histone H4 specific acetylation, potentiating mechano-transduction pathways and membrane elasticity in ovarian cancer cells. *Nanomedicine: Nanotechnology, Biology, and Medicine*, *17*, pp. 254–265.

Rahmani, F., Hasanzadeh, M., Hassanian, S.M., Khazaei, M., Esmaily, H., Naghipour, A., Ferns, G.A., and Avan, A., 2020. Association of a genetic variant in the angiopoietin-like protein 4 gene with cervical cancer. *Pathology-Research and Practice*, *216*(7), p. 153011.

Rajkumar, S.V., and Kumar, S., 2016, January. Multiple Myeloma: Diagnosis and Treatment. *In: Mayo Clinic Proceedings* (Vol. *91*(1), pp. 101–119). Elsevier.

Ramakrishnan, N., and Jialal, I., 2019. Bence-Jones Protein. In *StatPearls [Internet]*. StatPearls Publishing.

Ramlawi, B., Leja, M.J., Abu Saleh, W.K., Al Jabbari, O., Benjamin, R., Ravi, V., Shapira, O.M., Blackmon, S.H., Bruckner, B.A., and Reardon, M.J., 2016. Surgical treatment of primary cardiac sarcomas: Review of a single-institution experience. *Annals of Thoracic Surgery*, *101*(2), pp. 698–702.

Rapisuwon, Suthee, Eveline E. Vietsch, and Anton Wellstein, 2016. Circulating biomarkers to monitor cancer progression and treatment. *Computational and Structural Biotechnology Journal 14*, pp. 211–222.

Ribatti, D., Tamma, R., and Annese, T., 2020. Epithelial-mesenchymal transition in cancer: A historical overview. *Translational oncology*, *13*(6), p. 100773.

Robinson, D.R., Wu, Y.M., Lonigro, R.J., Vats, P., Cobain, E., Everett, J., Cao, X., Rabban, E., Kumar-Sinha, C., Raymond, V., and Schuetze, S., 2017. Integrative clinical genomics of metastatic cancer. *Nature*, *548*(7667), pp. 297–303.

Rohan, T.E., Henson, D.E., Franco, E.L., and Albores-Saavedra, J., 2006. Cancer Precursors. In *Cancer Prevention and Early Detection Third Edition* (pp. 21-46). Oxford University Press, New York.

Rojas-Puentes, L., Cardona, A.F., Carranza, H., Vargas, C., Jaramillo, L.F., Zea, D., Cetina, L., Wills, B., Ruiz-Garcia, E., and Arrieta, O. 2016. Epithelial-mesenchymal transition, proliferation, and angiogenesis in locally advanced cervical cancer treated with chemoradiotherapy. *Cancer Medicine*, *5*(8), pp. 1989–1999.

Saji, R.S., Prasana, J.C., Muthu, S., and George, J., 2021. Experimental and theoretical spectroscopic (FT-IR, FT-Raman, UV-VIS) analysis, natural bonding orbitals and molecular docking studies on 2-bromo-6-methoxynaphthalene: A potential anti-cancer drug. *Heliyon*, *7*(6), pp. e07213.

Sakthivel, K.M., and Guruvayoorappan, C., 2018. Targeted inhibition of tumor survival, metastasis and angiogenesis by Acacia ferruginea mediated regulation of VEGF, inflammatory mediators, cytokine profile and inhibition of transcription factor activation. *Regulatory Toxicology and Pharmacology*, *95*, pp. 400–411.

Sandgren, K., Nilsson, E., Keeratijarut Lindberg, A., Strandberg, S., Blomqvist, L., Bergh, A., Friedrich, B., Axelsson, J., Ögren, M., Ögren, M., Widmark, A., Thellenberg Karlsson, C., Söderkvist, K., Riklund, K., Jonsson, J., and Nyholm, T., 2021. Registration of histopathology to magnetic resonance imaging of prostate cancer. *Physics and Imaging in Radiation Oncology*, *18*, pp. 19–25.

Sarin, S.K., Choudhury, A., Sharma, M.K., Maiwall, R., Al Mahtab, M., Rahman, S., Saigal, S., Saraf, N., Soin, A.S., Devarbhavi, H., and Kim, D.J., 2019. Acute-on-chronic liver failure: Consensus recommendations of the Asian Pacific association for the study of the liver (APASL): an update. *Hepatology international, 13*(4), pp. 353–390.

Sato, A., Isono, M., Asano, T., Okubo, K., and Asano, T., 2017. Panobinostat and ixazomib inhibit bladder cancer growth synergistically by increasing histone acetylation and inducing endoplasmic reticulum stress. *European Urology Supplements, 16*(3), pp. e1468–e1469.

Savas, S., and Skardasi, G., 2018. The SWI/SNF complex subunit genes: Their functions, variations, and links to risk and survival outcomes in human cancers. *Critical Reviews in Oncology/Hematology, 123,* pp.114–131.

Saxena, S., and Gyanchandani, M., 2020. Machine learning methods for computer-aided breast cancer diagnosis using histopathology: A narrative review. *Journal of Medical Imaging and Radiation Sciences, 51*(1), pp.182–193.

Schardt, J.A., Meyer, M., Hartmann, C.H., Schubert, F., Schmidt-Kittler, O., Fuhrmann, C., Polzer, B., Petronio, M., Eils, R., and Klein, C.A., 2005. Genomic analysis of single cytokeratin-positive cells from bone marrow reveals early mutational events in breast cancer. *Cancer Cell, 8*(3), pp. 227–239.

Schiffman, Joshua D., Paul G. Fisher, and Peter Gibbs (2015) Early detection of cancer: Past, present, and future. American Society of Clinical Oncology educational book/ASCO. *American Society of Clinical Oncology. Meeting. American Society of Clinical Oncology.*

Seierstad, T., Hole, K.H., Tulipan, A.J., Strømme, H., Lilleby, W., Revheim, M.E., and Hernes, E., 2021. 18F-Fluciclovine PET for Assessment of prostate cancer with histopathology as reference standard: A systematic review. *PET Clinics, 16*(2), pp.167–176.

Selivanova, G., 2004. p53: fighting cancer. *Current Cancer Drug Targets, 4*(5), pp. 385–402.

Sevenich, L., and Joyce, J.A., 2014. Pericellular proteolysis in cancer. *Genes & Development, 28*(21), pp. 2331–2347.

Seyfried, T.N., and Huysentruyt, L.C., 2013. On the origin of cancer metastasis. *Critical Reviews in Oncogenesis, 18*(1-2), p. 43.

Shahid, A., Huang, M., Yeung, S., Parsa, C., Orlando, R., Andresen, B.T., Travers, J.B., and Huang, Y., 2021. 061 Absence of Adrb2 minimally affects UV-induced immunosuppression and skin cancer development. *Journal of Investigative Dermatology, 141*(5), pp. S11.

Shen, H., Sun, C.C., Kang, L., Tan, X., Shi, P., Wang, L., Liu, E., and Gong, J., 2021. Low-dose salinomycin inhibits breast cancer metastasis by repolarizing tumor hijacked macrophages toward the M1 phenotype. *European Journal of Pharmaceutical Sciences, 157*, p. 105629.

Shen, Q., and Reedijk, M., 2021. Notch signaling and the breast cancer microenvironment. In *Notch Signaling in Embryology and Cancer* (pp. 183–200). Springer, Cham.

Sheng, W., Shi, X., Lin, Y., Tang, J., Jia, C., Cao, R., Sun, J., Wang, G., Zhou, L., and Dong, M., 2020. Musashi2 promotes EGF-induced EMT in pancreatic cancer via ZEB1-ERK/MAPK signaling. *Journal of Experimental & Clinical Cancer Research, 39*(1), pp. 1–15.

Sinha, S., Chatterjee, S.S., Biswas, M., Nag, A., Banerjee, D., De, R., and Sengupta, A., 2018. SWI/SNF subunit expression heterogeneity in human aplastic anemia stem/progenitors. *Experimental Hematology, 62,* pp. 39–44.e2.

Skvortsova, K., Masle-Farquhar, E., Luu, P.L., Song, J.Z., Qu, W., Zotenko, E., Gould, C.M., Du, Q., Peters, T.J., Colino-Sanguino, Y., Pidsley, R., Nair, S.S., Khoury, A., Smith, G.C., Miosge, L.A., Reed, J.H., Kench, J.G., Rubin, M.A., Horvath, L., Bogdanovic, O., Lim, S.M., Polo, J.M., Goodnow, C.C., Stirzaker, C., and Clark, S.J., 2019. DNA hypermethylation encroachment at CpG Island borders in cancer is predisposed by H3K4 monomethylation patterns. *Cancer Cell, 35*(2), pp. 297–314.e8.

Smeda, M., Przyborowski, K., Stojak, M., and Chlopicki, S., 2020. The endothelial barrier and cancer metastasis: Does the protective facet of platelet function matter? *Biochemical Pharmacology, 176,* pp.113886.

Smith, H.G., Thomas, J.M., Smith, M.J.F., Hayes, A.J., and Strauss, D.C., 2016. Multivisceral resection of retroperitoneal sarcomas in the elderly. *European Journal of Cancer, 69*, pp. 119–126.

Spencer, D.H., Russler-Germain, D.A., Ketkar, S., Helton, N.M., Lamprecht, T.L., Fulton, R.S., Fronick, C.C., O'Laughlin, M., Heath, S.E., Shinawi, M., Westervelt, P., Payton, J.E., Wartman, L.D., Welch, J.S., Wilson, R.K., Walter, M.J., Link, D.C., DiPersio, J.F., and Ley, T.J., 2017. CpG Island hypermethylation mediated by DNMT3A is a consequence of AML progression. *Cell, 168*(5), pp. 801–816.e13.

Spill, F., Reynolds, D.S., Kamm, R.D., and Zaman, M.H., 2016. Impact of the physical microenvironment on tumor progression and metastasis. *Current Opinion in Biotechnology, 40*, pp. 41–48.

Steeg, P.S., 2007. Micromanagement of metastasis. *Nature*, *449*(7163), pp. 671–673.

Stephens, S., Tollesson, G., Robertson, T., and Campbell, R., 2019. Diffuse midline glioma metastasis to the peritoneal cavity via ventriculo-peritoneal shunt: Case report and review of literature. *Journal of Clinical Neuroscience*, *67*, pp. 288–293.

Stuart, C.E., Singh, R.G., Ramos, G.C.A., Priya, S., Ko, J., DeSouza, S.V., Cho, J., and Petrov, M.S., 2020. Relationship of pancreas volume to tobacco smoking and alcohol consumption following pancreatitis. *Pancreatology*, *20*(1), pp. 60–67.

Subbotin, V.M., 2018. Privileged portal metastasis of hepatocellular carcinoma in light of the coevolution of a visceral portal system and liver in the chordate lineage: A search for therapeutic targets. *Drug Discovery Today*. doi: 10.1016/j.drudis.2018.01.020.

Subbotin, V.M., 2019. A hypothesis on paradoxical privileged portal vein metastasis of hepatocellular carcinoma. Can organ evolution shed light on patterns of human pathology, and vice versa? *Medical Hypotheses*, *126*, pp. 109–128.

Sundaram, P., Hultine, S., Smith, L.M., Dews, M., Fox, J.L., Biyashev, D., Schelter, J.M., Huang, Q., Cleary, M.A., Volpert, O.V., and Thomas-Tikhonenko, A., 2011. p53-responsive miR-194 inhibits thrombospondin-1 and promotes angiogenesis in colon cancers. *Cancer Research*, *71*(24), pp. 7490–7501.

Szymanski, M., Sarnowska, E., Ornoch, A., Rusetska, N., Abramowicz, S., Chrzan, A., Ligaj, M., Maassen, A., Siedlecki, J., and Sarnowski, T., 2017. Loss of SWI/SNF chromatin remodelling complex is linked to advanced urinary bladder cancer. *Annals of Oncology*, *28*, pp. v602.

Taieb, S., Saada-Bouzid, E., Tresch, E., Ryckewaert, T., Bompas, E., Italiano, A., Guillemet, C., Peugniez, C., Piperno-Neumann, S., Thyss, A., Maynou, C., Clisant, S., and Penel, N., 2015. Comparison of response evaluation criteria in solid tumours and Choi criteria for response evaluation in patients with advanced soft tissue sarcoma treated with trabectedin: A retrospective analysis. *European Journal of Cancer*. pp. 202–209. doi: 10.1016/j.ejca.2014.11.008.

Talamini, R., Polesel, J., Gallus, S., Dal Maso, L., Zucchetto, A., Negri, E., Bosetti, C., Lucenteforte, E., Boz, G., Franceschi, S., Serraino, D., and La Vecchia, C., 2010. Tobacco smoking, alcohol consumption and pancreatic cancer risk: A case-control study in Italy. *European Journal of Cancer*, *46*(2), pp. 370–376.

Tang, T., Abu-Sbeih, H., Richards, D.M., and Wang, Y., 2019. Tu1695 – anti-angiogenesis cancer therapy-related gastrointestinal injury can lead to serious adverse consequences in a small proportion of patients. *Gastroenterology*, *156*(6), pp. S-1088–S-1089.

Tape, C.J., 2016. Systems biology analysis of heterocellular signaling. *Trends in Biotechnology*, *34*(8), pp. 627–637.

Tape, C.J., Ling, S., Dimitriadi, M., McMahon, K.M., Worboys, J.D., Leong, H.S., Norrie, I.C., Miller, C.J., Poulogiannis, G., Lauffenburger, D.A., and Jørgensen, C., 2016. Oncogenic KRAS regulates tumor cell signaling via stromal reciprocation. *Cell*, *165*(4), pp. 910–920.

Tappero, S., Parodi, S., Mantica, G., Dotta, F., Ndrevataj, D., Pacchetti, A., Testino, N., Caviglia, A., Beverini, M., Malinaric, R., Ambrosini, F., Guano, G., Chierigo, F., Rebuffo, S., Traverso, P., Borghesi, M., Suardi, N., and Terrone, C., 2020. The quality of bladder resection improves the histological characterization of bladder cancer. an analysis based on rare variant histotypes. *European Urology Open Science*, *20*, pp. S159.

Tarver, T., 2012. *Cancer Facts & Figures 2012*. American Cancer Society (ACS). Atlanta, GA.

Taylor, W.R., and Stark, G.R., 2001. Regulation of the G2/M transition by p53. *Oncogene*, *20*(15), pp. 1803–1815.

Teleanu, R.I., Chircov, C., Grumezescu, A.M., and Teleanu, D.M., 2020. Tumor angiogenesis and anti-angiogenic strategies for cancer treatment. *Journal of Clinical Medicine*, *9*(1), p. 84.

Terpos, E. 2018. MultipleMyeloma: Clinical updates from the American Society of Hematology Annual Meeting, 2017. *Clinical Lymphoma, Myeloma and Leukemia*, *18*(5), pp. 321–334.

Tey, J.C., Soon, Y.Y., Vellayappan, B.A., and Ho, F., 2019. Palliative gastric radiotherapy with or without chemotherapy versus non-radiotherapy approaches for locally advanced or metastatic (or both) gastric cancer. *Cochrane Database of Systematic Reviews*, (10). doi: 10.1002/14651858.CD013450.

Thomas, R.K., Baker, A.C., DeBiasi, R.M., Winckler, W., LaFramboise, T., Lin, W.M., Wang, M., Feng, W., Zander, T., MacConaill, L.E., and Lee, J.C., 2007. High-throughput oncogene mutation profiling in human cancer. *Nature Genetics*, *39*(3), pp. 347–351.

Thomas, S.K., Lee, J., and Beatty, G.L., 2020. Paracrine and cell autonomous signalling in pancreatic cancer progression and metastasis. *EBioMedicine*, *53*, p. 102662.

Tie, J., Kinde, I., Wang, Y., Wong, H.L., Roebert, J., Christie, M., Tacey, M., Wong, R., Singh, M., Karapetis, C.S., Desai, J., Tran, B., Strausberg, R.L., Diaz, L.A. Jr, Papadopoulos, N., Kinzler, K.W.,

Vogelstein, B., and Gibbs, P. 2015. Circulating tumor DNA as an early marker of therapeutic response in patients with metastatic colorectal cancer. *Annals of Oncology 26*(8), pp. 1715–1722.

Triki, W., Kacem, A., Itami, A., Baraket, O., Rebai, M.H., and Bouchoucha, S., 2019. Penile metastasis of colon carcinoma: A rare case report. *Urology Case Reports, 24*. doi: 10.1016/j.eucr.2019.100875.

Ustrell-Borràs, M., Traboulsi-Garet, B., and Gay-Escoda, C., 2020. Alcohol-based mouthwash as a risk factor of oral cancer: A systematic review. *Medicina Oral, Patología Oral y Cirugía Bucal, 25*(1), p. e1.

Vähäkangas, K., 2003. Molecular epidemiology of human cancer risk. *Lung Cancer*, pp. 43–59. doi: 10.1002/(SICI)1096-9896(199901)187.

Van Bilsen, W.P., Kovaleva, A., Bleeker, M.C., King, A.J., Bruisten, S.M., Brokking, W., De Vries, H.J., Meijer, C.J., and Van Der Loeff, M.F.S., 2019. HPV infections and flat penile lesions of the penis in men who have sex with men. *Papillomavirus Research, 8*, p. 100173.

Vega, S., Morales, A.V., Ocaña, O.H., Valdés, F., Fabregat, I., and Nieto, M.A., 2004. Snail blocks the cell cycle and confers resistance to cell death. *Genes & Development, 18*(10), pp. 1131–1143.

Vega-Arroyo, M., Ramos-Peek, M.Á., Álvarez-Gamiño, C.T. de J., Meza-Berlanga, C., Kerik-Rotenberg, N.E., and Tena-Suck, M.L., 2019. Medulloblastoma with supratentorial and massive extraneural metastasis: Literature review in a case documented with 18-FDG PET. *Interdisciplinary Neurosurgery: Advanced Techniques and Case Management, 16*, pp. 117–122.

Viale, P.H., 2020. The American Cancer Society's facts & figures: 2020 edition. *Journal of the Advanced Practitioner in Oncology, 11*(2), p. 135.

Vishwakarma, R., and McManus, K.J., 2020. Chromosome instability; implications in cancer development, progression, and clinical outcomes. *Cancers, 12*(4), p. 824.

Vogelstein, B., Fearon, E.R., Hamilton, S.R., Kern, S.E., Preisinger, A.C., Leppert, M., Smits, A.M., and Bos, J.L., 1988. Genetic alterations during colorectal-tumor development. *New England Journal of Medicine, 319*(9), pp. 525–532.

Vogelstein, B., and Kinzler, K.W., 2004. Cancer genes and the pathways they control. *Nature Medicine, 10*(8), pp. 789–799.

Wang, B., Kohli, J., and Demaria, M., 2020. Senescent cells in cancer therapy: Friends or foes? *Trends in Cancer, 6*(10), pp. 838–857.

Wang, C., Mu, Z., Chervoneva, I., Austin, L., Ye, Z., Rossi, G., Palazzo, J.P., Sun, C., Abu-Khalaf, M., Myers, R.E., and Zhu, Z., 2017. Longitudinally collected CTCs and CTC-clusters and clinical outcomes of metastatic breast cancer. *Breast Cancer Research and Treatment, 161*(1), pp. 83–94.

Wang, D., Xin, Y., Tian, Y., Li, W., Sun, D., and Yang, Y., 2017. Pseudolaric acid B inhibits gastric cancer cell metastasis in vitro and in haematogenous dissemination model through PI3K/AKT, ERK1/2 and mitochondria-mediated apoptosis pathways. *Experimental Cell Research, 352*(1), pp. 34–44.

Wang, P., Lin, C., Smith, E.R., Guo, H., Sanderson, B.W., Wu, M., Gogol, M., Alexander, T., Seidel, C., Wiedemann, L.M., and Ge, K., 2009. Global analysis of H3K4 methylation defines MLL family member targets and points to a role for MLL1-mediated H3K4 methylation in the regulation of transcriptional initiation by RNA polymerase II. *Molecular and Cellular Biology, 29*(22), pp. 6074–6085.

Wang, W., Zhang, X.H., Li, M., Hao, C.H., and Liang, H.P., 2020. Association between vaginal infections and the types and viral loads of human papillomavirus: A clinical study based on 4,449 cases of gynecologic outpatients. *Canadian Journal of Infectious Diseases and Medical Microbiology, 2020*. doi: 10.1155/2020/9172908.

Ward, Y., Lake, R., Faraji, F., Sperger, J., Martin, P., Gilliard, C., Ku, K.P., Rodems, T., Niles, D., Tillman, H., Yin, J.J., Hunter, K., Sowalsky, A.G., Lang, J., and Kelly, K., 2018. Platelets promote metastasis via binding tumor CD97 leading to bidirectional signaling that coordinates transendothelial migration. *Cell Reports, 23*(3), pp. 808–822.

Wilson, A.S., Power, B.E., and Molloy, P.L., 2007. DNA hypomethylation and human diseases. *Biochimica et Biophysica Acta (BBA)-Reviews on Cancer, 1775*(1), pp. 138–162.

Witsch, E., Sela, M., and Yarden, Y., 2010. Roles for growth factors in cancer progression. *Physiology, 25*(2), pp. 85–101.

Wong, A.D., Russell, L.M., Katt, M.E., & Searson, P.C. (2018). Chemotherapeutic drug delivery and quantitative analysis of proliferation, apoptosis, and migration in a tissue-engineered three-dimensional microvessel model of the tumor microenvironment. *ACS Biomaterials Science & Engineering, 5*(2), pp. 633–643.

Wrenn, E.D., Yamamoto, A., Moore, B.M., Huang, Y., McBirney, M., Thomas, A.J., Greenwood, E., Rabena, Y.F., Rahbar, H., Partridge, S.C., and Cheung, K.J., 2020. Regulation of collective metastasis by nanolumenal signaling. *Cell, 183*(2), pp. 395–410.

Wu, James T., 1999. Review of circulating tumor markers: from enzyme, carcinoembryonic protein to oncogene and suppressor gene. *Annals of Clinical & Laboratory Science 29*(2), pp. 106–111.

Wu, L., Yao, C., Xiong, Z., Zhang, R., Wang, Z., Wu, Y., Qin, Q., and Hua, Y., 2016. The effects of a picosecond pulsed electric field on angiogenesis in the cervical cancer xenograft models. *Gynecologic Oncology*, *141*(1), pp. 175–181.

Wu, X., Liu, L., and Zhang, H., 2020. miR-802 inhibits the epithelial-mesenchymal transition, migration and invasion of cervical cancer by regulating BTF3. *Molecular Medicine Reports*, *22*(3), pp. 1883–1891.

Xi, Z., Zinman, L., Moreno, D., Schymick, J., Liang, Y., Sato, C., Zheng, Y., Ghani, M., Dib, S., Keith, J., Robertson, J., and Rogaeva, E., 2013. Hypermethylation of the CpG island near the G4C2 repeat in ALS with a C9orf72 expansion. *American Journal of Human Genetics*, *92*(6), pp. 981–989.

Xia, H., Zhao, Y.N., Yu, C.H., Zhao, Y.L., and Liu, Y., 2016. Inhibition of metabotropic glutamate receptor 1 suppresses tumor growth and angiogenesis in experimental non-small cell lung cancer. *European Journal of Pharmacology*, *783*, pp. 103–111.

Xue, C., Wyckoff, J., Liang, F., Sidani, M., Violini, S., Tsai, K.L., Zhang, Z.Y., Sahai, E., Condeelis, J., and Segall, J.E., 2006. Epidermal growth factor receptor overexpression results in increased tumor cell motility in vivo coordinately with enhanced intravasation and metastasis. *Cancer Research*, *66*(1), pp. 192–197.

Yamada, Y., Simon-Keller, K., Belharazem-Vitacolonnna, D., Bohnenberger, H., Kriegsmann, M., Kriegsmann, K., Hamilton, G., Graeter, T., Preissler, G., Ott, G., Roessner, E.D., Dahmen, I., Thomas, R.K., Ströbel, P., and Marx, A., 2021. A tuft cell–like signature is highly prevalent in thymic squamous cell carcinoma and delineates new molecular subsets among the major lung cancer histotypes. *Journal of Thoracic Oncology*, *16*(6), pp. 1003–1016.

Yamaguchi, J., Fujino, T., Isa, R., Nishiyama, D., Kuwahara-Ota, S., Kawaji, Y., Tsukamoto, T., Chinen, Y., Shimura, Y., Kobayashi, T., and Horiike, S., 2019. Epstein-Barr virus-associated lymphoproliferative disease during imatinib mesylate treatment for chronic myeloid leukemia. *Haematologica*, *104*(8), p. e376.

Yamamoto, G., Sakakibara-Konishi, J., Ikari, T., Kitai, H., Mizugaki, H., Asahina, H., Kikuchi, E., and Shinagawa, N., 2019. Response of BRAF V600E-mutant lung adenocarcinoma with brain metastasis and leptomeningeal dissemination to Dabrafenib Plus trametinib treatment. *Journal of Thoracic Oncology*, *14*(5), no. e97–e99.

Yen, C.Y., Huang, H.W., Shu, C.W., Hou, M.F., Yuan, S.S.F., Wang, H.R., Chang, Y.T., Farooqi, A.A., Tang, J.Y., and Chang, H.W., 2016. DNA methylation, histone acetylation and methylation of epigenetic modifications as a therapeutic approach for cancers. *Cancer Letters*, *373*(2), pp.185–192.

Yoon, J.H., McArthur, M.J., Park, J., Basu, D., Wakamiya, M., Prakash, L., and Prakash, S., 2019. Error-prone replication through UV lesions by DNA polymerase θ protects against skin cancers. *Cell*, *176*(6), 1295–1309.e15.

Yoon, J.Y., Sigel, K., Martin, J., Jordan, R., Beasley, M.B., Smith, C., Kaufman, A., Wisnivesky, J., and Kim, M.K., 2019. Evaluation of the prognostic significance of TNM staging guidelines in lung carcinoid tumors. *Journal of Thoracic Oncology*, *14*(2), pp. 184–192.

Yoshimoto, T., Matsubara, D., and Niki, T., 2017. P1.02-041 mutation of SWI/SNF complex genes is frequent in poorly differentiated, mesenchymal-like lung cancer without major driver mutation. *Journal of Thoracic Oncology*, *12*(11), pp. S1939–S1940.

Zanotelli, M.R., Chada, N.C., Johnson, C.A., and Reinhart-King, C.A., 2021. The Physical Microenvironment of Tumors: Characterization and Clinical Impact. In: *The Physics of Cancer: Research Advances* (pp. 165–195). World Scientific Publishing Co Pte Ltd.

Zefferino, R., Piccoli, C., Di Gioia, S., Capitanio, N., and Conese, M., 2019. Gap junction intercellular communication in the Carcinogenesis hallmarks: Is this a phenomenon or epiphenomenon?. *Cells*, *8*(8), p. 896.

Zhang, A., Xu, Y., Xu, H., Ren, J., Meng, T., Ni, Y., Zhu, Q., Zhang, W.B., Pan, Y.B., Jin, J., and Bi, Y., 2021. Lactate-induced M2 polarization of tumor-associated macrophages promotes the invasion of pituitary adenoma by secreting CCL17. *Theranostics*, *11*(8), p. 3839.

Zhang, G., Zhang, Y., He, F., Wu, H., Wang, C., and Fu, C., 2021. Preoperative controlling nutritional status (CONUT) score is a prognostic factor for early-stage cervical cancer patients with high-risk factors. *Gynecologic Oncology*, *162*(3), pp.763–769.

Zhang, L., and Li, S., 2020. Lactic acid promotes macrophage polarization through MCT-HIF1α signaling in gastric cancer. *Experimental Cell Research*, *388*(2), p. 111846.

Zhang, L., Zhang, S., Yao, J., Lowery, F.J., Zhang, Q., Huang, W.C., Li, P., Li, M., Wang, X., Zhang, C., and Wang, H., 2015. Microenvironment-induced PTEN loss by exosomal microRNA primes brain metastasis outgrowth. *Nature*, *527*(7576), pp. 100–104.

Zhao, H., Ahirwar, D.K., Oghumu, S., Wilkie, T., Powell, C.A., Nasser, M.W., Satoskar, A.R., Li, D.Y., and Ganju, R.K., 2016. Endothelial Robo4 suppresses breast cancer growth and metastasis through regulation of tumor angiogenesis. *Molecular Oncology*, *10*(2), pp. 272–281.

Zhao, T., Ding, X., Yan, C., and Du, H., 2017. Endothelial Rab7 GTPase mediates tumor growth and metastasis in lysosomal acid lipase– deficient mice. *Journal of Biological Chemistry*, *292*(47), pp. 19198–19208.

Zheng, X., Carstens, J.L., Kim, J., Scheible, M., Kaye, J., Sugimoto, H., Wu, C.C., LeBleu, V.S., and Kalluri, R., 2015. Epithelial-to-mesenchymal transition is dispensable for metastasis but induces chemoresistance in pancreatic cancer. *Nature*, *527*(7579), pp. 525–530.

Zhou, C., Jin, Y., Chen, Y., Huang, S., Huang, R., Wang, Y., Zhao, Y., Chen, Y., Guo, L., and Liao, J., 2021. Histopathology classification and localization of colorectal cancer using global labels by weakly supervised deep learning. *Computerized Medical Imaging and Graphics*, *88*. doi: 10.1016/j.compmedimag.2021.101861

Zubair, H., and Ahmad, A., 2017. Cancer metastasis: An introduction. In *Introduction to Cancer Metastasis* (pp. 3–12). Academic Press.

2 Cancer Incidence and Mortality: In India and Worldwide

2.1 INTRODUCTION

Globally, the major factor responsible for deaths is non-communicable diseases (NCDs). Cancer ranks at the number one position for causing death (Saei et al., 2018; Pinheiro et al., 2019; Brozek-Pluska, Dziki and Abramczyk, 2020; Chang et al., 2020; Kamal et al., 2020; Abd-Elnaby, Alfonse and Roushdy, 2021; Saab et al., 2021; Vanshika et al., 2021) and acts as a barrier for life expectancy in the world. According to the WHO Report of 2015, in 91 countries, cancer is the leading cause of death before the age of 70 years among 172 countries and ranks third in another 22 countries. The incidence and mortality rates of cancer are increasing rapidly worldwide. The reasons are multifaceted but actually, aging and population are the main reasons, as well as socioeconomic development (Gersten and Wilmoth, 2002; Bray, 2014). Periodic surveillance is the current displacement of cancer related to infection and poverty by highly frequent cancers in developed countries. These cancers are mostly due to the Westernization of lifestyle (Pingali, 2007; Colli, 2009; Tanaka et al., 2009; Bourke et al., 2011; Venier et al., 2011; Asadzadeh Vostakolaei et al., 2013; Shin et al., 2018; Murphy et al., 2019; Shankar et al., 2019; Maule and Merletti, 2002; Bray, 2014), yet the diverse geographical regions are also responsible for local risk factors. This is exemplified by the prominent rate differences of cancer associated with infections, including the cervix and stomach (Bray, 2014). According to GLOBOCAN 2018, cancer incidence and mortality estimates were provided by the International Agency for Research on Cancer (IARC) (Ferlay et al., 2018). It is also based primarily on the global cancer rate of incidence and mortality and geographical variability, as seen in earlier reports. The worldwide burden of cancer increased to approximately 14 million new cases per year in 2012, and is expected to rise with a rate of 22 million annually within the next two decades and it is predicted that cancer deaths are going to rise from 8.2 to 13 million annually. On the world level, the most common cancers diagnosed in 2012 were of the lung (1.8 million cases, 13.0% of the total), breast (1.7 million, 11.9%), and large bowel (1.4 million, 9.7%). Most deaths from cancer were of the lung (1.6 million, 19.4% of the total), liver (0.8 million, 9.1%), and stomach (0.7 million, 8.8%) (GLOBOCAN 2012).

2.2 SOURCES AND METHODS OF DATA

Global Cancer Observatory, available online at *gco.iarc.fr*, has been used for estimating the cancer incidence and mortality rate in GLOBOCAN 2018. The Global Cancer Observatory online portal provides the graphical conception and tabulation of GLOBOCAN databank of 185 countries and 36 types of cancers accordingly combined by age and sex. The global profile of cancer is made up in GLOBOCAN by exploring the best available cancer incidence and mortality sources in a given country. The cancer sites have been extended to 36 types in GLOBOCAN 2018, with one major addition of non-melanoma skin cancer (NMSC) (Ferlay et al., 2018). Therefore, overall cancer estimates are provided at a combined level of 185 countries worldwide categorized into sex and age.

Cancer incidence is the new number of cases arising in a definite period and geographical area, taken either in the form of an absolute number of cases per year or as a rate according to

DOI: 10.1201/9781003201946-2

100,000 individuals per year. Incidence data has been generated by population-based cancer registries (PBCRs) (Mohan and Lando, 2015; Soto-Perez-de-Celis and Chavarri-Guerra, 2016; Feliciano et al., 2019; Jabbaripour et al., 2019; Gupta et al., 2020; Tian et al., 2020; Ganguly et al., 2021; Piñeros et al., 2021). However, PBCRs may include national populations, and cover subnational areas, like particular urban areas, specifically in developing countries. According to IARC's Cancer incidence (Volume IX) in Five Continents, in 2010, around 15% of the world population was explored by different cancer registries, with the lowest rates in South America, 7.5%; 6.5% in Asia; and Africa at 1%. Mortality is produced by incidence and fatality rate. Therefore, the mortality rate given in this repository measures the average death risk of a specific cancer in a population (Bray, 2014).

2.3 WORLDWIDE DISTRIBUTION OF CASES AND DEATHS BY CANCER TYPES

Worldwide, a total of 18.1 million new cases and 9.6 million cancer deaths were estimated in 2018. Collectively for both sexes, around one-half of the new cases and more than one-half of the deaths due to cancer were projected in Asia in the year of 2018, representing 60% of the world's population (Eniu et al., 2019; Lojanapiwat et al., 2019; Arbyn et al., 2020; Atun et al., 2020; Ward, Scott, et al., 2020; Shen et al., 2021). Europe records 23.4% of new cases and 20.3% of deaths due to cancer, despite representing only 9% of the total population, followed by 21% incidence and 14.4% mortality rate of America. From the graphs in Figures 2.1 and 2.2, the worldwide number of new cases and deaths, combined by sex, have been shown in the year 2018. In the case of both sexes cumulative, the highest percentage of cases was found in the case of lung cancer (11.6% of total) and is also responsible for the maximum number of deaths, i.e., 18.4% of total deaths, closely followed with cases of female breast cancer (11.6%) and 10.2% (colorectal cancer) in incidence, and colorectal cancer (9.2%) and liver cancer (8.2%) for deaths (Bray, 2014).

2.4 WORLDWIDE CANCER PATTERNS BASED ON SEX

Globally, the combined incidence rate of all cancers was approximately 20% higher in men in comparison to women. For the sex-based categories (Table 2.1 and Figure 2.3), in the case of males, lung cancer was found to be responsible for the highest number of deaths (22%) as well as

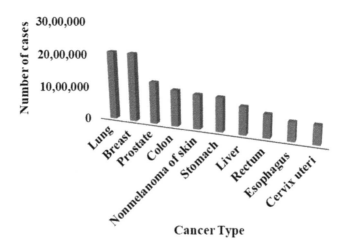

FIGURE 2.1 Graph showing the number of new cases of different cancer types (both sexes) worldwide in the year 2018 (see Bray et al., 2018).

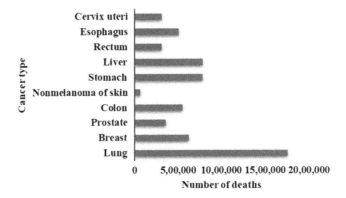

FIGURE 2.2   Graph showing number of deaths that occurred in the year 2018 due to different types of cancer (both sexes) worldwide (see Bray et al., 2018).

TABLE 2.1

Worldwide Highest Percentage of Different Cancer Types Based on Sex in the year 2018 (see Bray et al., 2018)

	Males			Females		
S. No.	Type of Cancer	Incidence (%)	Mortality (%)	Type of Cancer	Incidence (%)	Mortality (%)
1.	**Lung**	14.5	22.0	**Breast**	24.2	15.0
2.	Prostate	13.5	6.7	Colorectal	9.5	9.5
3.	Colorectal	10.9	9.0	Lung	8.4	13.8
4.	Stomach	7.2	9.5	Cervix uteri	6.6	7.5
5.	Liver	6.3	10.2	Stomach	4.1	6.5
6.	Bladder	4.5	2.8	Liver	2.8	5.6
7.	Esophagus	4.2	6.6	Ovary	3.4	4.4
8.	Non-Hodgkin Lymphoma	3.0	2.7			
9.	Leukemia	2.6	3.3			

had shown the highest incidence (14.5%), followed by prostate (13.5% incidence and 6.7% mortality) and colorectal cancer with 10.9% incidence and 9.0% mortality. Among females, the most frequently diagnosed cancer was breast cancer (24.2%) and found to be the leading reason for death (15.0%), followed by colorectal cancer (incidence 9.5% and mortality 9.5%) and lung cancer (incidence 8.4% and mortality 13.8%), while cervical cancer was ranked fourth with 6.6% incidence and 7.5% mortality (Bray, 2014).

2.5   CANCER DISTRIBUTION IN INDIA

In the year 2018, according to the data source of GLOBOCAN (2018), for both sexes (Figure 2.4), a maximum number of incidence and mortality in India was estimated for breast cancer with 14% and 11.1%, respectively, followed by lip and oral cavity cancer (10.4% incidence and 9.3% mortality), cervical cancer (8.4% incidence and 7.7% mortality), and lung cancer (5.9% incidence and 8.1% mortality).

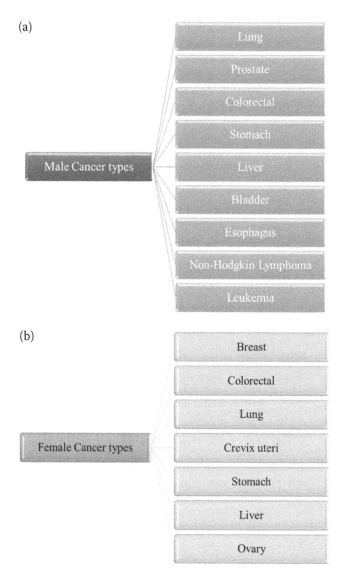

FIGURE 2.3   Different cancer types in (a) males and (b) females.

The sex-based distribution has shown that the maximum number of incidences among females (Figure 2.5) in 2018 was due to breast cancer (27.7% of total) followed by cervical (16.5% of total), ovary (6.2% of total), and lip and oral cavity cancer (4.8% of total). In the case of males (Figure 2.6), lip and oral cavity cancer (16.15 of total) ranks number one in incidence, followed by lung cancer (8.5% of total), stomach cancer (6.8% of total), and colorectal cancer (6.5% of total) (GLOBOCAN, 2018).

2.6   HUMAN DEVELOPMENT INDEX AND SITE-SPECIFIC CANCER

There is a strong relationship between the overall Human Development Index (HDI) and worldwide cancer burden, in the case of both sexes (Bray and Shield, 2016; Fidler, Bray, et al., 2017; Martínez-Mesa et al., 2017; Veisani et al., 2018; Znaor et al., 2018). This is mainly due to lifestyle and adaptation of behaviors in Western countries (Fidler et al., 2016). The incidence rate of the current cancer load is inordinate in most developed countries; and the countries, which

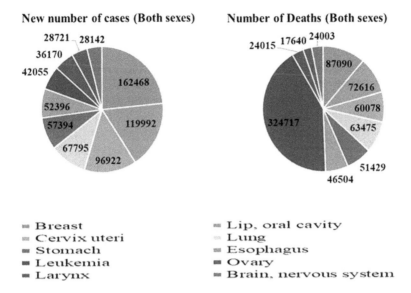

FIGURE 2.4   Pie chart showing the number of new cases and deaths of different cancer types (both sexes) occurred in India in the year 2018 (see GLOBOCAN, 2018).

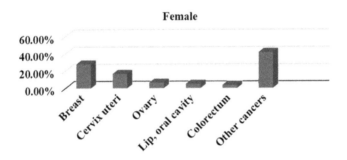

FIGURE 2.5   Graph showing the number of new cases in females due to different cancer types in India in the year 2018 (see GLOBOCAN, 2018).

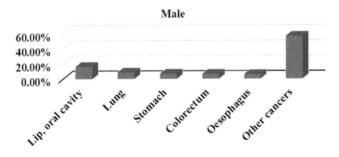

FIGURE 2.6   Graph showing the number of new cases in males due to different cancer types in India in the year 2018 (see GLOBOCAN, 2018).

are categorized into different levels of human development indices, experience the highest proportional upsurge in the future cancer load (Stewart and Wild, 2014). A particular estimation has predicted 100% and 81% cancer incidence increase in low and medium HDI countries until 2030 (Bray et al., 2012). Such estimated projections underline the necessity of targeted resource-dependent intrusions to reduce future cases, mainly in the developing world. The age-standardized mortality (Minicozzi et al., 2018; Bektur et al., 2019; Rutherford et al., 2020; Hessock et al., 2021; Kuzmickiene and Everatt, 2021; Lee et al., 2021; Molassiotis et al., 2021) was found to be comparable among the four HDI stages, which results in a higher proportion of the worldwide cancer mortality burden in medium and low HDI countries upon calculating the mortality to incidence ratio (Bray et al., 2012). These findings have suggested that there is huge inequality in cancer mortality and poor survival in the lower level of HDI countries due to lack of early diagnosis, cancer medicines, surgery, and radio- and chemotherapy facilities (Atun et al., 2015; Sullivan et al., 2015).

2.6.1 COLORECTAL CANCER

In previous decades, colorectal cancer has emerged as a human development marker, due to a parallel incidence increase in the HDI with time (Stewart and Wild, 2014; Arnold et al., 2017; Fidler, Bray, et al., 2017). In very high HDI countries, ASR (age-standardized incidence rate) was observed to be six times higher in comparison to low HDI countries in both sexes (Fidler, Bray, et al., 2017). Likewise, a positive mortality gradient was observed in spite of similar estimated death risk in high or very high HDI countries (Stewart and Wild, 2014; Arnold et al., 2017).

It was observed that a high prevalence of colorectal cancer is due to bad lifestyle factors, including Westernization, alcohol and red meat consumption, lack of physical activity, etc. (Asano and McLeod, 2002; Park et al., 2005; Larsson and Wolk, 2006; Moskal et al., 2007; Botteri, et al., 2008; Huncharek et al., 2008; Liang et al., 2009; Fedirko et al., 2011; Bagnardi et al., 2015; Bouvard et al., 2015).

2.6.2 BREAST CANCER

2.6.2.1 Incidence

Breast cancer has the second position in worldwide cancer and ranks first among women (Erlay et al., 2012; DeSantis et al., 2015; Ferlay et al., 2015; Torre et al., 2016). There is a chance of one in eight women developing breast cancer in their lifetime worldwide (DeSantis et al., 2017). In 2012, a global incidence of 1.67 million people with breast cancer was identified, accounting for 25% of all cancers (Erlay et al., 2012). Cancer can be found anywhere, irrespective of race, height, or country. However, the breast cancer incidence rate is higher in developed countries (Martínez-Mesa et al., 2017; Forjaz de Lacerda et al., 2018; Zhao et al., 2020; Gini et al., 2021; Wojtyla et al., 2021) and varies with ethnicity and race (DeSantis et al., 2014). The breast cancer incidence rate also varies in different parts of the world, with 27 per 100,000 in Africa and Asia and 92 per 100,000 in North America (Erlay et al., 2012; Ghoncheh et al., 2016). It is estimated that by 2050, the breast cancer incidence rate will increase to 3.2 million (Hortobagyi et al., 2005). In developed countries, breast cancer is increasing in older people (Hortobagyi et al., 2005; Tao et al., 2015; Li et al., 2016). In the United States, around 252,710 cases of invasive breast cancer and 6,341 in situ cases were diagnosed in 2017 (DeSantis et al., 2017). Approximately 24% of total breast cancer cases arise in the Asia-Pacific region with the most patients in China, Indonesia, and Japan (Youlden et al., 2014; Ghoncheh et al., 2016; Malvia et al., 2017).

An increasing trend of breast cancer cases was observed from 1998–2006 in Asian and American countries and Southeast Asia in 1988–2013, showing the highest prevalence of this deadly disease (Jemal et al., 2012; Gomez et al., 2017; Shoemaker et al., 2018). In 2012, a total of 277,054 new breast cancer cases were diagnosed in East Asia; in Southeast Asia, a total of 107,545 and in South-Central Asia 223,899 patients were estimated (Azubuike et al., 2018; Morey et al., 2019). Due to the advantageous screening and diagnosis methods, it became possible to increase the survival rate of breast cancer patients (Bourke et al., 2011). The 5-year survival rate was estimated at 89% (Rojas and Stuckey, 2016; Williams et al., 2018), while in European countries, this rate was 94.1% (Clèries et al., 2018;Momenimovahed and Salehiniya, 2019; Ruggeri et al., 2019). However, the survival rate scenario is entirely different in the African region due to lack of proper screening and treatment methods (Abdulrahman and Rahman, 2012; Balekouzou et al., 2016; Vanderpuye et al., 2017). The breast cancer incidence rate (age-standardized rate per 100,000) in different world regions is higher in developed regions, 74.1%, and 31.3% in less developed areas. In South-Central and Eastern Asia, the incidence rate of breast cancer was 28.2% and 27%, respectively (Bray et al., 2018; Torre et al., 2015; Torre et al., 2017).

2.6.2.2 Mortality

In 2012, breast cancer was ranked number five in leading cancer deaths, with 324,000 deaths in developing countries. However, in developed countries, death numbers reached 197,000, accounting for 15.4% of total cancer deaths (Erlay et al., 2012). It is estimated that higher mortality and breast cancer incidence rates will increase by 2020–21 in developed and developing countries (Grigoriadis, 2020). In the United States, 89% of deaths were observed in women above 50 years old in 2017 (DeSantis et al., 2017). Due to advancements in diagnostic and treatment methods, a significant reduction in incidence and mortality rates of breast cancer is observed in high-income countries (Carioli et al., 2017; Carioli et al., 2018). Breast cancer's age-standardized mortality rate (ASMR) is 12.9 (Kim et al., 2015) in the world. Africa has the highest ASMR (Azubuike et al., 2018). The mortality rate diverges from 6 in East Asia to 20 cases in Western Africa per 100,000 people (Erlay et al., 2012). In North America, the mortality-to-incidence rate ratio is 0.16, indicating a higher survival rate, while in Asia, it varies from 0.23–0.48 (Kim et al., 2015). Most Asian countries have low to middle income; hence, breast cancer causes most of the mortality in these countries (Sharma et al., 2011; Fan et al., 2015; Giri et al., 2018; Mubarik et al., 2020). The breast cancer mortality (age-standardized rate per 100,000) around the world is as follows: 14.9 in higher developed and 11.5 in less developed regions such as 16.2, 4.8, 16.4, . 13.5, and 6.1 in Western Europe, Northern America, Australia/New Zealand, South Central Asia and Eastern Asia, respectively (Torre et al., 2015).

In females, breast cancer has achieved the top rank in frequency across all the HDI levels as highest in prevalence and incidence (Parkin et al., 2011; Schütze et al., 2011; Bray et al., 2012; Stewart and Wild, 2014; Arnold et al., 2015). However, when comparing the ASR, a three times higher rate was observed in very high HDI countries than low HDI countries (Fidler, Gupta, et al., 2017). Hormonal and reproductive alterations and increased life expectancy are mainly responsible for the higher incidence of breast cancer in high HDI countries, along with Westernization, alcohol consumption, obesity, etc. (Parkin et al., 2011; Schütze et al., 2011; Arnold et al., 2015). However, the mortality gradient was found to be inconsistent in countries with high and very high HDI than in the medium HDI countries (Stewart and Wild, 2014).

2.6.3 PROSTATE CANCER

2.6.3.1 Incidence

There is a significant variation in prostate cancer incidence rates in different regions and populations (Ferlay et al., 2018). In 2018, 1,276,106 new prostate cancer cases were identified in the world, accounting for 7.1% of all cancers found in men (Bray et al., 2018). The incidence ASR was estimated to be the maximum in Oceania, that is, 79.1 per 100,000 people, followed by North America (73.7) and Europe (62.1). On the contrary, in Asia and Africa, the prostate cancer incidence rate is lower than the developed countries, that is, 26.6 and 11.5 (Bray et al., 2018). Differences of 190-fold were observed among the highest and lowest rate populations. Prostate cancer cases increase with age (Bray et al., 2018; C et al., 2021; Chao and Lepor, 2021; Cook et al., 2021; Deniz et al. 2021; Dhillon et al. 2021). However, it is predicted that only 1 out of 350 men will be diagnosed with this disease under 50 years of age (Perdana et al., 2017; Rawla, 2019). The incidence rate of prostate cancer upsurges to 1 in 52 men in the 50–59 age bracket. In men above

65 years, the incidence percentage rate is around 60% (Howlader et al., 2013). These variations could occur due to differences in PSA testing among the countries worldwide (Quinn and Babb, 2002). For example, in 2018, prostate cancer was the most frequently diagnosed cancer in Europe, with 450,000 new cases; 24% of all new cancers diagnosed (Crocetti, 2015). However, in the USA, it is the second-highest cancer analyzed commonly among men, with 164,690 new cases in 2018 (Cancer Stat Facts: SEER 2018). According to different research studies, about 20–40% of the prostate cancer cases in Europe and the USA arise due to overdiagnosis, i.e., comprehensive PSA testing (Draisma et al., 2009; Etzioni et al., 2002; Loeb et al., 2014; Butler et al., 2020; Jain and Sapra, 2020; Remmers and Roobol, 2020).

2.6.3.2 Mortality

Mortality rates of prostate cancer differ worldwide (Ferlay et al., 2018). In 2018, the highest rate of mortality per 100,000 was observed in Central America (10.7), followed by Australia (10.2) and Western Europe (10.1) (Ferlay et al., 2018). The lowest mortality rate was observed in South-Central (3.3) and Eastern Asia (4.7) and Northern Africa (5.8) (Ferlay et al., 2018). However, one-third of the prostate cancer deaths was reported in Asia with 33%, followed by Europe with 29.9% deaths. The mortality rate due to prostate cancer increases with age (Nelen, 2007; Vanacore et al., 2017; Pishgar et al., 2018; Taitt, 2018) and approximately 55% of all deaths occur over 65 years of age (Ferlay et al., 2018). It is observed that the highest proportion of prostate cancer mortality and incidence occurs in African-American men. This could be due to African-American men who have specific genes that are vulnerable to mutations that cause prostate cancer (Chakravarty et al., 2020; Herberts et al., 2020; Hosseinzadeh et al., 2020; Kheirkhah et al., 2020; Marciscano and Barbieri, 2020; Merrick et al., 2020; Nguyen et al., 2020; Peitzsch et al., 2020; van der Doelen et al., 2020; L. Zhang et al., 2020; Kumaraswamy et al., 2021). In the case of prostate cancer, a significant correlation was seen between the incidence rate and HDI with lower ASR in medium HDI countries than in low HDI countries (Coccia 2015; Pakzad et al. 2015; Hassanipour-Azgomi et al. 2016; Wong et al. 2016; Fidler, Gupta, et al., 2017; Luzzati et al. 2018; Huang et al. 2021). This data has shown greater susceptibility in black populations for prostate cancer (Delongchamps et al., 2007). On the contrary, higher ASR of high HDI countries corresponds to the highly prevalent usage of prostate-specific antigen testing (Baade et al., 2009). As this testing has a greater influence on incidence than mortality, therefore lesser variation has been observed in mortality rates worldwide. As a result, an insignificant correlation was observed between HDI and mortality.

2.6.4 CERVICAL CANCER

2.6.4.1 Incidence

This cancer is one of the leading causes of death due to cancer in women (Mathers, 2020; Mattiuzzi and Lippi, 2020; S. Zhang et al., 2020; Keykhaei et al., 2021). In the past 30 years, young women's proportion of cervical cancer has increased from 10–40% (Song et al., 2017).

WHO and IARC (International Agency for Research on Cancer) reported 529,000 new cervical cancer cases in the year 2008. In less developed countries, a total of 452,000 new cases of cervical cancer were estimated and ranked second among female malignancies cases (Ferlay et al., 2010). On the contrary, in developed countries, a total of 77,000 new cervical cancer cases were estimated, which was tenth among female malignancies. In 2018, cervical cancer ranked fourth for the highest number of cancer cases diagnosed and deaths, totaling 570,000 new cases and 311,000 deaths estimated in the world (Bray et al., 2018).

Though around 85% of the cervical cancer cases worldwide are from developing and low-income countries, the death rate is also 18 times higher in these countries than in developed countries (Prabhu and Eckert, 2016).

Cervical cancer ranks second, following breast cancer incidence and mortality rate in lower HDI settings. It is the highest diagnosed cancer among 28 countries and the leading cause of death in 42 countries, mainly in Sub-Saharan Africa and South-Eastern Asia (Ferlay et al., 2019). On comparing the regional incidence and mortality rates worldwide, the highest rate is seen in Africa. While in North America, Australia/New Zealand, and Western Asia, incidence and mortality rates are 7–10 times lower than in Africa (Small et al., 2017; Olorunfemi et al., 2018; Ferlay et al., 2019).

Inspite of being the most avoidable cancer type, the worldwide burden of cervical cancer varies across all the HDI levels widely. Particularly, the incidence rate of cervical cancer is negatively related to HDI (Fidler, Bray, et al., 2017). The biggest burden was found in low HDI countries, accounting for one-third of the total neoplastic cases diagnosed among both sexes (Sarker et al., 2016). It was observed that cervical cancer mortality and incidence could decrease down by 33% and 20% with 0.2 HDI increment. Moreover, the high mortality rate in low HDI sets is due to the late diagnosis and poor survival and this is directly linked to fragile health infrastructures and constrained accessibility of cancer services (Matsuo et al., 2018; Bjurberg et al., 2019; Huynh-Le et al., 2019; Ward, Grover, et al., 2020; Purohit et al., 2020; Abbas et al., 2021; Han et al., 2021; Pacífico de Carvalho et al., 2021; Qin et al., 2021; Zafar et al., 2021).

2.6.5 LUNG CANCER

2.6.5.1 Incidence

The incidence rate of lung cancer has been declining since the mid-1980s in men and since the 2000s in women due to gender variances in the historical form of smoking uptake and ending. The lung cancer incidence rate has been steadily decreasing by 2% per year, with a faster pace in men than women since the mid-2000s (American Cancer Society: Cancer Facts & Figures 2021).

Since 1990, the mortality trend has declined in lung cancer by 54% in men and by 30% in women since 2002 due to smoking reductions. This trend and the mortality rate decreased by more than 5% per year in men and 4% per year in women (American Cancer Society: Cancer Facts & Figures 2021).

2.6.5.2 Risk Factors

In the United States, the primary risk factor for lung cancer is cigarette smoking, which is responsible for 80% of deaths due to lung cancer. In addition, the duration and quantity of tobacco also increased the risk of lung cancer (Sung et al., 2021; Triplette et al., 2021). Other essential factors responsible for lung cancer are redon gas (collecting in indoor air); secondhand smoke; certain heavy metals such as chromium, cadmium, and arsenic; asbestos; air pollution; diesel; chimney proofing; rubber industries and paint industries; etc. (Rudin et al., 2021).

2.6.6 OVARIAN CANCER

2.6.6.1 Incidence

In 2020, around 21,750 new cases and 13,940 deaths of ovarian cancer were reported, accounting for 1.2% of total cancer cases. The 5-year survival rate of ovarian cancer is projected to be 48.6%, among which 15.7% of ovarian cancers are locally staged and 58% are at the metastasis stage. If cancer is detected early, the 5-year survival rate slopes down to 30.2% from 92.6% (Modugno and Edwards, 2012; Momenimovahed et al., 2019; Mancari et al., 2020).

The incidence of age-standardized rate is 11.1 in the United States during 2012–2016. Most commonly, the ovarian cancer type is epithelial with a serous subtype. It is predicted that around 21,410 new cases will be diagnosed in 2021 in the United States and approximately 13,770 women will die due to this disease. From the 1980s to 2017, the incidence rate of ovarian cancer dipped by 1% to 2% per year. In addition, the death rate of ovarian cancer declined by 2% per year during the years 2009–2018 (Torre et al., 2018; Siegel et al., 2020; Arora et al., 2021).

2.6.6.2 Risk Factors

Age is the most vital factor for ovarian cancer, alongside the family history of this disease (Ridgeway et al., 2017; Forman et al., 2018; Sundar et al., 2018; Wittenberg et al., 2018; Bennetsen et al., 2020; Cabasag et al., 2020; Zheng et al., 2020; Hemmingsen et al., 2021). Women having inherited mutations of BRCA1 or BRCA2 genes or genes related to Lynch syndrome are found to be at higher risk (Shulman, 2010; Minion et al., 2015; Hall et al., 2016; Evans et al., 2018; Gockley et al., 2018; Samtani and Saksena, 2019; Rush et al., 2020; Cowan et al., 2021; Hickey et al., 2021; Wong et al., 2021). Other medical characteristics responsible for ovarian cancer are endometriosis, inflammatory disease of the pelvis region, and personal and family breast or ovarian cancer. Additionally, other factors responsible for ovarian cancer are menopausal hormonal therapy or hormonal replacement therapy. In the mucinous subtype, cigarette smoking is responsible. Use of oral contraceptives, pregnancy, and fallopian tube removal are the factors responsible in the case of lower risk (Pujade-Lauraine et al., 2017; Friedlander et al., 2018).

REFERENCES

Abbas, M., Mehdi, A., Khan, F.H., Verma, S., Ahmad, A., Khatoon, F., Raza, S.T., Afreen, S., Glynn, S.A., and Mahdi, F., 2021. Role of miRNAs in cervical cancer: A comprehensive novel approach from pathogenesis to therapy. *Journal of Gynecology Obstetrics and Human Reproduction*, *50*(9), pp. 102159.

Abd-Elnaby, M., Alfonse, M., and Roushdy, M., 2021. Classification of breast cancer using microarray gene expression data: A survey. *Journal of Biomedical Informatics*. doi: 10.1016/j.jbi.2021.103764.

Abdulrahman, G.O., and Rahman, G.A., 2012. Epidemiology of breast cancer in Europe and Africa. *Journal of Cancer Epidemiology*, *2012*. doi: 10.1155/2012/915610.

American Cancer Society. Cancer Facts & Figures 2021. American Cancer Society, Atlanta.

Arbyn, M., Weiderpass, E., Bruni, L., de Sanjosé, S., Saraiya, M., Ferlay, J., and Bray, F., 2020. Estimates of incidence and mortality of cervical cancer in 2018: A worldwide analysis. *The Lancet Global Health*, *8*(2), pp. e191–e203.

Arnold, M., Pandeya, N., Byrnes, G., Renehan, A.G., Stevens, G.A., Ezzati, M., Ferlay, J., Miranda, J.J., Romieu, I., Dikshit, R., and Forman, D., 2015. Global burden of cancer attributable to high body-mass index in 2012: a population-based study. *The Lancet Oncology*, *16*(1), pp. 36–46.

Arnold, M., Sierra, M.S., Laversanne, M., Soerjomataram, I., Jemal, A., and Bray, F., 2017. Global patterns and trends in colorectal cancer incidence and mortality. *Gut*, *66*(4), pp. 683–691.

Arora, T., Mullangi, S., and Lekkala, M.R., 2021. Ovarian Cancer. *StatPearls [Internet]*.

Asadzadeh Vostakolaei, F., Broeders, M.J.M., Mousavi, S.M., Kiemeney, L.A.L.M., and Verbeek, A.L.M., 2013. The effect of demographic and lifestyle changes on the burden of breast cancer in Iranian women: A projection to 2030. *Breast*, *22*(3), pp. 277–281.

Asano, T.K., and McLeod, R.S., 2002. Dietary fibre for the prevention of colorectal adenomas and carcinomas. *Cochrane Database of Systematic Reviews*, (1), pp. CD003430.

Atun, R., Bhakta, N., Denburg, A., Frazier, A.L., Friedrich, P., Gupta, S., Lam, C.G., Ward, Z.J., Yeh, J.M., Allemani, C., Coleman, M.P., Di Carlo, V., Loucaides, E., Fitchett, E., Girardi, F., Horton, S.E., Bray, F., Steliarova-Foucher, E., Sullivan, R., Aitken, J.F., Banavali, S., Binagwaho, A., Alcasabas, P., Antillon, F., Arora, R.S., Barr, R.D., Bouffet, E., Challinor, J., Fuentes-Alabi, S., Gross, T., Hagander, L., Hoffman, R.I., Herrera, C., Kutluk, T., Marcus, K.J., Moreira, C., Pritchard-Jones, K., Ramirez, O., Renner, L., Robison, L.L., Shalkow, J., Sung, L., Yeoh, A., and Rodriguez-Galindo, C., 2020. Sustainable care for children with cancer: A Lancet Oncology Commission. *The Lancet Oncology*, *21*(4), pp. e185–e224.

Atun, R., Jaffray, D.A., Barton, M.B., Bray, F., Baumann, M., Vikram, B., Hanna, T.P., Knaul, F.M., Lievens, Y., Lui, T.Y., and Milosevic, M., 2015. Expanding global access to radiotherapy. *The Lancet Oncology*, *16*(10), pp. 1153–1186.

Azubuike, S.O., Muirhead, C., Hayes, L., and McNally, R., 2018. Rising global burden of breast cancer: the case of sub-Saharan Africa (with emphasis on Nigeria) and implications for regional development: A review. *World Journal of Surgical Oncology*, 16(1), pp. 1–13.

Baade, P.D., Youlden, D.R., and Krnjacki, L.J., 2009. International epidemiology of prostate cancer: geographical distribution and secular trends. *Molecular Nutrition & Food Research*, 53(2), pp. 171–184.

Bagnardi, V., Rota, M., Botteri, E., Tramacere, I., Islami, F., Fedirko, V., Scotti, L., Jenab, M., Turati, F., Pasquali, E., and Pelucchi, C., 2015. Alcohol consumption and site-specific cancer risk: A comprehensive dose–response meta-analysis. *British Journal Of Cancer*, 112(3), pp. 580–593.

Balekouzou, A., Yin, P., Pamatika, C.M., Bishwajit, G., Nambei, S.W., Djeintote, M., Ouansaba, B.E., Shu, C., Yin, M., Fu, Z., and Qing, T., 2016. Epidemiology of breast cancer: Retrospective study in the Central African Republic. *BMC Public Health*, 16(1), pp. 1–10.

Bektur, C., Dushimova, Z., Kaidarova, D., and Pennington, M.W., 2019. PCN171 ovarian cancer in kazakhstan: trends in age-standardised incidence and mortality rates. *Value in Health*, 22, S88.

Bennetsen, A.K.K., Baandrup, L., Aalborg, G.L., and Kjaer, S.K., 2020. Non-epithelial ovarian cancer in Denmark – incidence and survival over nearly 40 years. *Gynecologic Oncology*, 157(3), 693–699.

Bjurberg, M., Holmberg, E., Borgfeldt, C., Flöter-Rådestad, A., Dahm-Kähler, P., Hjerpe, E., Högberg, T., Kjølhede, P., Marcickiewicz, J., Rosenberg, P., Stålberg, K., Tholander, B., Hellman, K., and Åvall-Lundqvist, E., 2019. Primary treatment patterns and survival of cervical cancer in Sweden: A population-based Swedish Gynecologic Cancer Group Study. *Gynecologic Oncology*, 155(2), pp. 229–236.

Botteri, E., Iodice, S., Bagnardi, V., Raimondi, S., Lowenfels, A.B., and Maisonneuve, P., 2008. Smoking and colorectal cancer: a meta-analysis. *Jama*, 300(23), pp. 2765–2778.

Bourke, L., Thompson, G., Gibson, D.J., Daley, A., Crank, H., Adam, I., Shorthouse, A., and Saxton, J., 2011. Pragmatic lifestyle intervention in patients recovering from colon cancer: A randomized controlled pilot study. *Archives of Physical Medicine and Rehabilitation*, 92(5), pp. 749–755.

Bouvard, V., Loomis, D., Guyton, K.Z., Grosse, Y., Ghissassi, F.E., Benbrahim-Tallaa, L., Guha, N., Mattock, H., Straif, K., and Corpet, D., 2015. Carcinogenicity of consumption of red and processed meat. *The Lancet Oncology*, 16(16), pp. 1599–1600.

Bray, F., 2014. Transitions in human development and the global cancer burden. In *World Cancer Report 2014*, edited by Bernard W. Stewart, and Christopher P. Wild, pp. 54–68. Lyon: Naturaprint.

Bray, F., Ferlay, J., Soerjomataram, I., Siegel, R.L., Torre, L.A. and Jemal, A., 2018. Global cancer statistics 2018: GLOBOCAN estimates of incidence and mortality worldwide for 36 cancers in 185 countries. *CA: A Cancer Journal for Clinicians*, 68(6), pp. 394–424.

Bray, F., Jemal, A., Grey, N., Ferlay, J., and Forman, D., 2012. Global cancer transitions according to the Human Development Index (2008–2030): A population-based study. *The Lancet Oncology*, 13(8), pp. 790–801.

Bray, F., and Shield, K.D., 2016. Cancer: Global Burden, Trends, and Projections. *In: International Encyclopedia of Public Health* (pp. 347–368). Elsevier Inc.

Brozek-Pluska, B., Dziki, A., and Abramczyk, H., 2020. Virtual spectral histopathology of colon cancer - biomedical applications of Raman spectroscopy and imaging. *Journal of Molecular Liquids*, 303. doi: 10.1016/j.molliq.2020.112676.

Butler, E.N., Kelly, S.P., Coupland, V.H., Rosenberg, P.S., and Cook, M.B., 2020. Fatal prostate cancer incidence trends in the United States and England by race, stage, and treatment. *British Journal of Cancer*, 123(3), pp. 487–494.

C, J., Y, P., SF, B., and RJ, B., 2021. "More men die with prostate cancer than because of it" – an old adage that still holds true in the 21st century. *Cancer Treatment and Research Communications*, 26, p.100225.

Cabasag, C.J., Butler, J., Arnold, M., Rutherford, M., Bardot, A., Ferlay, J., Morgan, E., Møller, B., Gavin, A., Norell, C.H., Harrison, S., Saint-Jacques, N., Eden, M., Rous, B., Nordin, A., Hanna, L., Kwon, J., Cohen, P.A., Altman, A.D., Shack, L., Kozie, S., Engholm, G., De, P., Sykes, P., Porter, G., Ferguson, S., Walsh, P., Trevithick, R., Tervonen, H., O'Connell, D., Bray, F., and Soerjomataram, I., 2020. Exploring variations in ovarian cancer survival by age and stage (ICBP SurvMark-2): A population-based study. *Gynecologic Oncology*, 157(1), pp. 234–244.

Cancer Stat Facts: Prostate Cancer [Internet]. SEER, 2018. Available from: https://seer.cancer.gov/statfacts/html/prost.html.

Carioli, G., Malvezzi, M., Rodriguez, T., Bertuccio, P., Negri, E., and La Vecchia, C., 2017. Trends and predictions to 2020 in breast cancer mortality in Europe. *The Breast*, 36, pp. 89–95.

Carioli, G., Malvezzi, M., Rodriguez, T., Bertuccio, P., Negri, E., and La Vecchia, C., 2018. Trends and predictions to 2020 in breast cancer mortality: Americas and Australasia. *The Breast, 37,* pp. 163–169.

Chakravarty, D., Huang, L., Kahn, M., and Tewari, A.K., 2020. Immunotherapy for metastatic prostate cancer: Current and emerging treatment options. *Urologic Clinics of North America, 47*(4), pp. 487–510.

Chang, Y.W., Hsu, C.L., Tang, C.W., Chen, X.J., Huang, H.C., and Juan, H.F., 2020. Multiomics reveals ectopic ATP synthase blockade induces cancer cell death via a lncrna-mediated phospho-signaling network. *Molecular and Cellular Proteomics, 19* (11), pp. 1805–1825.

Chao, B., and Lepor, H., 2021. 5-year outcomes following focal laser ablation of prostate cancer. *Urology.* doi: 10.1016/j.urology.2021.03.054.

Clèries, R., Rooney, R.M., Vilardell, M., Espinàs, J.A., Dyba, T., and Borras, J.M., 2018. Assessing predicted age-specific breast cancer mortality rates in 27 European countries by 2020. *Clinical and Translational Oncology, 20*(3), pp. 313–321.

Coccia, M., 2015. The Nexus between technological performances of countries and incidence of cancers in society. *Technology in Society, 42,* pp. 61–70.

Colli, J., 2009. Can men reduce the risk of prostate cancer through lifestyle changes? *Journal of Urology, 182*(5), pp. 2101–2102.

Cook, M.B., Hurwitz, L.M., Geczik, A.M., and Butler, E.N., 2021. An up-to-date assessment of US prostate cancer incidence rates by stage and race: A novel approach combining multiple imputation with age and delay adjustment. *European Urology, 79* (1), pp. 33–41.

Cowan, R., Nobre, S.P., Pradhan, N., Yasukawa, M., Zhou, Q.C., Iasonos, A., Soslow, R.A., Arnold, A.G., Trottier, M., Catchings, A., Roche, K.L., Gardner, G., Robson, M., Abu Rustum, N.R., Aghajanian, C., and Cadoo, K., 2021. Outcomes of incidentally detected ovarian cancers diagnosed at time of risk-reducing salpingo-oophorectomy in BRCA mutation carriers. *Gynecologic Oncology, 161*(2), pp. 521–526.

Crocetti, E., 2015. *Epidemiology of Prostate Cancer in Europe.* Centre for Parliamentary Studies, European Union.

Delongchamps, N.B., Singh, A., and Haas, G.P., 2007. Epidemiology of prostate cancer in Africa: Another step in the understanding of the disease?. *Current Problems in Cancer, 31*(3), p. 226.

Deniz, M., Zengerling, F., Gundelach, T., Moreno-Villanueva, M., Bürkle, A., Janni, W., Bolenz, C., Kostezka, S., Marienfeld, R., Benckendorff, J., Friedl, T.W.P., Wiesmüller, L., and Rall-Scharpf, M., 2021. Age-related activity of Poly (ADP-Ribose) Polymerase (PARP) in men with localized prostate cancer. *Mechanisms of Ageing and Development, 196.* doi: 10.1016/j.mad.2021.111494.

DeSantis, C., Ma, J., Bryan, L., and Jemal, A., 2014. Breast cancer statistics, 2013. *CA: A Cancer Journal for Clinicians, 64*(1), pp. 52–62.

DeSantis, C.E., Bray, F., Ferlay, J., Lortet-Tieulent, J., Anderson, B.O., and Jemal, A., 2015. International variation in female breast cancer incidence and mortality rates. *Cancer Epidemiology and Prevention Biomarkers, 24*(10), pp. 1495–1506.

DeSantis, C.E., Ma, J., Goding Sauer, A., Newman, L.A., and Jemal, A., 2017. Breast cancer statistics, 2017, racial disparity in mortality by state. *CA: A Cancer Journal for Clinicians, 67*(6), pp. 439–448.

Dhillon, V.S., Deo, P., Bonassi, S., and Fenech, M., 2021. Lymphocyte micronuclei frequencies in skin, haematological, prostate, colorectal and esophageal cancer cases: A systematic review and meta-analysis. *Mutation Research – Reviews in Mutation Research.* doi: 10.1016/j.mrrev.2021.108372.

Draisma, G., Etzioni, R., Tsodikov, A., Mariotto, A., Wever, E., Gulati, R., Feuer, E., and De Koning, H., 2009. Lead time and overdiagnosis in prostate-specific antigen screening: importance of methods and context. *Journal of the National Cancer Institute, 101*(6), pp. 374–383.

Eniu, A., Cherny, N.I., Bertram, M., Thongprasert, S., Douillard, J.Y., Bricalli, G., Vyas, M., and Trapani, D., 2019. Cancer medicines in Asia and Asia-Pacific: What is available, and is it effective enough? *ESMO Open, 4*(4), pp. e000483.

Erlay, J., Ervik, M., Dikshit, R., Eser, S., and Mathers, C., 2012. Cancer incidence and mortality worldwide: IARC CancerBase No. 11. GLOBOCAN 2012 *v1.*

Etzioni, R., Penson, D.F., Legler, J.M., Di Tommaso, D., Boer, R., Gann, P.H., and Feuer, E.J., 2002. Overdiagnosis due to prostate-specific antigen screening: Lessons from US prostate cancer incidence trends. *Journal of the National Cancer Institute, 94*(13), pp. 981–990.

Evans, D.G.R., van Veen, E.M., Byers, H.J., Wallace, A.J., Ellingford, J.M., Beaman, G., Santoyo-Lopez, J., Aitman, T.J., Eccles, D.M., Lalloo, F.I., Smith, M.J., and Newman, W.G., 2018. A dominantly inherited 5′ UTR variant causing methylation-associated silencing of BRCA1 as a cause of breast and ovarian cancer. *American Journal of Human Genetics, 103*(2), pp. 213–220.

39

Fan, L., Goss, P.E., and Strasser-Weippl, K., 2015. Current status and future projections of breast cancer in Asia. *Breast Care*, *10*(6), pp. 372–378.

Fedirko, V., Tramacere, I., Bagnardi, V., Rota, M., Scotti, L., Islami, F., Negri, E., Straif, K., Romieu, I., La Vecchia, C., and Boffetta, P., 2011. Alcohol drinking and colorectal cancer risk: An overall and dose–response meta-analysis of published studies. *Annals Of Oncology*, *22*(9), pp. 1958–1972.

Feliciano, S.V.M., Santos, M. de O., Pombo-de-Oliveira, M.S., de Aquino, J.Â.P., de Aquino, T.A., Arregi, M.M.U., Antoniazzif, B.N., da Costa, A.M., Formigosa, L.A.C., Laporte, C.A., Lima, C.A., Machado, N.C., de Oliveira, J.C., Pereira, L.D.A., de Souza, A., dos Santos, C.M.A., de Souza, P.C.F., and Venezian, D.B., 2019. Incidence and mortality of myeloid malignancies in children, adolescents and Young adults in Brazil: A population-based study. *Cancer Epidemiology*, *62*. doi: 10.1016/j.canep.2019.101583.

Ferlay, J., Colombet, M., and Soerjomataram, I., 2018. Global and regional estimates of the incidence and mortality for 38 cancers. GLOBOCAN, 2018.

Ferlay, J., Colombet, M., Soerjomataram, I., Mathers, C., Parkin, D.M., Piñeros, M., Znaor, A., and Bray, F., 2019. Estimating the global cancer incidence and mortality in 2018: GLOBOCAN sources and methods. *International Journal of Cancer*, *144*(8), pp. 1941–1953.

Ferlay, J., Ervik, M., Lam, F., Colombet, M., Mery, L., Piñeros, M., Znaor, A., Soerjomataram, I., and Bray, F., 2018. *Global Cancer Observatory: Cancer Today*. International Agency for Research on Cancer, Lyon, France.

Ferlay, J., Shin, H.R., Bray, F., Forman, D., Mathers, C., and Parkin, D.M., 2010. Estimates of worldwide burden of cancer in 2008: GLOBOCAN 2008. *International Journal of Cancer*, *127*(12), pp. 2893–2917.

Ferlay, J., Soerjomataram, I., Dikshit, R., Eser, S., Mathers, C., Rebelo, M., Parkin, D.M., Forman, D., and Bray, F., 2015. Cancer incidence and mortality worldwide: sources, methods and major patterns in GLOBOCAN 2012. *International Journal of Cancer*, *136*(5), pp. E359–E386.

Ferlay, J., Soerjomataram, I., Ervik, M., Dikshit, R., Eser, S., Mathers, C., Rebelo, M., Parkin, D.M., Forman, D., & Bray, F. 2014. GLOBOCAN 2012 v1. 0, Cancer Incidence and Mortality Worldwide: IARC CancerBase No. 11 [Internet]. 2013. Lyon, France: International Agency for Research on Cancer. In globocan.iarc. fr/Default.aspx

Fidler, M.M., Bray, F., Vaccarella, S., and Soerjomataram, I., 2017. Assessing global transitions in human development and colorectal cancer incidence. *International Journal of Cancer*, *140*(12), pp. 2709–2715.

Fidler, M.M., Gupta, S., Soerjomataram, I., Ferlay, J., Steliarova-Foucher, E., and Bray, F., 2017. Cancer incidence and mortality among young adults aged 20–39 years worldwide in 2012: a population-based study. *The Lancet Oncology*, *18*(12), pp. 1579–1589.

Fidler, M.M., Soerjomataram, I., and Bray, F., 2016. A global view on cancer incidence and national levels of the human development index. *International Journal Of Cancer*, *139*(11), pp. 2436–2446.

Forjaz de Lacerda, G., Kelly, S.P., Bastos, J., Castro, C., Mayer, A., Mariotto, A.B., and Anderson, W.F., 2018. Breast cancer in Portugal: Temporal trends and age-specific incidence by geographic regions. *Cancer Epidemiology*, *54*, pp. 12–18.

Forman, D., Bauld, L., Bonanni, B., Brenner, H., Brown, K., Dillner, J., Kampman, E., Manczuk, M., Riboli, E., Steindorf, K., Storm, H., Espina, C., and Wild, C.P., 2018. Time for a European initiative for research to prevent cancer: A manifesto for Cancer Prevention Europe (CPE). *Journal of Cancer Policy*, *17*, pp. 15–23.

Friedlander, M., Gebski, V., Gibbs, E., Davies, L., Bloomfield, R., Hilpert, F., Wenzel, L.B., Eek, D., Rodrigues, M., Clamp, A., and Penson, R.T., 2018. Health-related quality of life and patient-centred outcomes with olaparib maintenance after chemotherapy in patients with platinum-sensitive, relapsed ovarian cancer and a BRCA1/2 mutation (SOLO2/ENGOT Ov-21): A placebo-controlled, phase 3 randomised trial. *The Lancet Oncology*, *19*(8), pp. 1126–1134.

Ganguly, S., Kinsey, S., and Bakhshi, S., 2021. Childhood cancer in India. *Cancer Epidemiology*, *71*, pp.101679.

Gersten, O., and Wilmoth, J.R., 2002. The cancer transition in Japan since 1951. *Demographic Research*, *7*, pp. 271–306.

Ghonceh, M., Pournamdar, Z., and Salehiniya, H., 2016. Incidence and mortality and epidemiology of breast cancer in the world. *Asian Pacific Journal of Cancer Prevention*, *17*(sup3), pp. 43–46.

Gini, A., van Ravesteyn, N.T., Jansen, E.E.L., Heijnsdijk, E.A.M., Senore, C., Anttila, A., Novak Mlakar, D., Veerus, P., Csanádi, M., Zielonke, N., Heinävaara, S., Széles, G., Segnan, N., de Koning, H.J., and Lansdorp-Vogelaar, I., 2021. The EU-TOPIA evaluation tool: An online modelling-based tool for informing breast, cervical, and colorectal cancer screening decisions in Europe. *Preventive Medicine Reports*, *22*, pp. 101392.

Giri, M., Giri, M., Thapa, R.J., Upreti, B., and Pariyar, B., 2018. Breast Cancer in Nepal: Current status and future directions. *Biomedical reports*, *8*(4), pp. 325–329.

GLOBOCAN, 2018: https://gco.iarc.fr/.

Gockley, A.A., Kolin, D.L., Awtrey, C.S., Lindeman, N.I., Matulonis, U.A., and Konstantinopoulos, P.A., 2018. Durable response in a woman with recurrent low-grade endometrioid endometrial cancer and a germline BRCA2 mutation treated with a PARP inhibitor. *Gynecologic Oncology*, *150*(2), pp. 219–226.

Gomez, S.L., Von Behren, J., McKinley, M., Clarke, C.A., Shariff-Marco, S., Cheng, I., Reynolds, P., and Glaser, S.L., 2017. Breast cancer in Asian Americans in California, 1988–2013: increasing incidence trends and recent data on breast cancer subtypes. Breast cancer research and treatment, *164*(1), pp. 139–147.

Grigoriadis, G., 2020. *Physiological Factors Impacting Fitness in Breast Cancer Survivors* (Doctoral dissertation, University of Illinois at Chicago).

Gupta, S., Aitken, J., Bartels, U., Bhakta, N., Bucurenci, M., Brierley, J.D., De Camargo, B., Chokunonga, E., Clymer, J., Coza, D., Fraser, C., Fuentes-Alabi, S., Gatta, G., Gross, T., Jakab, Z., Kohler, B., Kutluk, T., Moreno, F., Nakata, K., Nur, S., Parkin, D.M., Penberthy, L., Pole, J., Poynter, J.N., Pritchard-Jones, K., Ramirez, O., Renner, L., Steliarova-Foucher, E., Sullivan, M., Swaminathan, R., Van Eycken, L., Vora, T., and Frazier, A.L., 2020. Development of paediatric non-stage prognosticator guidelines for population-based cancer registries and updates to the 2014 Toronto Paediatric Cancer Stage Guidelines. *The Lancet Oncology*, *21*(9), pp. e444–e451.

Hall, M.J., Obeid, E.I., Schwartz, S.C., Mantia-Smaldone, G., Forman, A.D., and Daly, M.B., 2016. Genetic testing for hereditary cancer predisposition: BRCA1/2, Lynch syndrome, and beyond. *Gynecologic Oncology*, *140*(3), pp. 565–574.

Han, X., Liu, S., Yang, G., Hosseinifard, H., Imani, S., Yang, L., Maghsoudloo, M., Fu, S.Z., Wen, Q.L., and Liu, Q., 2021. Prognostic value of systemic hemato-immunological indices in uterine cervical cancer: A systemic review, meta-analysis, and meta-regression of observational studies. *Gynecologic Oncology*, *160*(1), pp. 351–360.

Hassanipour-Azgomi, S., Mohammadian-Hafshejani, A., Ghoncheh, M., Towhidi, F., Jamehshorani, S., and Salehiniya, H., 2016. Incidence and mortality of prostate cancer and their relationship with the Human Development Index worldwide. *Prostate International*, *4*(3), pp. 118–124.

Hemmingsen, C.H., Kjaer, S.K., Bennetsen, A.K.K., Dehlendorff, C., and Baandrup, L., 2021. The association of reproductive factors with risk of non-epithelial ovarian cancer and comparison with serous ovarian cancer. *Gynecologic Oncology*, *162*(2), pp. 469–474.

Herberts, C., Murtha, A.J., Fu, S., Wang, G., Schönlau, E., Xue, H., Lin, D., Gleave, A., Yip, S., Angeles, A., Hotte, S., Tran, B., North, S., Taavitsainen, S., Beja, K., Vandekerkhove, G., Ritch, E., Warner, E., Saad, F., Iqbal, N., Nykter, M., Gleave, M.E., Wang, Y., Annala, M., Chi, K.N., and Wyatt, A.W., 2020. Activating AKT1 and PIK3CA mutations in metastatic castration-resistant prostate cancer. *European Urology*, *78*(6), pp. 834–844.

Hessock, M., Brewer, T., Hutson, S., and Anderson, J., 2021. Use of a standardized tool to identify women at risk for hereditary breast and ovarian cancer. *Nursing for Women's Health*, *25*(3), pp. 187–197.

Hickey, I., Jha, S., and Wyld, L., 2021. The psychosexual effects of risk-reducing bilateral salpingo-oophorectomy in female BRCA1/2 mutation carriers: A systematic review of qualitative studies. *Gynecologic Oncology*. doi: 10.1016/j.ygyno.2020.12.001.

Hortobagyi, G.N., de la Garza Salazar, J., Pritchard, K., Amadori, D., Haidinger, R., Hudis, C.A., Khaled, H., Liu, M.C., Martin, M., Namer, M., and O'Shaughnessy, J.A., 2005. The global breast cancer burden: Variations in epidemiology and survival. *Clinical breast cancer*, *6*(5), pp. 391–401.

Hosseinzadeh, O., Hekmat, Z., Nekoufar, S., Ahmad, M., Mohammadzadeh, N., and Monfaredan, A., 2020. Evaluate the gene expression of TPT1, EDN3, and ANO7 in prostate cancer tissues and their relation with age, tumor stage and family history. *Meta Gene*, *24*. doi: 10.1016/j.mgene.2020.100671.

Howlader, N.N.A.K.M., Noone, A.M., Krapcho, M., Garshell, J., Neyman, N., Altekruse, S.F., Kosary, C.L., Yu, M., Ruhl, J., Tatalovich, Z., and Cho, H., 2013. SEER cancer statistics review, 1975–2010. *Bethesda, MD: National Cancer Institute*, *21*, p. 12.

Huang, J., Lok, V., Ngai, C.H., Zhang, L., Yuan, J., Lao, X.Q., Ng, K., Chong, C., Zheng, Z.J., and Wong, M.C.S., 2021. Worldwide burden of, risk factors for, and trends in pancreatic cancer. *Gastroenterology*, *160*(3), pp. 744–754.

Huncharek, M., Muscat, J., and Kupelnick, B., 2008. Colorectal cancer risk and dietary intake of calcium, vitamin D, and dairy products: A meta-analysis of 26,335 cases from 60 observational studies. *Nutrition and Cancer*, *61*(1), pp. 47–69.

Huynh-Le, M.P., Klapheke, A., Cress, R., Mell, L.K., Yashar, C.M., Einck, J.P., Mundt, A.J., and Mayadev, J.S., 2019. Impact of marital status on receipt of brachytherapy and survival outcomes in locally advanced cervical cancer. *Brachytherapy*, *18*(5), pp. 612–619.

Jabbaripour, P., Sanaat, Z., Dolatkhah, R., and Somi, M.H., 2019. Epidemiologic profile of gastric cancer in East Azerbaijan, Iran: 2 years population-based cancer registry results. *Annals of Oncology*, *30*, pp. ix62.

Jain, M.A., and Sapra, A., 2020. *Cancer Prostate Screening*. StatPearls [Internet].

Jemal, A., Bray, F., Forman, D., O'Brien, M., Ferlay, J., Center, M., and Parkin, D.M., 2012. Cancer burden in Africa and opportunities for prevention. *Cancer*, *118*(18), pp. 4372–4384.

Kamal, A., Mohsen, A., Kamal, A., and Siam, I., 2020. Study of the association between Urotensin 2 (p.T21M and p.S89N) variants and breast cancer in Egyptian patients. *Gene Reports*, *20*(6750), p.100789.

Keykhaei, M., Masinaei, M., Mohammadi, E., Azadnajafabad, S., Rezaei, N., Moghaddam, S.S., Rezaei, N., Nasserinejad, M., Abbasi-Kangevari, M., Malekpour, M.R., and Ghamari, S.H., 2021. A global, regional, and national survey on burden and Quality of Care Index (QCI) of hematologic malignancies; global burden of disease systematic analysis 1990–2017. *Experimental Hematology & Oncology*, *10*(1), pp. 1–15.

Kheirkhah, S., Javanzad, M., Hoseinzadeh, M., Hekmati Azar Mehrabani, Z., Mohammadzadeh, N., and Monfaredan, A., 2020. Monitoring prostate cancer (PCa) with appraise the gene expression of PRUNE2, NCAPD3 and ASPA and their connection with age, family history and tumor stage. *Gene Reports*, *21*(14), p.100840.

Kim, Y., Yoo, K.Y., and Goodman, M.T., 2015. Differences in incidence, mortality and survival of breast cancer by regions and countries in Asia and contributing factors. *Asian Pacific Journal of Cancer Prevention*, *16*(7), pp. 2857–2870.

Kumaraswamy, A., Welker Leng, K.R., Westbrook, T.C., Yates, J.A., Zhao, S.G., Evans, C.P., Feng, F.Y., Morgan, T.M., and Alumkal, J.J., 2021. Recent advances in epigenetic biomarkers and epigenetic targeting in prostate cancer. *European Urology*, *80*(1), pp. 71–81.

Kuzmickiene, I., and Everatt, R., 2021. Trends and age-period-cohort analysis of upper aerodigestive tract and stomach cancer mortality in Lithuania, 1987–2016. *Public Health*, *196*, pp. 62–68.

Larsson, S.C., and Wolk, A., 2006. Meat consumption and risk of colorectal cancer: a meta-analysis of prospective studies. *International Journal Of Cancer*, *119*(11), pp. 2657–2664.

Lee, C.L., Huang, K.G., Chua, P.T., Mendoza, M.C.V.R., Lee, P.S., and Lai, S.Y., 2021. Standardization and experience may influence the survival of laparoscopic radical hysterectomy for cervical cancer. *Taiwanese Journal of Obstetrics and Gynecology*, *60*(3), pp. 463–467.

Li, T., Mello-Thoms, C., and Brennan, P.C., 2016. Descriptive epidemiology of breast cancer in China: Incidence, mortality, survival and prevalence. *Breast Cancer Research and Treatment*, *159*(3), pp. 395–406.

Liang, P.S., Chen, T.Y., and Giovannucci, E., 2009. Cigarette smoking and colorectal cancer incidence and mortality: Systematic review and meta-analysis. *International Journal Of Cancer*, *124*(10), pp. 2406–2415.

Loeb, S., Bjurlin, M.A., Nicholson, J., Tammela, T.L., Penson, D.F., Carter, H.B., Carroll, P., and Etzioni, R., 2014. Overdiagnosis and overtreatment of prostate cancer. *European urology*, *65*(6), pp. 1046–1055.

Lojanapiwat, B., Lee, J.Y., Gang, Z., Kim, C.S., Fai, N.C., Hakim, L., Umbas, R., Ong, T.A., Lim, J., Letran, J.L., Chiong, E., Lee, S.H., Türkeri, L., Murphy, D.G., Moretti, K., Cooperberg, M., Carlile, R., Hinotsu, S., Hirao, Y., Kitamura, T., Horie, S., Onozawa, M., Kitagawa, Y., Namiki, M., Fukagai, T., Miyazaki, J., and Akaza, H., 2019. Report of the third Asian Prostate Cancer study meeting. *Prostate International*, *7*(2), pp. 60–67.

Luzzati, T., Parenti, A., and Rughi, T., 2018. Economic growth and cancer incidence. *Ecological Economics*, *146*, pp. 381–396.

Malvia, S., Bagadi, S.A., Dubey, U.S., and Saxena, S., 2017. Epidemiology of breast cancer in Indian women. *Asia-Pacific Journal of Clinical Oncology*, *13*(4), pp. 289–295.

Mancari, R., Cutillo, G., Bruno, V., Vincenzoni, C., Mancini, E., Baiocco, E., Bruni, S., Vocaturo, G., Chiofalo, B., and Vizza, E., 2020. Development of new medical treatment for epithelial ovarian cancer recurrence. *Gland Surgery*, *9*(4), p. 1149.

Marciscano, A.E., and Barbieri, C.E., 2020. CDK12 gene alterations in prostate cancer: Present, but clinically actionable? *European Urology*, *78*(5), pp. 680–681.

Martínez-Mesa, J., Werutsky, G., Stefan, M., Pereira Filho, C.A.S., Dueñas-González, A., Zarba, J.J., Mano, M., Villarreal-Garza, C., Gómez, H., and Barrios, C.H., 2017. Exploring disparities in incidence and mortality rates of breast and gynecologic cancers according to the Human Development Index in the Pan-American region. *Public Health, 149*, pp. 81–88.

Mathers, C.D., 2020. History of global burden of disease assessment at the World Health Organization. *Archives of Public Health, 78*(1), pp. 1–13.

Matsuo, K., Blake, E.A., Machida, H., Mandelbaum, R.S., Roman, L.D., and Wright, J.D., 2018. Incidences and risk factors of metachronous vulvar, vaginal, and anal cancers after cervical cancer diagnosis. *Gynecologic Oncology, 150*(3), pp. 501–508.

Mattiuzzi, C., and Lippi, G., 2020. Cancer statistics: a comparison between world health organization (WHO) and global burden of disease (GBD). *European journal of public health, 30*(5), pp. 1026–1027.

Maule, M., and Merletti, F., 2012. Cancer transition and priorities for cancer control. *The Lancet Oncology, 8*(13), pp. 745–746.

Merrick, B.A., Phadke, D.P., Bostrom, M.A., Shah, R.R., Wright, G.M., Wang, X., Gordon, O., Pelch, K.E., Auerbach, S.S., Paules, R.S., DeVito, M.J., Waalkes, M.P., and Tokar, E.J., 2020. KRAS-retroviral fusion transcripts and gene amplification in arsenic-transformed, human prostate CAsE-PE cancer cells. *Toxicology and Applied Pharmacology, 397*, p.115017.

Minicozzi, P., Cassetti, T., Vener, C., and Sant, M., 2018. Analysis of incidence, mortality and survival for pancreatic and biliary tract cancers across Europe, with assessment of influence of revised European age standardisation on estimates. *Cancer Epidemiology, 55*, pp. 52–60.

Minion, L.E., Dolinsky, J.S., Chase, D.M., Dunlop, C.L., Chao, E.C., and Monk, B.J., 2015. Hereditary predisposition to ovarian cancer, looking beyond BRCA1/BRCA2. *Gynecologic Oncology, 137*(1), pp. 86–92.

Modugno, F., and Edwards, R.P., 2012. Ovarian cancer: Prevention, detection, and treatment of the disease and its recurrence. Molecular mechanisms and personalized medicine meeting report. *International Journal of Gynecologic Cancer, 22*(Supp 2). doi: 10.1097/IGC.0b013e31826bd1f2.

Mohan, P., and Lando, H.A., 2015. Cancer registries in oral cancer control in India. *Journal of Cancer Policy, 4*, pp. 13–14.

Molassiotis, A., Tyrovolas, S., Giné-Vázquez, I., Yeo, W., Aapro, M., and Herrstedt, J., 2021. Organized breast cancer screening not only reduces mortality from breast cancer but also significantly decreases disability-adjusted life years: analysis of the Global Burden of Disease Study and screening programme availability in 130 countries. *ESMO Open, 6*(3), pp. 100111.

Momenimovahed, Z., and Salehiniya, H., 2019. Epidemiological characteristics of and risk factors for breast cancer in the world. *Breast Cancer: Targets and Therapy, 11*, p. 151.

Momenimovahed, Z., Tiznobaik, A., Taheri, S., and Salehiniya, H., 2019. Ovarian cancer in the world: epidemiology and risk factors. *International Journal of Women's Health, 11*, p. 287.

Morey, B.N., Gee, G.C., von Ehrenstein, O.S., Shariff-Marco, S., Canchola, A.J., Yang, J., Allen, L., Lee, S.S., Bautista, R., La Chica, T., and Tseng, W., 2019. Peer reviewed: Higher breast cancer risk among immigrant asian american women than among US-born Asian American women. *Preventing Chronic Disease, 16*. doi: 10.5888/pcd16.180221.

Moskal, A., Norat, T., Ferrari, P., and Riboli, E., 2007. Alcohol intake and colorectal cancer risk: A dose–response meta-analysis of published cohort studies. *International Journal Of Cancer, 120*(3), pp. 664–671.

Mubarik, S., Wang, F., Fawad, M., Wang, Y., Ahmad, I., and Yu, C., 2020. Trends and projections in breast cancer mortality among four Asian countries (1990–2017): Evidence from five stochastic mortality models. *Scientific reports, 10*(1), pp. 1–12.

Murphy, N., Moreno, V., Hughes, D.J., Vodicka, L., Vodicka, P., Aglago, E.K., Gunter, M.J., and Jenab, M., 2019. Lifestyle and dietary environmental factors in colorectal cancer susceptibility. *Molecular Aspects of Medicine, 69*, pp. 2–9.

Nelen, V., 2007. Epidemiology of prostate cancer. *Prostate Cancer, 10*(2), pp. 63–89.

Nguyen, B., Mota, J.M., Nandakumar, S., Stopsack, K.H., Weg, E., Rathkopf, D., Morris, M.J., Scher, H.I., Kantoff, P.W., Gopalan, A., Zamarin, D., Solit, D.B., Schultz, N., and Abida, W., 2020. Pan-cancer Analysis of CDK12 Alterations Identifies a Subset of Prostate Cancers with Distinct Genomic and Clinical Characteristics. *European Urology, 78*(5), pp. 671–679.

Olorunfemi, G., Ndlovu, N., Masukume, G., Chikandiwa, A., Pisa, P.T., and Singh, E., 2018. Temporal trends in the epidemiology of cervical cancer in South Africa (1994–2012). *International Journal of Cancer, 143*(9), pp. 2238–2249.

Pacífico de Carvalho, N., Pilecco, F.B., and Cherchiglia, M.L., 2021. Regional inequalities in cervical cancer survival in Minas Gerais State, Brazil. *Cancer Epidemiology, 71*. doi: 10.1016/j.canep.2021.101899.

Pakzad, R., Mohammadian-Hafshejani, A., Ghoncheh, M., Pakzad, I., and Salehiniya, H., 2015. The incidence and mortality of prostate cancer and its relationship with development in Asia. *Prostate International*, 3(4), pp. 135–140.

Park, Y., Hunter, D.J., Spiegelman, D., Bergkvist, L., Berrino, F., Van Den Brandt, P.A., Buring, J.E., Colditz, G.A., Freudenheim, J.L., Fuchs, C.S., and Giovannucci, E., 2005. Dietary fiber intake and risk of colorectal cancer: A pooled analysis of prospective cohort studies. *JAMA*, 294(22), pp. 2849–2857.

Parkin, D.M., Boyd, L., and Walker, L.C., 2011. 16. The fraction of cancer attributable to lifestyle and environmental factors in the UK in 2010. *British Journal Of Cancer*, 105(S2), pp. S77–S81.

Peitzsch, C., Gorodetska, I., Klusa, D., Shi, Q., Alves, T.C., Pantel, K., and Dubrovska, A., 2020. Metabolic regulation of prostate cancer heterogeneity and plasticity. *Seminars in Cancer Biology*.

Perdana, N.R., Mochtar, C.A., Umbas, R., and Hamid, A.R.A., 2017. The risk factors of prostate cancer and its prevention: A literature review. *Acta Medica Indonesiana*, 48(3), pp. 228–238.

Piñeros, M., Saraiya, M., Baussano, I., Bonjour, M., Chao, A., and Bray, F., 2021. The role and utility of population-based cancer registries in cervical cancer surveillance and control. *Preventive Medicine*, 144, p. 106237.

Pingali, P., 2007. Westernization of Asian diets and the transformation of food systems: Implications for research and policy. *Food Policy*, 32(3), pp. 281–298.

Pinheiro, P.S., Callahan, K.E., Jones, P.D., Morris, C., Ransdell, J.M., Kwon, D., Brown, C.P., and Kobetz, E.N., 2019. Liver cancer: A leading cause of cancer death in the United States and the role of the 1945–1965 birth cohort by ethnicity. *JHEP Reports*, 1(3), pp. 162–169.

Pishgar, F., Ebrahimi, H., Saeedi Moghaddam, S., Fitzmaurice, C., and Amini, E., 2018. Global, regional and national burden of prostate cancer, 1990 to 2015: Results from the global burden of disease study 2015. *The Journal of Urology*, 199(5), pp. 1224–1232.

Poveda, A., Floquet, A., Ledermann, J.A., Asher, R., Penson, R.T., Oza, A.M., Korach, J., Huzarski, T., Pignata, S., Friedlander, M., and Baldoni, A., 2021. Olaparib tablets as maintenance therapy in patients with platinum-sensitive relapsed ovarian cancer and a BRCA1/2 mutation (SOLO2/ENGOT-Ov21): A final analysis of a double-blind, randomised, placebo-controlled, phase 3 trial. *The Lancet Oncology*, 22(5), pp. 620–631.

Prabhu, M., and Eckert, L.O., 2016. Development of World Health Organization (WHO) recommendations for appropriate clinical trial endpoints for next-generation Human Papillomavirus (HPV) vaccines. *Papillomavirus Research*, 2, pp. 185–189.

Pujade-Lauraine, E., Ledermann, J.A., Selle, F., Gebski, V., Penson, R.T., Oza, A.M., Korach, J., Huzarski, T., Poveda, A., Pignata, S., and Friedlander, M., 2017. Olaparib tablets as maintenance therapy in patients with platinum-sensitive, relapsed ovarian cancer and a BRCA1/2 mutation (SOLO2/ENGOT-Ov21): a double-blind, randomised, placebo-controlled, phase 3 trial. *The Lancet Oncology*, 18(9), pp. 1274–1284.

Purohit, S., Ferris, D.G., Alvarez, M., Tran, P.M.H., Tran, L.K.H., Mysona, D.P., Hopkins, D., Zhi, W., Dun, B., Wallbillich, J.J., Cummings, R.D., Wang, P.G., and She, J.X., 2020. Better survival is observed in cervical cancer patients positive for specific anti-glycan antibodies and receiving brachytherapy. *Gynecologic Oncology*, 157(1), pp. 181–187.

Qin, F., Pang, H., Ma, J., Zhao, M., Jiang, X., Tong, R., Yu, T., Luo, Y., and Dong, Y., 2021. Combined dynamic contrast enhanced MRI parameter with clinical factors predict the survival of concurrent chemo-radiotherapy in patients with 2018 FIGO IIICr stage cervical cancer. *European Journal of Radiology*, 141, 109787.

Quinn, M., and Babb, P., 2002. Patterns and trends in prostate cancer incidence, survival, prevalence and mortality. Part I: international comparisons. *BJU international*, 90(2), pp. 162–173.

Rawla, P., 2019. Epidemiology of prostate cancer. *World Journal of Oncology*, 10(2), p. 63.

Remmers, S., and Roobol, M.J., 2020. Personalized strategies in population screening for prostate cancer. *International Journal of Cancer*, 147(11), pp. 2977–2987.

Ridgeway, J.L., Asiedu, G.B., Carroll, K., Tenney, M., Jatoi, A., and Radecki Breitkopf, C., 2017. Patient and family member perspectives on searching for cancer clinical trials: A qualitative interview study. *Patient Education and Counseling*, 100(2), pp. 349–354.

Rojas, K., and Stuckey, A., 2016. Breast cancer epidemiology and risk factors. *Clinical Obstetrics and Gynecology*, 59(4), pp. 651–672.

Rudin, C.M., Brambilla, E., Faivre-Finn, C., and Sage, J., 2021. Small-cell lung cancer. *Nature Reviews Disease Primers*, 7(1), pp. 1–20.

Ruggeri, M., Pagan, E., Bagnardi, V., Bianco, N., Gallerani, E., Buser, K., Giordano, M., Gianni, L., Rabaglio, M., Freschi, A., and Cretella, E., 2019. Fertility concerns, preservation strategies and quality of life in young women with breast cancer: Baseline results from an ongoing prospective cohort study in selected European Centers. *The Breast*, *47*, pp. 85–92.

Rush, S.K., Swisher, E.M., Garcia, R.L., Pennington, K.P., Agnew, K.J., Kilgore, M.R., and Norquist, B.M., 2020. Pathologic findings and clinical outcomes in women undergoing risk-reducing surgery to prevent ovarian and fallopian tube carcinoma: A large prospective single institution experience. *Gynecologic Oncology*, *157*(2), pp. 514–520.

Rutherford, M.J., Dickman, P.W., Coviello, E., and Lambert, P.C., 2020. Estimation of age-standardized net survival, even when age-specific data are sparse. *Cancer Epidemiology*, *67*. doi: 10.1016/j.canep.2020.101745.

Saab, M.M., Noonan, B., Kilty, C., FitzGerald, S., Collins, A., Lyng, A., Kennedy, U., O'Brien, M., and Hegarty, J., 2021. Awareness and help-seeking for early signs and symptoms of lung cancer: A qualitative study with high-risk individuals. *European Journal of Oncology Nursing*, *50*. doi: 10.1016/j.ejon.2020.101880.

Saei, A.A., Sabatier, P., Tokat, Ü.G., Chernobrovkin, A., Pirmoradian, M., and Zubarev, R.A., 2018. Comparative proteomics of dying and surviving cancer cells improves the identification of drug targets and sheds light on cell life/death decisions. *Molecular and Cellular Proteomics*, *17*(6), pp. 1144–1155.

Samtani, R., and Saksena, D., 2019. BRCA gene mutations: A population based review. *Gene Reports*. doi: 10.1016/j.genrep.2019.100380.

Sarker, M., Krishnan, S., Parham, G., Bray, F., Ginsburg, O., Sullivan, R., Blas, M.M., Kotha, S.R., Oomman, N., Dvaladze, A., and Vanderpuye, V., 2016. The global burden of women's cancers: A grand challenge in global health. *Lancet*. doi: 10.1016/S0140-6736(16)31392-7.

Schütze, M., Boeing, H., Pischon, T., Rehm, J., Kehoe, T., Gmel, G., Olsen, A., Tjønneland, A.M., Dahm, C.C., Overvad, K., and Clavel-Chapelon, F., 2011. Alcohol attributable burden of incidence of cancer in eight European countries based on results from prospective cohort study. *BMJ*, *342*. doi: 10.1136/bmj.d1584.

Shankar, E., Gupta, K., and Gupta, S., 2019. Dietary and Lifestyle Factors in Epigenetic Regulation of Cancer. *In Epigenetics of Cancer Prevention* (pp. 361–394). Elsevier.

Sharma, V., Kerr, S.H., Kawar, Z., and Kerr, D.J., 2011. Challenges of cancer control in developing countries: current status and future perspective. *Future Oncology*, *7*(10), pp. 1213–1222.

Shen, B.J., Lo, W.C., and Lin, H.H., 2021. Global burden of tuberculosis attributable to cancer in 2019: Global, regional, and national estimates. *Journal of Microbiology, Immunology and Infection*. doi: 10.1016/j.jmii.2021.02.005.

Shin, S., Saito, E., Sawada, N., Ishihara, J., Takachi, R., Nanri, A., Shimazu, T., Yamaji, T., Iwasaki, M., Sasazuki, S., Inoue, M., Tsugane, S., Tsugane, S., Sawada, N., Iwasaki, M., Sasazuki, S., Yamaji, T., Shimazu, T., Hanaoka, T., Ogata, J., Baba, S., Mannami, T., Okayama, A., Kokubo, Y., Miyakawa, K., Saito, F., Koizumi, A., Sano, Y., Hashimoto, I., Ikuta, T., Tanaba, Y., Sato, H., Roppongi, Y., Takashima, T., Suzuki, H., Miyajima, Y., Suzuki, N., Nagasawa, S., Furusugi, Y., Nagai, N., Ito, Y., Komatsu, S., Minamizono, T., Sanada, H., Hatayama, Y., Kobayashi, F., Uchino, H., Shirai, Y., Kondo, T., Sasaki, R., Watanabe, Y., Miyagawa, Y., Kobayashi, Y., Machida, M., Kobayashi, K., Tsukada, M., Kishimoto, Y., Takara, E., Fukuyama, T., Kinjo, M., Irei, M., Sakiyama, H., Imoto, K., Yazawa, H., Seo, T., Seiko, A., Ito, F., Shoji, F., Saito, R., Murata, A., Minato, K., Motegi, K., Fujieda, T., Yamato, S., Matsui, K., Abe, T., Katagiri, M., Suzuki, M., Doi, M., Terao, A., Ishikawa, Y., Tagami, T., Sueta, H., Doi, H., Urata, M., Okamoto, N., Ide, F., Goto, H., Fujita, R., Onga, N., Takaesu, H., Uehara, M., Nakasone, T., Yamakawa, M., Horii, F., Asano, I., Yamaguchi, H., Aoki, K., Maruyama, S., Ichii, M., Takano, M., Tsubono, Y., Suzuki, K., Honda, Y., Yamagishi, K., Sakurai, S., Tsuchiya, N., Kabuto, M., Yamaguchi, M., Matsumura, Y., Sasaki, S., Watanabe, S., Akabane, M., Kadowaki, T., Inoue, M., Noda, M., Mizoue, T., Kawaguchi, Y., Takashima, Y., Yoshida, Y., Nakamura, K., Takachi, R., Ishihara, J., Matsushima, S., Natsukawa, S., Shimizu, H., Sugimura, H., Tominaga, S., Hamajima, N., Iso, H., Sobue, T., Iida, M., Ajiki, W., Ioka, A., Sato, S., Maruyama, E., Konishi, M., Okada, K., Saito, I., Yasuda, N., Kono, S., Akiba, S., Isobe, T., and Sato, Y., 2018. Dietary patterns and colorectal cancer risk in middle-aged adults: A large population-based prospective cohort study. *Clinical Nutrition*, *37*(3), pp. 1019–1026.

Shoemaker, M.L., White, M.C., Wu, M., Weir, H.K., and Romieu, I., 2018. Differences in breast cancer incidence among young women aged 20–49 years by stage and tumor characteristics, age, race, and ethnicity, 2004–2013. *Breast Cancer Research and Treatment*, *169*(3), pp. 595–606.

Shulman, L.P., 2010. Hereditary Breast and Ovarian Cancer (HBOC): Clinical features and counseling for BRCA1 and BRCA2, Lynch syndrome, Cowden syndrome, and Li-Fraumeni syndrome. *Obstetrics and Gynecology Clinics of North America, 37*(1), pp. 109–33.

Siegel, R. L., Miller, K. D., and Jemal, A., 2020. Cancer statistics, 2020. *CA: A Cancer Journal for Clinicians.* doi: 10.3322/caac.21590.

Small Jr, W., Bacon, M.A., Bajaj, A., Chuang, L.T., Fisher, B.J., Harkenrider, M.M., Jhingran, A., Kitchener, H.C., Mileshkin, L.R., Viswanathan, A.N., and Gaffney, D.K., 2017. Cervical cancer: A global health crisis. *Cancer, 123*(13), pp. 2404–2412.

Song, R., Cong, L., Ni, G., Chen, M., Sun, H., Sun, Y., and Chen, M. 2017. MicroRNA-195 inhibits the behavior of cervical cancer tumors by directly targeting HDGF. *Oncology Letters, 14*(1), pp. 767–775.

Soto-Perez-de-Celis, E., and Chavarri-Guerra, Y., 2016. National and regional breast cancer incidence and mortality trends in Mexico 2001-2011: Analysis of a population-based database. *Cancer Epidemiology, 41*, pp. 24–33.

Stewart, B.W., and Wild, C.P. eds., 2014. *World Cancer Report 2012.* International Agency for Research on Cancer, Lyon.

Sullivan, R., Alatise, O.I., Anderson, B.O., Audisio, R., Autier, P., Aggarwal, A., Balch, C., Brennan, M.F., Dare, A., D'Cruz, A., and Eggermont, A.M., 2015. Global cancer surgery: delivering safe, affordable, and timely cancer surgery. *The Lancet Oncology, 16*(11), pp. 1193–1224.

Sundar, S., Khetrapal-Singh, P., Frampton, J., Trimble, E., Rajaraman, P., Mehrotra, R., Hariprasad, R., Maitra, A., Gill, P., Suri, V., Srinivasan, R., Singh, G., Thakur, J.S., Dhillon, P., and Cazier, J.B., 2018. Harnessing genomics to improve outcomes for women with cancer in India: key priorities for research. *The Lancet Oncology, 19*(2), pp. e102–e112.

Sung, W.W., Au, K.K., Wu, H.R., Yu, C.Y., and Wang, Y.C., 2021. Improved trends of lung cancer mortality-to-incidence ratios in countries with high healthcare expenditure. *Thoracic Cancer.* doi: 10.1111/1759-7714.13912.

Taitt, H.E., 2018. Global trends and prostate cancer: A review of incidence, detection, and mortality as influenced by race, ethnicity, and geographic location. *American Journal of Men's Health, 12*(6), pp. 1807–1823.

Tanaka, S., Yamamoto, S., Inoue, M., Iwasaki, M., Sasazuki, S., Iso, H., and Tsugane, S., 2009. Projecting the probability of survival free from cancer and cardiovascular incidence through lifestyle modification in Japan. *Preventive Medicine, 48*(2), pp. 128–133.

Tao, Z., Shi, A., Lu, C., Song, T., Zhang, Z., and Zhao, J., 2015. Breast cancer: Epidemiology and etiology. *Cell Biochemistry and Biophysics, 72*(2), pp. 333–338.

Tian, H., Yang, W., Hu, Y., Liu, Z., Chen, L., Lei, L., Zhang, F., Cai, F., Xu, H., Liu, M., Guo, C., Chen, Y., Xiao, P., Chen, J., Ji, P., Fang, Z., Liu, F., Liu, Y., Pan, Y., dos-Santos-Silva, I., He, Z., and Ke, Y., 2020. Estimating cancer incidence based on claims data from medical insurance systems in two areas lacking cancer registries in China. *EClinicalMedicine, 20.* doi: 10.1016/j.eclinm.2020.100312.

Torre, L.A., Bray, F., Siegel, R.L., Ferlay, J., Lortet-Tieulent, J., and Jemal, A., 2015. Global cancer statistics, 2012. *CA: A Cancer Journal for Clinicians, 65*(2), pp. 87–108.

Torre, L.A., Islami, F., Siegel, R.L., Ward, E.M., and Jemal, A., 2017. Global cancer in women: burden and trends. *Cancer Epidemiology and Prevention Biomarkers, 26*(4), pp. 444–457.

Torre, L.A., Siegel, R.L., Ward, E.M., and Jemal, A., 2016. Global cancer incidence and mortality rates and trends – an update. *Cancer Epidemiology and Prevention Biomarkers, 25*(1), pp. 16–27.

Torre, L.A., Trabert, B., DeSantis, C.E., Miller, K.D., Samimi, G., Runowicz, C.D., Gaudet, M.M., Jemal, A., and Siegel, R.L., 2018. Ovarian cancer statistics, 2018. *CA: A Cancer Journal for Clinicians, 68*(4), pp. 284–296.

Triplette, M., Thayer, J.H., Kross, E.K., Cole, A.M., Wenger, D., Farjah, F., Nair, V.S., and Crothers, K., 2021. The impact of smoking and screening results on adherence to follow-up in an academic multisite lung cancer screening program. *Annals of the American Thoracic Society, 18*(3), pp. 545–547.

van der Doelen, M.J., Isaacsson Velho, P., Slootbeek, P.H.J., Pamidimarri Naga, S., Bormann, M., van Helvert, S., Kroeze, L.I., van Oort, I.M., Gerritsen, W.R., Antonarakis, E.S., and Mehra, N., 2020. Impact of DNA damage repair defects on response to radium-223 and overall survival in metastatic castration-resistant prostate cancer. *European Journal of Cancer, 136*, 16–24.

Vanacore, D., Boccellino, M., Rossetti, S., Cavaliere, C., D'Aniello, C., Di Franco, R., Romano, F.J., Montanari, M., La Mantia, E., Piscitelli, R., and Nocerino, F., 2017. Micrornas in prostate cancer: An overview. *Oncotarget, 8*(30), p. 50240.

Vanderpuye, V., Grover, S., Hammad, N., Simonds, H., Olopade, F., and Stefan, D.C., 2017. An update on the management of breast cancer in Africa. *Infectious Agents and Cancer, 12*(1), pp. 1–12.

Vanshika, S., Preeti, A., Sumaira, Q., Vijay, K., Shikha, T., Shivanjali, R., Shankar, S.U., and Mati, G.M., 2021. Incidence OF HPV and EBV in oral cancer and their clinico-pathological correlation– a pilot study of 108 cases. *Journal of Oral Biology and Craniofacial Research, 11*(2), pp. 180–184.

Veisani, Y., Jenabi, E., Khazaei, S., and Nematollahi, S., 2018. Global incidence and mortality rates in pancreatic cancer and the association with the Human Development Index: Decomposition approach. *Public Health, 156*, pp. 87–91.

Venier, N., Vandersluis, A., Colquhoun, A., Kiss, A., Fleshner, N., Klotz, L., and Venkateswaran, V., 2011. Sustained aerobic exercise alone does not counteract the tumor-promoting effects of a westernized diet on prostate cancer progression in vivo. *Journal of Urology, 185*(4S). doi: 10.1016/j.juro.2011.02 .1725.

Ward, Z.J., Grover, S., Scott, A.M., Woo, S., Salama, D.H., Jones, E.C., El-Diasty, T., Pieters, B.R., Trimble, E.L., Vargas, H.A., Hricak, H., and Atun, R., 2020. The role and contribution of treatment and imaging modalities in global cervical cancer management: survival estimates from a simulation-based analysis. *The Lancet Oncology, 21*(8), pp. 1089–1098.

Ward, Z.J., Scott, A.M., Hricak, H., Abdel-Wahab, M., Paez, D., Lette, M.M., Vargas, H.A., Kingham, T.P., and Atun, R., 2020. Estimating the impact of treatment and imaging modalities on 5-year net survival of 11 cancers in 200 countries: A simulation-based analysis. *The Lancet Oncology, 21*(8), pp. 1077–1088.

Williams, L.J., Fletcher, E., Douglas, A., Anderson, E.D., McCallum, A., Simpson, C.R., Smith, J., Moger, T.A., Peltola, M., Mihalicza, P., and Sveréus, S., 2018. Retrospective cohort study of breast cancer incidence, health service use and outcomes in Europe: A study of feasibility. *The European Journal of Public Health, 28*(2), pp. 327–332.

Wittenberg, E., Reb, A., and Kanter, E., 2018. Communicating with patients and families around difficult topics in cancer care using the COMFORT communication curriculum. *Seminars in Oncology Nursing, 34*(3), pp.264–273.

Wojtyla, C., Bertuccio, P., Wojtyla, A., and La Vecchia, C., 2021. European trends in breast cancer mortality, 1980–2017 and predictions to 2025. *European Journal of Cancer, 152*, pp. 4–17.

Wong, M.C.S., Goggins, W.B., Wang, H.H.X., Fung, F.D.H., Leung, C., Wong, S.Y.S., Ng, C.F., and Sung, J.J.Y., 2016. Global incidence and mortality for prostate cancer: Analysis of temporal patterns and trends in 36 countries. *European Urology, 70*(5), pp. 862–874.

Wong, M.C.S., Huang, J., Lok, V., Wang, J., Fung, F., Ding, H., and Zheng, Z.J., 2021. Differences in incidence and mortality trends of colorectal cancer worldwide based on sex, age, and anatomic location. *Clinical Gastroenterology and Hepatology, 19* (5), pp. 955–966.e61.

Youlden, D.R., Cramb, S.M., Yip, C.H., and Baade, P.D., 2014. Incidence and mortality of female breast cancer in the Asia-Pacific region. *Cancer Biology & Medicine, 11*(2), p. 101.

Zafar, A., Alruwaili, N.K., Imam, S.S., Alharbi, K.S., Afzal, M., Alotaibi, N.H., Yasir, M., Elmowafy, M., and Alshehri, S., 2021. Novel nanotechnology approaches for diagnosis and therapy of breast, ovarian and cervical cancer in female: A review. *Journal of Drug Delivery Science and Technology, 61*. doi: 10.1016/j.jddst.2020.102198.

Zhang, L., Meng, X., Pan, C., Qu, F., Gan, W., Xiang, Z., Han, X., and Li, D., 2020. piR-31470 epigenetically suppresses the expression of glutathione S-transferase pi 1 in prostate cancer via DNA methylation. *Cellular Signalling, 67*. doi: 10.1016/j.cellsig.2019.109501.

Zhang, S., Xu, H., Zhang, L., and Qiao, Y., 2020. Cervical cancer: Epidemiology, risk factors and screening. *Chinese Journal of Cancer Research, 32*(6), p. 720.

Zhao, T., Cui, Z., McClellan, M.G., Yu, D., Amy Sang, Q.X., and Zhang, J., 2020. Identifying county-level factors for female breast cancer incidence rate through a large-scale population study. *Applied Geography, 125*. doi: 10.1016/j.apgeog.2020.102324.

Zheng, L., Cui, C., Shi, O., Lu, X., Li, Y. Kun, Wang, W., Li, Y., and Wang, Q., 2020. Incidence and mortality of ovarian cancer at the global, regional, and national levels, 1990–2017. *Gynecologic Oncology, 159*(1), pp. 239–247.

Znaor, A., Eser, S., Anton-Culver, H., Fadhil, I., Ryzhov, A., Silverman, B.G., Bendahou, K., Demetriou, A., Nimri, O., Yakut, C., and Bray, F., 2018. Cancer surveillance in northern Africa, and central and western Asia: Challenges and strategies in support of developing cancer registries. *The Lancet Oncology, 19*(2), pp. e85–e92.

3 Trends in Cancer Screening: Different Diagnostic Approaches

3.1 INTRODUCTION

Cancer deaths are avoidable by early detection. Various detection methods are being used for the diagnosis of cancer. To check the growth of this disease, precise screening is required. There are different epidemiologic concepts to understand biomarkers and other tests used for earlier detection of new or recurrent cancer (Al-Shaheri et al., 2021; de Kock et al., 2021; Henning, Barashi and Smith, 2021; Jalil, Pandey and Kumar, 2021; Mollasalehi and Shajari, 2021; Özgür et al., 2021; Vandghanooni et al., 2021). In simple words, screening is a method of detecting cancer at its early or late stage among individuals who have a higher chance of getting cancer. Sensitivity is the ability of the test to indicate those who have cancer among the population with cancer, whereas specificity defines the ability of the test to recognize those who do not have cancer among the population without cancer. Several screening tests have been shown to detect early cancer and can reduce the number of deaths from cancer. Some tests are used only for early detection but cannot reduce the deaths from cancer (Rapisuwon et al., 2016). Currently, the use of circulating markers is considered as the most robust and effective method for detection (Lin, Huang and Chang, 2011; Marcuello et al., 2019; Alizadeh Savareh et al., 2020; Azad et al., 2020; Valihrach, Androvic and Kubista, 2020; Vandghanooni et al., 2021; Vasantharajan et al., 2021). The development of circulating nucleic acid markers allows the use of minimally invasive serial blood samples to assess the mutational status (Chiacchiarini et al., 2021; Goodman and Speers, 2021; Hulstaert et al., 2021; Lin and Chang, 2021; Palmela Leitão et al., 2021). ctDNA (circulating tumor DNA) is found to be present at higher concentrations in blood, independent of the concentration of CTCs (circulating tumor cells). It exhibits a response to chemotherapy or molecular targeted therapy (Rapisuwon et al., 2016). Circulating RNAs are potent cancer markers. As a result of cell proliferation and stromal remodeling, different RNA species are found to be deregulated. Circulating RNA markers show greater precision than other markers. So, circulating blood-based marker screening tests for cancer are feasible regarding sensitivity and specificity (Schiffman et al., 2015).

3.2 EXISTING DIAGNOSTIC TECHNIQUES

Screening tests that have been utilized to minimize the death from cancer: In Figure 3.1, different screening methods have been shown.

a. **Colonoscopy, sigmoidoscopy, and high sensitivity fecal occult blood tests (FOBTs):** To reduce the death from colorectal cancer, these tests are shown to be useful (Chen et al., 2019; Grobbee et al., 2020; Randel et al., 2021). Colonoscopy and sigmoidoscopy can detect abnormal growth of colorectal cells, which can be removed before it develops into a tumor, so these tests can also prevent cancer. People in the age range of 50–75 are at a higher risk of colorectal cancer.

b. **Mammography:** This technique is used to screen and reduce the number of deaths from

DOI: 10.1201/9781003201946-3

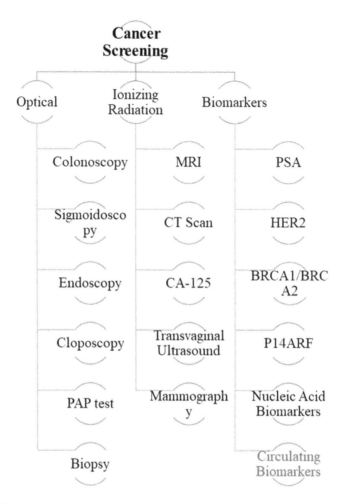

FIGURE 3.1 Different cancer screening methods.

breast cancer among women of the age group 40–74 (Conlon, Lester and Friedewald, 2019; Hadjipanteli et al., 2019; Magario et al., 2019).

c. **Pap test and human papillomavirus testing (HPV) testing:** These tests are used for the identification of abnormal cells of of the cervix and also reduce deaths from cervical cancer (Song, 2018; Kilic et al., 2020; Kim, Jun and Maeng, 2020; Lefeuvre et al., 2020; Padhy et al., 2020).

3.2.1 OTHER SCREENING TESTS

a. **Alpha-fetoprotein blood test:** This test used for early detection of liver cancer (Zakaria et al., 2020; Patil et al., 2021). Sometimes it is performed along with the ultrasound of the liver.

b. **Breast MRI:** This is an imaging technique. Women who have a mutation in the *BRCA1* and *BRCA2* gene at higher risk of breast cancer. This test used to detect mutation in these genes (Brown et al., 2021; Thompson and Wright, 2021).

c. **CA-125 test:** To detect ovarian cancer early, this test is performed along with the trans-vaginal ultrasound. It is also useful in diagnosing the recurrence of cancer in women who have been diagnosed previously with ovarian cancer (Wenstrom et al., 1997; Aisaka et al., 2001; Aisaka et al., 2002; Stone et al., 2014; Burki, 2015).

d. **PSA test:** This blood test is used to detect colorectal cancer, performed along with the digital rectal examination. But this test is not very recommendable because it is found in different studies that it could lead to overdiagnosis and overtreatment and not very effective for cancer death prevention (Alam et al., 2017; Pérez-Ibave, Burciaga-Flores and Elizondo-Riojas, 2018).

e. **Transvaginal ultrasound:** This imaging test is used to detect mutation in *BRCA1* and *BRCA2* responsible for ovarian cancer and also used for endometrial cancer (Dietrich et al., 2020; F et al., 2021; González-Timoneda et al., 2021).

3.3   MARKERS-BASED DIAGNOSIS OF CANCER

A biomarker is best defined as "a biological molecule that signifies a disease or condition, can be found in blood or tissues resulting in a normal or abnormal process" (Cheng and Meiser, 2019; Ding et al., 2019; Iwasaki, Shimura and Kataoka, 2019; Muinao, Deka Boruah and Pal, 2019; Xiong et al., 2019; Solanki, Venkatesulu and Efstathiou, 2021). It is used to differentiate normal ones with the diseased person. These differences could arise due to germline or somatic mutations, transcriptional changes, and post-translational modifications. Different varieties of biomarkers are categorized including proteins, nucleic acid, peptides antibodies, etc. Conditions such as changes in gene expression, proteomic, and metabolomic profiling can also work as a biomarker (Cai et al., 2010; Tayanloo-Beik et al., 2020). It can be found in blood or other body fluids that can be assessed easily or maybe tissue-derived for which biopsy detection method is needed. By using different techniques such as gene expression arrays, mass spectroscopy, and high-throughput sequencing, biomarkers can be identified easily.

It is important to focus on the analytical validity of a newly developed biomarker which describes the technical aspects of a biomarker assay to determine robustness, sensitivity, and specificity and it must be accurate and reproducible.

Along with that, the clinical validity of the biomarker must also be described to know whether it has biological relevance or not (Henry and Hayes, 2012).

3.3.1   Tumor Markers Currently in Practice for Cancer Detection

Following are the different markers currently being used in detection of different types of cancer (Figure 3.2).

3.3.1.1   Tissue-Derived Tumor Markers

For the detection of these types of markers, biopsy removal from patients is required. Common tumor markers that are analyzed in malignant tissues are the following (Romano, 2015).

3.3.1.2   21-Gene Signature (Oncotype *DX*)

This set of markers is used to analyze the reappearance of tumors in breast cancer patients. The 21-gene signature includes estrogen receptors, *HER2*, cell proliferation, cellular invasion, macrophage marker CD68, anti-apoptotic genes (*BAG1* and *GSTM1*), and five housekeeping reference genes (β-actin, *GAPDH*, *RPLPO*, *GUS*, and *TFRC*) (Solin et al., 2013). Oncotype *DX* is a method of profiling for gene expression (Dobbe et al., 2008).

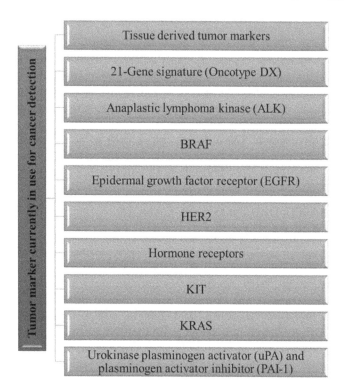

FIGURE 3.2 Different markers currently in use for cancer detection.

3.3.1.3   Anaplastic Lymphoma Kinase (*ALK*)

This gene is used to analyze the presence of the mutation in NSCLC (non-small cell lung cancer) patient by using techniques fluorescent in situ hybridization (FISH), immunohistochemistry (IHC), and polymerase chain reaction (PCR) (Shackelford et al., 2014).

3.3.1.4   *BRAF*

Mutations in the *BRAF* gene can be present in melanoma (Glitza and Davies, 2014), colorectal, and thyroid cancer. Various assays are used for the detection of this mutation such as pyrosequencing (Romano, 2015).

3.3.1.5   Epidermal Growth Factor Receptor (EGFR)

EGFR is used to detect the cancer stage and diagnostic approaches for breast cancer, head and neck, and colon cancer (Centuori and Martinez, 2014). Mutations in this gene are analyzed through real-time PCR (Romano, 2015).

3.3.1.6 *HER2*

This cellular receptor also called *HER2/neu, erbB-2, or EGFR2* helps in tumor growth. Overexpression of this receptor molecule has been detected in the breast (Advani et al., 2015), stomach, and esophageal cancers. The status of *HER2* can be seen through FISH analysis by using probes of different kits (Romano, 2015).

3.3.1.7   Hormone Receptors

Estrogen and progesterone receptors can proliferate breast cancer. Breast cancer patients are tested for these two hormones through biopsy. The ER expression level can be detected through IHC, PET scan, etc. (van Kruchten et al., 2013).

3.3.1.8 *KIT*

KIT is a proto-oncogene, also called CD117 (Jones et al., 2015). It is associated with the gastrointestinal stromal tumor and mucosal melanoma and can be detected through IHC (Romano, 2015).

3.3.1.9 *KRAS*

Mutation in the *KRAS* gene is responsible for increasing chemoresistance in colorectal cancer (Westwood et al., 2014). This gene mutation can be detected through real-time PCR (Romano, 2015).

3.3.1.10 Urokinase Plasminogen Activator (*uPA*) and Plasminogen Activator Inhibitor (*PAI-1*)

These two markers are used to detect the level of malignancy and treatment criteria for breast cancer (Harbeck et al., 2014). *uPA* level can be detected through the ELISA kit.

3.3.1.11 Tumor Markers Detectable in Blood or Fluids

Following are the different protein-based markers for different types of cancer present in blood (Figure 3.3).

3.3.1.11.1 5-Protein Signature (Ova1)

Ova1 consists of five different proteins, CA 125, β-2-microglobulin (B2M), ApoA1, transthyretin (TT), and transferrin (TF) (Nolen and Lokshin, 2012). Detection of this marker in the blood is used to diagnose ovarian cancer.

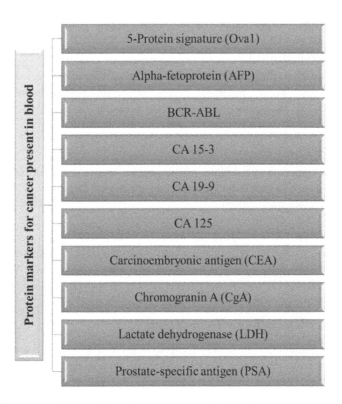

FIGURE 3.3 Different protein biomarkers of cancer present in blood.

3.3.1.11.2 Alpha-fetoprotein (AFP)

Overexpression of AFP is found in hepatocellular carcinoma (Y.J. Zhao 2013Zhao et al., 2013) and germ cell tumors. It is also used to diagnose liver cancer (Y.J. Zhao et al., 2013). AFP serum detection can be analyzed through the ELISA kit (Romano, 2015).

3.3.1.11.3 BCR-ABL

This marker is specifically present in chronic myeloid leukemia cells that are produced through specific chromosomal translocation resulting in the Philadelphia chromosome (Pemmaraju and Cortes, 2014). It can be analyzed through a real-time PCR kit.

3.3.1.11.4 CA 15-3

It is Carcinoma Antigen 15-3. It is used to diagnose and treatment for breast cancer (Donepudi et al., 2014). An ELISA kit is used to detect this marker in the blood.

3.3.1.11.5 CA 19-9

It is an acronym for carbohydrate antigen 19-9. Detection of CA19-9 in the blood is used to diagnose colorectal, pancreas (Ballehaninna and Chamberlain, 2011), and bladder cancer which can be measured through the ELISA kit.

3.3.1.11.6 CA 125

It stands for cancer antigen 125 or carbohydrate antigen 125. It is associated with epithelial ovarian cancer (Musto et al., 2014), primary peritoneal cancer (Togo Peraza et al., 2014), and fallopian tube cancer. Detection of this marker is based on the ELISA kit.

3.3.1.11.7 Carcinoembryonic Antigen (CEA)

For diagnosis and screening of breast, colorectal, and lung cancer (Swiderska et al., 2014), the level of CEA in the blood is measured. Detection can be done with the ELISA kit.

3.3.1.11.8 Chromogranin A (CgA)

This blood marker is used to diagnose neuroendocrine tumors like neuroblastoma, lung cancer, and carcinoma (D'Herbomez et al., 2010). It is also associated with advanced prostate cancer (Conteduca et al., 2014). Its expression level can be measured in blood via the ELISA method (Romano, 2015).

3.3.1.11.9 Lactate Dehydrogenase (LDH)

This marker has been detected in the blood of lymphoma (Hsu et al., 2014), melanoma, and germ cell tumors. It is also used to check the recurrence of the tumor. Detection of this marker in blood carried out by ELISA (Romano, 2015).

3.3.1.11.10 Prostate-Specific Antigen (PSA)

This marker is detected in the blood of prostate cancer patients. It is used to diagnose as well as the clinical course of prostate cancer (Stephan et al., 2014). The serum level of PSA can be detected through the ELISA kit.

3.3.2 MOLECULAR MARKERS-BASED TECHNOLOGIES

Any change in gene sequences, its level of expression, or change in structure and function of proteins can be used as a "molecular marker" which is used to detect cancer progression at an early stage (Sidransky, 2002; Edmondson et al., 2017; Arend et al., 2018; Nair et al., 2018; Bronkhorst et al., 2019; Moroney et al., 2019; Amelot et al., 2020; Doello et al., 2020; Rao et al.,

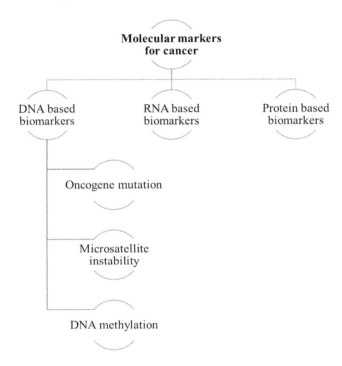

FIGURE 3.4 Different types of molecular markers for cancer.

2020; Mohanty et al., 2021). Many anatomical methods are designed to detect tumor masses such as mammography, physical examination, endoscopy, etc. Different molecular markers (Figure 3.4), on the other hand, independently rely on molecular signals present in the blood or other body fluids despite the presence of tumor mass. These markers are categorized into four classes (Baron, 2012):

1. **Carcinogenesis biomarker:** These molecules are shed by tumor cells as a result of the carcinogenic process in form of methylated DNA or mutated DNA.
2. **Response biomarker:** These molecules are produced by host cells in response to cancer. Examples include protein, antibodies, etc.
3. **Released biomarker:** These are produced in response to damage caused by cancer like blood in the stool and PSA in serum.
4. **Risk biomarker:** These are the factors associated with carcinogenic signals in the body like a high level of estradiol in breast cancer.

3.3.2.1 Nucleic-Acid-Based Markers

Many studies related to cancer have reached a point, i.e., association of genetic modification with cancer, which could be used to detect a certain type of cancer. Discovery of free DNA in serum or blood (Li et al., 2019; Sinha et al., 2019; van der Pol and Mouliere, 2019; Thakur et al., 2019; Weng et al., 2019; Bustamante Alvarez et al., 2020; Fettke et al., 2020; Liu et al., 2020; Mo et al., 2020; Ritter et al., 2020; Suryavanshi et al., 2020; Zvereva et al., 2020;) opened a large scope for the screening of cancer, serum, or plasma from cancer patients used to detect tumor markers such as oncogene mutations, MICROSATELLITE instability, hypermethylation of promoter regions (Anker et al., 2001).

a. **Oncogene mutation:** Highly sensitive polymerase chain reaction is required to detect mutation in individual DNA fragments (Mills et al., 1995). Different studies have shown the use of this approach to detect genetic mutation in the *RAS* gene and *p53* in stool samples of patients with colorectal cancer (Dong et al., 2001; Traverso et al., 2002). Another method such as mass spectroscopy has also been used for RAS mutation in patients of lung cancer and hepatocellular carcinoma (Jackson et al., 2001).

b. **Microsatellite instability:** Expansion or deletion of microsatellites in cancer cells is called microsatellite instability. Seventy-six percent of small-cell lung cancers (SCLC) and plasma samples have shown the alteration in microsatellite. Another means of detecting cancer cells is via SNPs (single nucleotide polymorphisms) (Sidransky, 2002).

c. **DNA methylation:** Hypermethylation of promoter regions of genes that are associated with cancer has been proved a very encouraging marker. To detect a methylated region of DNA accurately, PCR assays have been used (Herman et al., 1996). This method was used in cancer cells in the saliva of patients with oral cancer, methylation of genes that encode INK4A, methyl O-guanine methyltransferase involved in DNA repair. Methylation of these genes has been reported in serum samples of patients of the lung and head and neck cancer (Sanchez-Cespedes et al., 2000). For quantification of the number of methylated alleles (in a single region) among wild-type DNA, REAL-TIME PCR-based approaches have been used using the Taq man probe (Usadel et al., 2002).

d. **RNA-based approaches:** For the detection of cancer cells, several mRNA-based methods have been formulated for the identification and quantification of mRNA levels in tissue samples by reverse transcriptase PCR (Krismann et al., 1995). mRNA levels of markers such as tyrosinase, present in melanoma cells can be measured through this assay. Reverse transcriptase PCR is also used in the detection of other types of markers such as CEA in colorectal cancer and PSA in prostate cancer.

3.3.2.2   Protein Markers

Several protein-based assays have been proposed for cancer cell detection. Measuring the high serum levels of PSA, associated with prostate cancer is the most common protein-based approach for cancer detection. Along with that high serum levels of oncofetal proteins such as the human glycoprotein hormone (β-HCG) and α-fetoprotein (AFP) are used in the diagnosis of testicular cancer and hepatocellular cancer, whereas CEA is commonly used in the diagnosis of colon cancer for determination of response to therapy biomarkers, such as PSA for prostate cancer and cancer antigen (CA)125 for ovarian cancer have been used widely (Canizares et al., 2001).

3.3.2.2.1 Advantages of Molecular Screening

Molecular screening is a very high-quality method for cancer detection. By using this method it is possible to separate patients who can skip anatomical assessment from those who are more susceptible to adapt to cancer. This method is more accurate, safe, and convenient because it requires only drawing of blood, urine, or stool and it does not require any type of radiation or other unfriendly procedures such as colonoscopy for assessment of cancer. Hence, it has made detection methods more affordable, specific, sensitive, and also increased the chances of detecting cancer at an early age. Another advantage of it is that it can be combined with mathematical methods easily for enhancing the specificity and sensitivity of detection (Baron, 2012).

3.3.2.3   Robust Circulating Markers

Circulating markers for cancer diagnosis and treatment can be easily measured by simply drawing the blood, which made it a very comfortable method of analysis. The discovery of circulating markers have been progressed from the enzyme, protein products, hormones, and oncogenes to DNA and RNA to increase the sensitivity and specificity of the detection method (Wu, 1999).

3.3.2.3.1 Enzymes and Serum Proteins

Earlier most commonly used tumor markers were enzymes, hormones, and serum proteins, among which enzymes were most commonly used for cancer detection that rely upon the enzymatic activity. Some of the enzymes that are used as tumor markers are prostatic acid phosphatase in prostate cancer, lactate dehydrogenase in lung and breast cancer, fucosyltransferase in multiple malignant tumors, etc.

But the use of isoenzymes made the detection method more specific. Several tissue-specific isozymes have been used like lactate dehydrogenase, alkaline phosphatase, creatine phosphokinase, etc. (Rotenberg et al., 1988).

3.3.2.3.2 Carcinoembryonic Proteins

The more intense tumor marker that has been discovered was CEA (carcinoembryonic antigens) because their expression is switched off during normal fetal development and become reactivated during tumor condition (Wu and James T., 1999).

3.3.2.3.3 Circulating Nucleic Acid Markers

Advances in specificity and sensitivity of tumor increased by the discovery of circulating DNA and RNA to be used as a biomarker. Each organ that has been affected by cancer shed mutant DNA and RNA in the blood.

3.3.2.3.4 Circulating DNA

Mutation in oncogenes like Ras, p53, and other cancer-associated genes and hypermethylation of tumor suppressor genes promoter region is associated with several types of cancer including colon, breast, lung, melanoma, etc. (Bettegowda et al., 2014). Detection of circulating tumor DNA is associated with the burden of a tumor.

3.3.2.3.5 Detection Methods and Sensitivity

The detection of ctDNA (circulating tumor DNA) is not very easy. The main obstacle comes due to the low abundance of it into the blood. The new method of detections like next-generation sequencing and quantitative PCR (qPCR) has lowered the lower limit of detection to approximately 1–2% from the conventional methods like pyrosequencing that has a lower limit of detection at 10% of ctDNA. In a study of 69 patients of colorectal cancers with detectable *KRAS* ctDNA, it was found that the prevalence of ctDNA in non-metastatic have been found low approximately varies between 49–78%, which was much lower than metastatic forms of 86–100% (Rapisuwon et al., 2016).

But several limitations are there in detecting ctDNA like it is not always present in peripheral blood. Additionally, ctDNA detection is highly dependent on pre-analytical sampling. Some studies have shown the inconsistency in levels of ctDNA in plasma and serum due to the loss of DNA during purification (Rapisuwon et al., 2016).

3.3.2.3.6 Circulating RNA

mRNAs are known to play a critical role in intracellular protein translation and that status of the intracellular process is reflected by the extracellular process and are possibly potential biomarkers for cancer diagnosis (Rapisuwon et al., 2016). Circulating mRNA was first demarcated in the 1990s in cancer patients of the stomach, pancreas, lung, blood, etc. (Funaki et al., 1996).

3.3.2.3.7 Types of Circulating Cell-Free RNA: Non-Coding RNA

- **Piwi-interacting RNAs (piRNA):** They are 26–31 nucleotide long ss RNAs. The function of these RNAs is to repress transposons and target mRNA, this process is mediated by PIWI proteins that belong to the Argaunate proteins subfamily (Siomi et al., 2011). They are present in human plasma in ample quantity. It was found that plasma levels of PiR-019825 were deregulated in colorectal cancer patients, whereas piR-016658 and piR-020496 were found to be related with prostate cancer patients, and plasma levels of piR-001311 and piR-016658 were found to associated with pancreatic cancer (Yuan et al., 2016).

- **Long non-coding RNAs (lncRNA):** They are of > 200 nucleotides in length and can be categorized into five subclasses, intergenic, intronic, sense overlapping, antisense, and bi-directional lncRNAs (Archer et al., 2015). Many RNA transcripts have been recognized as lncRNAs that regulate growth, metabolism and cancer metastasis (Silva et al., 2015). In patients with gastric cancer, circulating levels of lncRNA H19 were found to be elevated in comparison to healthy ones.
- **microRNA:** Mature microRNAs (miRNA) are derivative of hairpin precursor transcripts. They are short strands of non-coding RNA and extremely conserved. The formation of 21–24 nucleotides are long, double-stranded mature miRNAs that take place in two steps: firstly by the cleavage of primary microRNA (pri-miRNA) with the help of enzyme complex Drosha which takes place inside the nucleus and then through DICER1 in the cytoplasm (Lin and Gregory, 2015). One of the mature miRNA strands, by binding primarily to the 3'un-translated region (UTR) region of mRNA regulates the protein translation process. Moreover, it can repress or activate translational efficacy by binding to the open reading frame (ORF) or 5'UTR of target mRNAs also (Portnoy et al., 2011). miRNAs are known to be involved in a varied range of biological processes such as growth, cell proliferation, body metabolism, and cell signaling (Swanton, 2012). It is noted that miRNAs are stable in the peripheral circulation (Rapisuwon et al., 2016).

3.3.2.3.8 Relation of miRNAs Alterations to Cancer

Different studies are showing the specific microRNAs dysregulation in cancer biology. These tumor suppressors and onco miRs signatures have been proven as a hallmark of cancer (Detassis et al., 2017; Precazzini et al., 2021). miRNAs are found to be involved in a different phase of tumorigenesis, supporting the proliferation signaling and empowering the replicative immortality. In colorectal cancer (CRC), a decreased expression level of miR-545 was found in contrast to normal tissues. At the same time, the overexpression of this gene resulted in a lower proliferation of cells (Huang and Lu, 2017; Cantile et al., 2021). miR-130b-301b regulates the mechanism of senescence. In prostate cancer cells, it is hypermethylated. After restoring the senescence me-chanism via ectopic expression, it was seen that it reduced the prostate cancer cells' malignant phenotype (Ramalho-Carvalho et al., 2017, Fort et al., 2018). miRNAs are also known to be involved in the dysregulation of cellular metabolism, a hallmark of cancer. Reactive oxygen species (ROS) regulate several cellular and physiological functions leading to often cancer. A higher level of ROS results in oxidized DNA and RNA, responsible for genome lesions and cancer progression (Fimognari, 2015; de Araújo et al., 2016; Van Houten et al., 2018; Costantini, 2019). In colorectal cancer, cell lines have a lower expression of miR-1 compared to normal epithelial cells. After restoring its activity, the proliferation of cancer cells decreased via dipping down the level of glucose uptake, aerobic glycolysis production and targeting the HIF-1 (Du et al., 2021; Xu et al., 2017). Several studies have found that miRNA expression patterns could be used to dif-ferentiate different tumor subtypes in the recent era. Lu and colleagues have developed a bead-based flow cytometry for miRNA expression profiling. Through this method, miRNA fingerprints have been generated to classify the origin of tumors and identify the degree of differentiation (Lu et al., 2005). For differentiation of lung adenocarcinoma and squamous cell carcinoma, miRNA signatures can be used for diagnosis purposes (Yanaihara et al., 2006; Yanaihara 2013; Inamura, 2017).

Additionally, the combination of miR-21 and miR-205 was identified as potential biomarkers to differentiate lung adenocarcinoma and squamous cell carcinoma (Lebanony et al., 2009; Jiang et al., 2013). In a study of Del Vescovo et al., the role of miR-205 was identified for discrimination of ADC from SSC (Vescovo et al., 2011). A combination of 63 miRNAs and ability to distinguish between basal and luminal subtypes in muscle-invasive bladder cancer (MIBC) was validated in a study. Further, this study also showed that basal subtype tumor patients had less survival (Ochoa et al., 2016). Zhang et al. have published a work in which miRNAs and mRNAs microarray experiments were executed on similar samples. To identify miRNA candidates in lung cancer, the regulatory network of miRNA-mRNA was obtained by integrating the gene expression and target predictions of miRNA. Through the microarray, miRNA-mRNA network and TCGA studies, 28 miRNAs were identified in lung cancer. These miRNAs are proven as promising biomarkers for the early detection of lung cancer (Zhang et al., 2017). Similarly, another miRNA, that is, miR-223, identifies as a potential biomarker for early-stage detection of non-small-cell lung cancer in 2019 by droplet digital PCR (ddPCR) technology (D'Antona et al., 2019). In FFPE samples (biopsy tissues) of lung tumors, conducted by Detassis and colleagues, it was demonstrated that miR375-3p could differentiate between low-grade neuroendocrine and non-neuroendocrine tumors (Detassis et al., 2020). A correlation study was performed in a study conducted by Mjelle et al. in 2019. In this study, profiling of small RNA and gene expression array analyses of cancer tissue, peritumoral tissues, soluble exosomes, and hepatocellular carcinoma serum samples and serum samples of non-cancer patients as controls were performed. After this study, a correlation was found between dysregulated miRNAs in tumor samples and circulating miRNAs levels (Mjelle et al., 2019). Similarly, a profiling study of exosomal miRNA was performed on lung cancer adenocarcinoma individuals before and after surgery by Xue et al. before surgery level of miR-484 was significantly increased compared to control and got decreased after surgery (Xue et al., 2020). However, the circulating miRNAs and miRNA expression in tumor cells are not so clear in some cases. For example, expression of miR-223 occurs in hematopoietic cells while secreted in microvesicles which further influence the invasive activity of tumor cells (Yang et al., 2011). These studies confirm the potential of miRNA as a non-invasive biomarker.

The expression of miRNA in the circulation was firstly reported in 2008 where the detection of four placental miRNAs (miR-141, miR-149, miR-299-5p, and miR-135b) have been reported in maternal plasma in the course of pregnancy and after the delivery level was decreased (Chim et al., 2008). Patients with diffuse large B-cell lymphoma had shown increased levels of circulating miR-21, miR-155, and miR-210 expression in comparison to healthy control in a study (Lawrie et al., 2008). Mitchell et al. also showed that patients with advanced prostate cancer can be distinguished from healthy panels by an increased level of miR-141 (Mitchell et al., 2008).

3.3.2.3.9 Circulating miRNAs: Diagnostic and Predictive Markers for Cancers

Currently available blood-based tumor markers for the diagnosis of cancer have drawbacks of low sensitivity, particularly concerning early tumor screening. From different studies, tissue miRNA signs have proven more reliable than the regulatory mRNA expression profiles for the detection of cancer (Cheng, 2015). The circulating miRNAs have been found to display a higher stability in body fluids and easier to extract from body fluids non-invasively. Several miRNA studies concluded that due to existence in exosomes and small double lipid layer vesicles, they show a high resistance towards RNases, so circulating miRNAs are extremely stable. Its level in plasma and serum is shown to be unaffected by different treatments such as high pH, freeze-thaw cycles, long storage, etc. (Tavallaie et al., 2015).

miRNAs have the properties to sustain the extreme temperature, pH, and harsh cell processings such as formalin-fixed paraffin-embedding (FFPE), making them highly stable biomarkers. In a study by Kakimoto and collaborators, a steady level of miRNAs is detected in postmortem tissue samples and can be further analyzed by qPCR after formalin fixation (Kakimoto et al., 2015). Deep sequencing of miRNA has been accomplished on FFPE samples and it was found that results are in good correlation with matched frozen samples results (Meng et al., 2013; Hedegaard et al., 2014). c-miRNA is considered a potential biomarker due to its high stability and potential to sustain the presence of ribonucleases and other harsh physiological conditions in body fluids. These properties increase the possibility of their use for clinical purposes. miRNAs level and their composition can change due to tumor conditions or diseases in the body (Li and Kowdley, 2012). Several research

TABLE 3.1

Examples of miRNA found in Serum/Plasma Associated with Different Types of Cancer

Type of Cancer	miRNA Upregulated	miRNA Downregulated	Sample	References
Non-small cell lung cancer (NSCLC)	miR-24, miR-223 miRNA-1, miR-30d, miR-499, miR-486 miR-21, miR-182, miR-210 miR-221, miR-660, miR-486-5p, miR-28-3p, miR-197, miR-106a, miR-451, miR-140-5p, miR-16, miR-125b, miR-198	miR-92a, miR-484, miR-328, miR-191, miR-376a, miR-342-3p, miR-331-2p, miR-30c, miR-28-5p, miR-98, miR-17, miR-26b, miR-374a, miR-30b, miR-26a, miR-134-3p, miR-103, miR-126, let-7a, let-7d	Serum/ plasma	Chen et al., 2005; Hu et al., 2010; Petriella et al., 2016
Breast cancer	miR-505-5p, miR-96-5p, miR-10b, miR-21, miR-125b, miR- 145, miR-155, miR-191, miR-10b, miRNA-373, miR-148b, miR-409-3p, miR-801, miR-484	miR-141, miR-144, miR-193b, miR- 200a, miR-200b, miR-200c, miR-203, miR-210, miR-215, miR-365, miR-375, miR-429, miR-486-5p, miR-801, miR-1260, miR-1274a	Serum/ plasma	Ng et al., 2013; Matamala et al., 2015; Cuk et al., 2013
Colorectal cancer (CRC)	let-7a, miR-1229, miR-1246, miR- 150, miR-21, miR-223, miR-23a miR-135b, miR-95, miR-222, miR-17-3p, miR-92, miR-193a-3p, miR-338-5p	miR-126, miR-21, miR-34a	Serum/ plasma	Ogata-Kawata et al., 2014; Ng et al., 2009; Wang & Gu, 2012
Pancreatic (PDAC)	miR-21, miR-210, miR-155, miR-196a,miR-20a, miR-21, miR-24, miR-25, miR-99a, miR-185, miR-191, miR-21, miR-483-3p, miR-21, miR-34a, miR-18a	miR-141	Serum/ plasma	Wang et al., 2009; R. Liu et al., 2012; X.G. Liu et al., 2012; G. Zhao et al., 2013

studies have already proven that miRNAs are sensitive towards the physiological or pathological states of the body and thus act as a reliable indicator of disease incidence and can be helpful for clinical diagnosis (De Guire et al., 2013; Ménard et al., 2016). Besides cancer, miRNAs have also identified as useful diagnostic biomarkers in the case of neurodegenerative diseases (Leidinger et al., 2013; Kumar et al., 2013; Grasso et al., 2014), cardiovascular illness (Schulte and Zeller, 2015; Orlicka-Płocka et al., 2016), and endometriosis, etc. (Esquela-Kerscher and Slack, 2006; Volinia et al., 2006; Huang et al., 2010; Wang et al., 2012, Bottani et al., 2020; Moretti et al., 2017; Agrawal et al., 2018). All these aspects together can reach an inference that circulating miRNAs epitomize an ideal biomarker class for cancer diagnosis and prognosis.

In different studies, several miRNAs (see Table 3.1) are found to be associated with different cancers, including lung, colorectal, pancreatic, breast, etc.

1. **Lung Cancer:** The first observational reports of involvement of circulating miRNAs in serum samples came from the study of Chen et al. in NSCLC patients, where the level of miR-25 and miR-223 found to be elevated in comparison to healthy panels (Cheng, 2015). In some studies, miRNA-1, miR-30d, miR-499, and miR-486 were underlined as serum-based

prognostic markers (Hu et al., 2010). In another study, serum levels of miR-486-5p were found to be prognostic of survival in NSCLC (Petriella et al., 2016).

2. **Breast Cancer:** The concentrations of miR-505-5p and miR-96-5p were found to be increased in the serum of early-stage breast cancer patients versus healthy panels, and after undergoing treatment it was found that the level of miR-505-5p was decreased (Matamala et al., 2015). The levels of several miRNAs (miR-148b, miR-376c, miR-409-3p, and miR-801) were found to be increased in plasma of breast cancer patients in comparison to healthy controls (Cuk et al., 2013). A panel of circulating miRNAs related to the inclusive survival of patients with metastatic breast cancer was recognized by Madhavan et al. (Ghai and Wang, 2016). Several members of this panel, like the miR-200 family, augmented in exosomes (Meng et al., 2016) are involved in metastasis.

3. **Colorectal Cancer:** Changes in circulating miRNA as a potent marker for colorectal cancer has been observed. In the serum of CRC patients, the increased concentration of miR-21 has been reported regularly in various studies. It was also reported that isolated serum miR-21 were exosomes enriched (Ogata-Kawata et al., 2014). Additionally, many other studies employed on the plasma of CRC patients have shown the raised levels of miR-135b, miR-95, miR- 222, miR-17-3p, and miR-92 in the plasma of CRC patients (Ng et al., 2009). Another potent marker for CRC is circulating miR-141, as elevated miR-141 is found to be associated with metastatic CRC patients (Cheng et al., 2011; Wang & Gu, 2012).

4. **Pancreatic Cancer:** The level of four miRNAs (miR-21, miR-155, miR-210, and miR-196a) in the plasma of pancreatic ductal adenocarcinoma (PDAC) is increased in comparison to healthy controls (Wang et al., 2009). Wang et al. have found the increased levels of miR-205, miR-210, miR-492, and miR-1247 in pancreatic juice of PDAC patients. This finding can differentiate PDAC patients with controls and found to give better sensitivity in combination with CA 19-9 (Wang et al., 2014). Cote et al. (2014) studied several miRNAs that are known to be associated with PDAC and elevation in the level of five miRNAs (miR-10b, miR-155, miR-106b, miR-30c, and miR-212) in the bile of PDAC patients has been reported, compared to healthy panels.

Hence, from the analysis of miRNA, it is indicated that, in comparison to other biomarkers, miRNAs are more stable, sensitive, and specific.

3.3.2.3.10 miRNA Expression Profiles in Cancer Tissues

1. **Hepatocellular Carcinoma:** Several studies have been performed to identify biomarkers and other therapeutic agents by evaluating the differential expression of miRNAs among the hepatocellular carcinoma tissue and the conforming nontumor liver tissue. Andrés-León et al. (2017) studied the miRNAs in The Cancer Genome Atlas (TCGA) data for HCC and other tumor types. In 2018, a study unified the miRNA expression database TCGA, GSE31384 (Wong et al., 2011), and GSE6857 (Gu et al., 2014) and evaluated their expression levels in liver cancer and normal samples to examine the difference in miRNAs' expression allied with carcinogenesis (Nagy et al., 2018). The researchers observed that hsa-miR-149, hsa-miR-139, hsa-miR-3677, hsa-miR-550a, and hsa-miR-212 were explicitly correlated with the overall survival of HCC patients (Li et al., 2016). The researchers also commenced the integrated analysis of several miRNA profiling studies (Murakami et al., 2009; Wang et al., 2013; Shi et al., 2015; Mou et al., 2017; Lou et al., 2019) in HCC samples for detection of differentially expressed miRNAs. The results suggested that miR- 221/222, miR-195, and miR-199a were reliably and differentially expressed in multiple independent studies. Though no overlapping expression for some of the miRNAs could be established in the existing literature and the available data, the miRNAs described most consistently as dysregulated, namely hsa-miR-18, hsa-miR-21, hsa-miR-106, hsa-miR-221/222, hsa-miR-

224, hsa-miR-99, hsa-miR-195, and hsa-miR-199, possess a significant role in the regulation of the hallmarks of HCC.

2. **Ovarian Cancer:** Tumors create a significant health concern and cause of death in humans. Their mechanisms of existence are complex with various factors, as many tumor types exhibit initiation of metastasis and recurrence. Also, multiple tumors are comparatively challenging to treat and have constantly been the focus and challenge for researchers. In recent years, the identification of miRNA has exhibited new insights to the study of tumors. Specifically, numerous miRNAs have been reported to have diverse roles in ovarian cancer. For example, the expression of some miRNAs is repressed in ovarian cancer, signifying that these can be considered tumor suppressor genes. Alternatively, other miRNAs are irrationally expressed in ovarian cancer and can be regarded as cancer-promoting genes. By associating the expression levels of miRNAs in ovarian cancer and normal tissues, Iorio et al. observed that the expression of miRNA-199a (miR-199a), miR-200a, miR-200b, and miR-200 had a noteworthy increase in its expression levels than that in normal tissues, whereas miR-140, miR-145, and miR-125b exhibited low expression in cancer tissues (Iorio et al., 2007). In turn, the miR-15 and miR-16 were down-regulated in ovarian cancer tissues, and miR-31 was expressed at low levels in ovarian cancer cells and tissues, signifying its role as a tumor suppressor gene in ovarian cancer (Aqeilan et al., 2010; Creighton et al., 2010). miR-200a, miR-200b, and miR-200c have also been reported to exhibit significantly higher levels in serous epithelial ovarian cancer in comparison to their expression levels in normal ovarian tissue (Kan et al., 2012). Because of the altered expression and roles of different miRNAs in ovarian cancer, specific miRNAs can be considered potential screening indicators as more specific molecular markers for the early diagnosis of ovarian cancer.

3.4 INTEGRATIVE TECHNIQUES

3.4.1 METHODS FOR DETECTION OF CIRCULATING MIRNA

3.4.1.1 Non-Sensor Techniques

For the detection of miRNA, most commonly employed methods are based on amplification techniques, including qRT-PCR (Ell et al., 2013), microarrays (Dong et al., 2013), and RNA sequencing in the laboratory. For the quantitative measurement of one constituent of a sample, it depends on the selective binding of the molecular probe.

For absolute miRNA quantification, the most commonly used method is qRT-PCR, which has an extensive product range with maximum accuracy (dilutions of synthetic miRNA oligonucleotides of known concentration are used to generate standard curves for detection).

miRNA microarrays are less costly but this method shows lower sensitivity and dynamic range. Therefore, it can be best used as discovery tools instead of using it as a quantitative method.

For RNA sequencing, for a given sample expression of miRNA quantification is used as a value relative to the total number of sequence reads (Tavallaie et al., 2015).

3.4.1.1.1 Quantitative PCR (qPCR)

qPCR is considered the most sensitive method of miRNA quantification among the "target amplification-based" methods. For differential gene expression analysis, qPCR is the most commonly used method.

However, miRNAs are such easy targets to quantify by PCR methods due to several reasons: (i) the small size of miRNA interferes with the optimal primer designing (ii) presence of stable hairpin structure in the pre-miRNA, (iii) omnipresence of different family members of miRNA that differ in only one or a few nucleotides, (iv) and most important, their low abundance in body fluids and interference in detection due to genomic contamination. Hence, genomic DNA contamination removal is required before the reverse transcription analysis. Furthermore, miniature miRNAs groups can be assessed with customized plates (e.g., Thermo Fisher) that are commercially available. Two main approaches combining the retro transcription and amplification steps are presently adopted to detect miRNA from qPCR. The first method ((Qiagen, Exiqon) comprises the use of poly(A) extension for cDNA synthesis before the reverse transcription process in a single step, disabling the restriction of short primer designing in qPCR. The assay is performed with double-stranded DNA binding dyes or SYBR green. In the second method (TaqMan technology), specific stem-loop primers are used to synthesize cDNA to enhance amplification specificity. Afterward, qPCR is performed by the fluorescent FAM method. RT-qPCR method of Exiquon is considered a more sensitive and specific miRNA detection method than the TaqMan Technology (Chugh and Dittmer, 2012; Leshkowitz et al., 2013; Ruiz-Tagle et al., 2020).

3.4.1.1.2 Digital PCR (dPCR)

Among the existing PCR methods, dPCR is recently considered the most capable method for absolute identification. The digital droplet PCR method is based on a sample division into thousands of micro-reactions of fixed volume (Pinheiro et al., 2012). Fluorescence measurement of each droplet quantifies the nucleic acid sample and ddPCR. These fluorescent signals (droplets) can be positive or negative and comprise the target nucleic acid. This permits the quantification of reaction molecules, assuming the Poisson distribution. It provides copies according to the reaction volume with confidence intervals as an experiment result (Hindson et al., 2011; Huggett, 2020). Hence, in this method, reference genes or a calibration curve are not required for quantification. Higher precision is obtained via this method compared to qPCR for quantification of miRNA (Campomenosi et al., 2016). Other advantages of ddPCR over qPCR include higher sensitivity for quantification and detection of fewer abundant targets (Pritchard et al., 2012; Boeckel et al., 2013; Brunetto et al., 2014; Zhao et al., 2015) and abridged sensitivity towards the PCR inhibitors (Rački et al., 2014; Doi et al., 2015). ddPCR has proven to be useful for miRNA detection as a diagnostic and prognostic method (Ma et al., 2013; Giraldez, Chevillet, et al., 2018; Tavano et al., 2018; Zhao et al., 2018; Choi et al., 2019; D'Antona et al., 2019; Zhao et al., 2019).

Some drawbacks of qPCR have not been solved by ddPCR as it also requires the extraction of miRNA and reverses transcription. ddPCR application for miRNA measurement is not commonly spread due to the high cost of instruments and reagents, relatively extended procedure, and need for highly trained operators.

3.4.1.1.3 miRNA Microarrays

Microarrays signify a high-throughput technique that can simultaneously detect the presence and alterations in expression levels of various miRNAs in an experiment. The principle of microarray involves nucleic acid hybridization of target molecules and their complementary probes. The probes have 50amine-modification in microarray analysis. It is synthesized in this form for immobilization upon solid support via covalent crosslinking. Various fluorescent dyes are available such as Alexa Fluor 546/647, Cy3, etc., to label miRNAs. Due to fluorescently labeled miRNAs hybridized to microarrays, specific binding occurs between the labeled miRNAs and the complementary probes. By analyzing the signal intensity of fluorescence emission, relative miRNA quantities can be evaluated in a specific experimental set (Li and Ruan, 2009; de Planell-Saguer and Rodicio, 2013). Some microarray platforms allow comparing two different samples using two types of fluorophores (Li and Ruan, 2009; de Planell-Saguer and Rodicio, 2013). In recent times, various alternatives have been forwarded to detect miRNAs using microarray technology. These technical variations include immobilization chemistry, labeling of samples, probe design, and chip-based microarray signal detection technique (Liu et al., 2008). Also, several commercial microarray platforms exist to detect and quantify the miRNA, and their ability to measure differential expression of miRNAs has been estimated. Different studies have assessed the other platforms for microarray and found significant variations in quantification and differential expression of miRNAs (Del Vescovo et al., 2013; Mestdagh et al., 2014). These variations could arise at the technical steps such as during enzymatic reactions, amplification process or microarray probe designing, microarray, or the algorithms used for data analyses. In the case of mRNA microarrays, identifying several housekeeping genes is possible, but in miRNA microarrays, it is not easy to locate invariable noncoding RNAs. For the normalization of data, mathematical models can be used as hundreds of non-coding RNA probes are used in microarray simultaneously.

These approaches depend on the assumptions that (1) differential expression of less quantity of miRNAs occur in samples under study against control samples and (2) upregulated downregulated miRNAs unevenly balance miRNAs. For reducing the differences in expression values of miRNAs, median and quantile normalization techniques have been used in the case of identical tissue samples (Rao et al., 2008). Overall, miRNA microarrays are most useful for initial screening but not sufficiently sensitive for absolute quantification.

3.4.1.1.4 Small RNA Sequencing (sRNA-Seq)

Due to recent advancements in next-generation sequencing technology, miRNA profiling becomes reliable and flexible. sRNA-Seq mainly comprises the following steps: isolation of total RNA, size fractionation, adapter ligation at 3' and 5' ends of small RNAs, reverse transcription and amplification via PCR. At the completion stage of sequencing, final reads are denoted by unique reads to a map upon the modern annotated genome or the mature miRNA sequences found on miRbase (www.mirbase.org). Library construction is achieved by employing different kits that depend upon the sequencing platforms. There are two strategies for sRNA-Seq: (i) with adapters having invariant ends (e.g., TruSeq (Illumina) or CleanTag (Trilink Biotech) and (ii) with adapters having four degenerative nucleotides at the ligation site (e.g., NEXTflex (Bioo Scientific)) (Dard-Dascot et al., 2018; Giraldez, Spengler, et al., 2018). Due to the ligation of adapters, serious bias can also be known as "sequence bias." This condition develops because of the structural effects of RNA that allows preferential ligation of specific sRNAs with an adapter sequence, making it a disadvantage to others. In comparison to long RNA-Seq, sRNA-Seq exhibits more sequence bias (Dard-Dascot et al., 2018). Researchers have tried to reduce the bias by optimizing the reaction conditions through the thermostable DNA/RNA ligase or changing ligation temperature (Giraldez, Spengler, et al., 2018).

3.4.1.2 Biosensor-Based Techniques

Recent advancements in nanotechnology for point-of-care diagnosis could lead to novel nano biosensors development with a high potential. Usually, a biosensor is a combination of the receptor (i.e., bioreceptor/biorecognition molecule) for the specific recognition of target analyte and the transducer (signal-converter and enhancer) for recognizing biomolecular interaction and conversion of this interaction into an assessable signal (Tavallaie et al., 2015; Islam, Yadav, et al., 2017). It is a portable analytical device and possible to achieve rapid and accurate results by producing biosensors even in a highly complex medium such as blood and urine by using this combination of biorecognition species with a sensitive transducer without the use of polymerase-based amplification steps. With these smart features, a new detection method based on biosensors for circulating miRNA assays has been developed (Tavallaie et al., 2015). Several novel biosensors comprising nanopore, optical, and electrochemical read-out techniques have widely been established to quantify and analyze miRNAs (Kong et al., 2018). Unfortunately, though, sensitivity, selectivity, cost, and response time of a developing biosensor continue to be issues (Wu et al., 2018).

3.4.1.2.1 Electrochemical Biosensors

Electrochemical methods for the detection of miRNA detection specifically rely on the hybridization of target RNA sequences to the complementary receptor probes bound on the surface (commonly the DNA oligonucleotides) of the electrode. The signal transduction of an electrochemical process relies on intrinsic and extrinsic including the electroactivity of nucleobases, redox indicators (e.g., methylene blue), covalently bound redox labels (e.g., nanoparticles), reporter enzymes (e.g., phosphatases, peroxidases), etc. (Palecek and Bartosik, 2012; Hartman et al., 2013). Electrochemical sensing systems are highly sensitive, specific, cost-effective, and straightforward to operate compared to other methods. In addition, due to their portability and amenability to miniaturization, they possess the potential for the development of point-of-care testing devices. Due to their high sensitivity (sub-fM detection limits), these electrochemical systems can be highly suitable for liquid biopsy-based analysis of low-frequency biomolecules such as circulating disease-specific miRNAs (Hossain et al., 2017).

For circulating miRNA detection, previously several electrochemical approaches that have already been used up to now are simply modifications of DNA hybridization biosensors, i.e., the changes in target sequence from DNA to miRNA has been employed. This method can improve detection limits also. It is found that by using 3D conductive materials that have been made by increasing the electroactive zone and decreasing the electrical resistance of the working electrodes, circulating miRNA in the diluted serum detection has been improved to a 10 fM concentration detection limit. Two auxiliary probes are another approach that has the capability to self-assemble to form one-dimensional DNA concatamers (Hong et al., 2013). It is used to amplify the signal and it is known that any hybridization-based method is dependent on probable hybridization of the probe with other RNA, which is a limitation of this approach. To overcome this problem, one inventive technique has been developed by using the p19 RNA-binding protein. Labib et al. have developed a striking way by taking advantage of this unique binding property of the p19 protein to the small double-stranded RNA. The three sensing types combined in this sensor link have high sensitivity, a wide potential range of measured concentrations (11 orders of magnitude, from 10 aM to 1 μM), triple confirmation of the miRNA concentration, and two different miRNAs are sequentially analyzed on one electrode. This method has been confirmed by qRT-PCR and used positively for direct detection and profiling of endogenous miRNAs in human serum, stated as biomarkers for colorectal, prostate, and liver tumors (Labib et al., 2013).

Electrochemical detection of miRNAs is generally accomplished via voltammetric, amperometric, and impedimetric methods (Johnson and Mutharasan, 2014). One such example includes the direct oxidation-based identification of circulating miRNA bases, reported by Lusi et al. (2009). In this study, the miR-122 was hybridized with its inosine substitute capture probe. Carbon-based nano-electrode and electroactive polymers were employed to increase the electroactive area and reduce electrical resistance on the electrode surface. During the RNA-capture probe hybridization process, direct oxidation of guanine on the surface of the electrode exerted an electrical signal, which was recorded by using differential pulse voltammetry (DPV). The reported electrode possessed a low limit of detection (LOD) with a value of 10 fM. Also, another highly sensitive (LOD = 100 aM) sensor was reported, in which the two auxiliary probes were allowed to self-assemble to form one-dimensional DNA concatemers. The hairpin capture probe was conjugated on the surface of the screen-printed gold electrode. The results depicted that in the absence of target (miR-21) on the sensor surface, the hairpin probe constituted its loop structure, exhibiting no binding site for the DNA concatemers and resulting in minimal electrochemical signals.

However, in the presence of a target miRNA, the stem-loop structure of the probe was disturbed, allowing hybridization with DNA concatemers. This resulted in a positively charged RuHex reporter molecule interacting with an anionic target-probe, thus generating a noteworthy increase in the electrochemical signal. As reported earlier, one of the highly concerning problems with miRNA-based biosensing is the probability of interference from RNA sequences of closely related sequences (e.g., different intermediates of RNA synthesis pathway including pri-miRNA, pre-miRNA and rRNA, dsRNA, miRNA from the same family, etc.). In 2013, Kilic et al. reported a specific method to overcome the reported issue (Kilic et al., 2013). One particular type of protein known as p19 was used due to its role as a molecular caliper for small, double-stranded RNA (21–23 base pairs) and isolated miRNAs in a size-dependent and sequence-independent manner. Due to its high specificity for miRNAs, the p19 protein does not interact with other nucleotide molecules such as ssRNA, rRNA, mRNA, ssDNA, or dsDNA (Khan et al., 2011). Hence, the addition of p19 in the sensor system results in the reduction of the non-specific detection. A simple and highly sensitive system for electrochemical detection of miRNA has also been reported (Boriachek et al., 2018). The sensor employs a target miRNA-specific probe altered with gold-coated magnetic nanoparticles as the capturing entities for isolation of target miRNA from the raw blood. Subsequent assembly of nanoparticles on the surface of a gold electrode is followed by electric field-induced reconfiguration. Hybridization of miRNA to the target probe overwhelms the change in current due to its augmented distance between the DNA and Au@MNPs and by translocating the hybridized nanosensors in the vicinity of the electrode surface where they act as blockades for the current tunneling. The method resulted in the highly sensitive detection of miR-21 and a broad concentration range (10 aM to 1 nM). Though this method was only established in a clinically relevant xenograft mouse model of human lung cancer, the adaptability of the system and the prime role of miR-21 in ovarian cancer suggest that the platform can be employed for detection and monitoring of ovarian cancer.

3.4.1.2.2 Nanopore-Based Biosensors

Nanopore sensors are being developed for the detection of the position and conformation of a single molecule that is present within the pore lumen (Bayley & Jayasinghe, 2004). Numerous nanopore sensors have been developed, including the next-generation DNA sequencing technology. A nanopore detection approach has been developed by Wang et al. using the translocation of single-stranded oligonucleotides through the 2-nm pore containing an α-hemolysin protein pore (Meller et al., 2001). A programmable oligonucleotide probe is being used for high selectivity and sensitivity in circulating miRNA detection from plasma RNA extracts. Detection of miRNA in blood with a detection limit of 100 fM has been reported via this method without the need for amplification of miRNA.

Nanopores, nano-scale pore structures, are among the most protuberant single-molecule sensors used in miRNA analysis (Wanunu et al., 2010; Wang et al., 2011;; Gu et al., 2012; Henley et al., 2016). Characteristically, in the existence of a conducting liquid, the nanopores generate current at constant potential due to the transport of charges in the holes. The induced current is a function of the size and physical parameters of the pore. Contingent on the existence of target molecules, including miRNAs in the pore, the alterations in the current can be measured, thus resulting in the detection of the molecule (Olasagasti et al., 2010; Clarke et al., 2009). Wang et al. developed a specific nanopore-based sensor to detect miRNAs employing a hemolysin protein pore (Wang et al., 2011). The sensor relies on the movement of single-stranded oligonucleotides through the 2-nm sized pore consisting of a programmable oligonucleotide probe. The sensor also has a high detection limit of 100 fM of miRNA in the blood sample. Also, the sensor was effectively tested to distinguish the relative levels of miR-155 in cancer patients.

3.4.1.2.3 Optical Biosensors

Optical biosensing methods rely on the energy transfer phenomenon among the two light-sensitive molecules. After reaching its highest excitation state, the donor chromophore transfers its energy to the acceptor molecule (chromophore) via dipole-dipole interactions (Orazem and Tribollet, 2008). The energy transfer mechanism depends on the donor and acceptor's distance and orientation (Orazem and Tribollet, 2008; Chang and Park, 2010; Balaji and Zhang, 2017). There are other energy transfer mechanisms for optical sensors, such as fluorescence, resonance, forster resonance, or electronic energy transfer mechanism (Chang and Park, 2010; Balaji and Zhang, 2017).

For the analysis of miRNAs associated with ovarian cancer, SPR (surface plasmon resonance) and SERS (surface-enhanced raman spectroscopy)-based optical sensors are used

(Carrascosa et al., 2016). SPR detects the refractive index change after the binding of biomolecules and the probe is immobilized on a metal surface. This technique is beautiful for real-time measurement of binding kinetics, in situ, label, and enzyme-free (Homola, 2008; Islam, Masud, et al., 2017). The main drawback of this method is low sensitivity. Different signal amplification approaches have been employed to improve the sensitivity, such as using metallic or metallic oxide nanoparticles (Wang et al., 2016; Liu et al., 2017; Wang et al., 2019), use of a DNA super sandwich and HCR (hybridization chain reaction), etc. In a study by He et al., it has been shown that gold nanoparticles can increase the sensitivity of SPR-based optical sensors by more than 1,000 times (He et al., 2000). In another study by Corn et al., an approach was used for miRNA detection with a LOD (limit of detection) value of 10 fM using SPR imaging. In this method, hybridization of miRNA and single-stranded locked nucleic acid (LNA) microarray was performed. This step was trailed by enzymatic polyadenylation of miRNA-3' end. Then, hybridization of T30-coated gold nanoparticles to a poly(A) tail amplified the SPRI signal. The only disadvantages of this method are its long procedure time in hybridization and multiple steps (Fang et al., 2006).

In a recent study by Hu et al., a sensitive SPRI approach was used to detect miRNA-15a with a LOD value of 0.56 fM in serum samples of humans (Hu et al., 2017). In another study, Vaisocherová et al. have simultaneously detected multiple miRNAs samples in a SPRI biosensor. For example, 0.5 pM of miRNAs were detected in an hour from the erythrocyte lysate (Vaisocherová et al., 2015). Similarly, in a study by Sipova et al., for detection of miR-122, a label-free portable sensor based on SPR technology was developed. It is a rapid, sensitive method and takes only 35 minutes for detection (Wang et al., 2016). In the same step, additional amplification was provided by recognition of an miRNA with antibody. This step lowered the limit of detection to 2 pM. One major problem of SPR is the production of steric hindrance due to biomolecule tags used in the amplification step. SERS-based biosensors rely on interactions between the LSPRs (localized surface plasmon resonance) of metallic nanostructures and electromagnetic fields generated through the emission of the molecules from the metal surface.

The SERS detection method is divided into direct and indirect ways based on the fabrication of plasmonic nanostructures. Through the immediate process, the essential spectrum of the target molecule is developed, while in the indirect approach, signals are generated from the Raman reporters associated with the target. However, in the direct process, there is a lack of sensitivity and reproducibility. Hence, nucleic acid biosensors rely on the indirect method (Garcia-Rico et al., 2018).

However, recently, a direct system in SERS has been recognized with greater sensitivity and reproducibility. In a study by Lee et al., silver nanocrystals were deposited inside the gold nanobowls electrochemically. The nanogap among these two, i.e., silver nanocrystals and gold nanobowls, enhances SERS and signal homogeneity. Furthermore, using this method lower than 1 fM, miR-34a was detected in the gastric cancer cell line (Lee et al., 2018).

Forster resonance energy transfer (FRET) is an optical detection method that depends on radiation-less electronic energy transmission from the donor to the acceptor molecule via dipole-dipole interactions (Andryushchenko and Chekmarev, 2018). This method was best demonstrated in the study of Qiu et al. (2018; Zhou et al., 2018). They established a single detection step method via isothermal amplification of miR-146a, miR-21, and miR-132. It was based on FRET among Tb donors and dye acceptors, with a detection limit of 4.2 ± 0.5 attomoles (Qiu et al., 2018; Zhou et al., 2018). However, the main disadvantage of this method is that it was verified on only a small group of samples (Qiu et al., 2018).

REFERENCES

Advani, P.P., Crozier, J.A., and Perez, E.A., 2015. HER2 testing and its predictive utility in anti-HER2 breast cancer therapy. *Biomarkers in Medicine*, 9(1), pp. 35–49.

Agrawal, S., Tapmeier, T.T., Rahmioglu, N., Kirtley, S., Zondervan, K.T., and Becker, C.M., 2018. The miRNA mirage: how close are we to finding a non-invasive diagnostic biomarker in endometriosis? A systematic review. *International Journal of Molecular Sciences*, 19(2), p. 599.

Aisaka, K., Obata, S., Koshino, T., Kaibara, M., and Mori, H., 2001. CA-125 dynamic test: a novel method for clinical diagnosis of endometriosis. *Fertility and Sterility*, 76(3), pp. S148–S149.

Aisaka, K., Takada, S.-I., and Obata, S., 2002. A novel method for clinical diagnosis of endometriosis: CA-125 dynamic test. *Fertility and Sterility*, 77, pp. S6–S7.

Alam, R., Tosoian, J.J., Okani, O., Ross, A.E., and Vuica-Ross, M., 2017. Metastatic prostate cancer diagnosed by bone marrow aspiration in an elderly man not undergoing PSA screening. *Urology Case Reports*, 11, pp. 7–8.

Alizadeh Savareh, B., Asadzadeh Aghdaie, H., Behmanesh, A., Bashiri, A., Sadeghi, A., Zali, M., and Shams, R., 2020. A machine learning approach identified a diagnostic model for pancreatic cancer through using circulating microRNA signatures. *Pancreatology*, 20(6), pp. 1195–1204.

Al-Shaheri, F.N., Alhamdani, M.S.S., Bauer, A.S., Giese, N., Büchler, M.W., Hackert, T., and Hoheisel, J.D., 2021. Blood biomarkers for differential diagnosis and early detection of pancreatic cancer. *Cancer Treatment Reviews*. doi: 10.1016/j.ctrv.2021.102193.

Amelot, A., Terrier, L.M., Cristini, J., Buffenoir, K., Pascal-Moussellard, H., Carpentier, A., Bonaccorsi, R., Le Nail, L.R., and Mathon, B., 2020. Survival in breast cancer patients with spine metastases: Prognostic assessment involving molecular markers. *European Journal of Surgical Oncology*, 46(6), pp. 1021–1027.

Andrés-León, E., Cases, I., Alonso, S., and Rojas, A.M., 2017. Novel miRNA-mRNA interactions conserved in essential cancer pathways. *Scientific reports*, 7(1), pp. 1–13.

Andryushchenko, V.A., and Chekmarev, S.F., 2018. Modeling of multicolor single-molecule förster resonance energy-transfer experiments on protein folding. *The Journal of Physical Chemistry B*, 122(47), pp. 10678–10685.

Anker, P., Lyautey, J., Lederrey, C., and Stroun, M., 2001. Circulating nucleic acids in plasma or serum. *Clinica Chimica Acta*, 313(1-2), pp. 143–146.

Aqeilan R.I., Calin G.A., and Croce C.M., 2010. miR-15a and miR-16-1 in cancer: Discovery, function and future perspectives. *Cell Death Differ*, 17, pp. 215–220. doi: 10.1038/cdd.2009.69.

Archer, K., Broskova, Z., Bayoumi, A.S., Teoh, J.P., Davila, A., Tang, Y., Su, H., and Kim, I.M., 2015. Long non-coding RNAs as master regulators in cardiovascular diseases. *International Journal of Molecular Sciences*, 16(10), pp. 23651–23667.

Arend, R.C., Jones, B.A., Martinez, A., and Goodfellow, P., 2018. Endometrial cancer: Molecular markers and management of advanced stage disease. *Gynecologic Oncology*. doi: 10.1016/j.ygyno.2018.05.015.

Azad, T.D., Chaudhuri, A.A., Fang, P., Qiao, Y., Esfahani, M.S., Chabon, J.J., Hamilton, E.G., Yang, Y.D., Lovejoy, A., Newman, A.M., Kurtz, D.M., Jin, M., Schroers-Martin, J., Stehr, H., Liu, C.L., Hui, A.B.Y., Patel, V., Maru, D., Lin, S.H., Alizadeh, A.A., and Diehn, M., 2020. Circulating tumor DNA analysis for detection of minimal residual disease after chemoradiotherapy for localized esophageal cancer. *Gastroenterology*, 158(3), pp. 494–505.e6.

Balaji, A., and Zhang, J., 2017. Electrochemical and optical biosensors for early-stage cancer diagnosis by using graphene and graphene oxide. *Cancer Nanotechnology*, 8(1), pp. 1–12.

Ballehaninna, U.K., and Chamberlain, R.S. 2011. Serum CA 19-9 as a biomarker for pancreatic cancer—a comprehensive review. *Indian Journal of Surgical Oncology*, 2(2), pp. 88–100.

Baron, J.A., 2012. Screening for cancer with molecular markers: Progress comes with potential problems. *Nature Reviews Cancer*, 12(5), pp. 368–371.

Bayley, H., and Jayasinghe, L., 2004. Functional engineered channels and pores. *Molecular Membrane Biology*, 21(4), pp. 209–220.

Bettegowda, C., Sausen, M., Leary, R.J., Kinde, I., Wang, Y., Agrawal, N., Bartlett, B.R., Wang, H., Luber, B., Alani, R.M., and Antonarakis, E.S., 2014. Detection of circulating tumor DNA in early-and late-stage human malignancies. *Science Translational Medicine*, 6(224), pp. 224ra24–224ra24.

Boeckel, J.N., Thomé, C.E., Leistner, D., Zeiher, A.M., Fichtlscherer, S., and Dimmeler, S., 2013. Heparin selectively affects the quantification of microRNAs in human blood samples. *Clinical Chemistry*, 59(7), pp. 1125–1127.

Boriachek, K., Umer, M., Islam, M.N., Gopalan, V., Lam, A.K., Nguyen, N.T., and Shiddiky, M.J., 2018. An amplification-free electrochemical detection of exosomal miRNA-21 in serum samples. *Analyst*, 143(7), pp. 1662–1669.

Bottani, M., Banfi, G., and Lombardi, G., 2020. The clinical potential of circulating miRNAs as biomarkers: present and future applications for diagnosis and prognosis of age-associated bone diseases. *Biomolecules*, *10*(4), p. 589.

Bronkhorst, A.J., Ungerer, V., and Holdenrieder, S., 2019. The emerging role of cell-free DNA as a molecular marker for cancer management. *Biomolecular Detection and Quantification*. doi: 10.1016/j.bdq.2019. 100087.

Brown, A.L., Jeong, J., Wahab, R.A., Zhang, B., and Mahoney, M.C., 2021. Diagnostic accuracy of MRI textural analysis in the classification of breast tumors. *Clinical Imaging*, *77*, pp. 86–91.

Brunetto, G.S., Massoud, R., Leibovitch, E.C., Caruso, B., Johnson, K., Ohayon, J., Fenton, K., Cortese, I., and Jacobson, S., 2014. Digital droplet PCR (ddPCR) for the precise quantification of human T-lymphotropic virus 1 proviral loads in peripheral blood and cerebrospinal fluid of HAM/TSP patients and identification of viral mutations. *Journal of Neurovirology*, *20*(4), pp. 341–351.

Burki, T.K. ha., 2015. CA-125 blood test in early detection of ovarian cancer. *The Lancet. Oncology*. doi: 10.1016/S1470-2045(15)70237-8.

Bustamante Alvarez, J.G., Janse, S., Owen, D.H., Kiourtsis, S., Bertino, E.M., He, K., Carbone, D.P., and Otterson, G.A., 2020. Treatment of non–small-cell lung cancer based on circulating cell-free DNA and impact of variation allele frequency. *Clinical Lung Cancer*. doi: 10.1016/j.cllc.2020.11.007.

Cai, Z., Zhao, J.S., Li, J.J., Peng, D.N., Wang, X.Y., Chen, T.L., Qiu, Y.P., Chen, P.P., Li, W.J., Xu, L.Y., Li, E.M., Tam, J.P.M., Qi, R.Z., Jia, W., and Xie, D., 2010. A combined proteomics and metabolomics profiling of gastric cardia cancer reveals characteristic dysregulations in glucose metabolism. *Molecular and Cellular Proteomics*, *9*(12), pp. 2617–2628.

Campomenosi, P., Gini, E., Noonan, D.M., Poli, A., D'Antona, P., Rotolo, N., Dominioni, L., and Imperatori, A., 2016. A comparison between quantitative PCR and droplet digital PCR technologies for circulating microRNA quantification in human lung cancer. *BMC Biotechnology*, *16*(1), pp. 1–10.

Canizares, F., Sola, J., Perez, M., Tovar, I., De Las Heras, M., Salinas, J., Penafiel, R., and Martinez, P., 2001. Preoperative values of CA 15-3 and CEA as prognostic factors in breast cancer: A multivariate analysis. *Tumor Biology*, *22*(5), p. 273.

Cantile, M., Di Bonito, M., Tracey De Bellis, M., and Botti, G., 2021. Functional interaction among lncRNA HOTAIR and MicroRNAs in cancer and other human diseases. *Cancers*, *13*, p. 570.

Carrascosa, L.G., Huertas, C.S., and Lechuga, L.M., 2016. Prospects of optical biosensors for emerging label-free RNA analysis. *Trac Trends in Analytical Chemistry*, *80*, pp. 177–189.

Centuori, S.M., and Martinez, J.D., 2014. Differential regulation of EGFR–MAPK signaling by deoxycholic acid (DCA) and ursodeoxycholic acid (UDCA) in colon cancer. *Digestive Diseases and Sciences*, *59*(10), pp. 2367–2380.

Chang, B.Y., and Park, S.M., 2010. Electrochemical impedance spectroscopy. *Annual Review of Analytical Chemistry*, *3*, pp. 207–229.

Chen, C., Ridzon, D.A., Broomer, A.J., Zhou, Z., Lee, D.H., Nguyen, J.T., Barbisin, M., Xu, N.L., Mahuvakar, V.R., Andersen, M.R., and Lao, K.Q., 2005. Real-time quantification of microRNAs by stem–loop RT–PCR. *Nucleic Acids Research*, *33*(20), pp. e179–e179.

Chen, C., Stock, C., Jansen, L., Chang-Claude, J., Hoffmeister, M., and Brenner, H., 2019. Trends in colonoscopy and fecal occult blood test use after the introduction of dual screening offers in Germany: Results from a large population-based study, 2003–2016. *Preventive Medicine*, *123*, pp. 333–340.

Cheng, G., 2015. Circulating miRNAs: roles in cancer diagnosis, prognosis and therapy. *Advanced Drug Delivery Reviews*, *81*, pp. 75–93.

Cheng, H., Zhang, L., Cogdell, D.E., Zheng, H., Schetter, A.J., Nykter, M., Harris, C.C., Chen, K., Hamilton, S.R., and Zhang, W., 2011. Circulating plasma MiR-141 is a novel biomarker for metastatic colon cancer and predicts poor prognosis. *PLoS One*, *6*(3). doi: 10.1371/journal.pone.0017745.

Cheng, L., and Meiser, B., 2019. The relationship between psychosocial factors and biomarkers in cancer patients: A systematic review of the literature. *European Journal of Oncology Nursing*. 10.1016/j.ejon.2019.06.002

Chiacchiarini, M., Trocchianesi, S., Besharat, Z.M., Po, A., and Ferretti, E., 2021. Role of tissue and circulating microRNAs and DNA as biomarkers in medullary thyroid cancer. *Pharmacology and Therapeutics*. doi: 10.1016/j.pharmthera.2020.107708.

Chim, S.S., Shing, T.K., Hung, E.C., Leung, T.Y., Lau, T.K., Chiu, R.W., and Dennis Lo, Y.M., 2008. Detection and characterization of placental microRNAs in maternal plasma. *Clinical Chemistry*, *54*(3), pp. 482–490.

Choi, R.S.Y., Lai, W.Y.X., Lee, L.T.C., Wong, W.L.C., Pei, X.M., Tsang, H.F., Leung, J.J., Cho, W.C.S.,

Chu, M.K.M., Wong, E.Y.L., and Wong, S.C.C., 2019. Current and future molecular diagnostics of gastric cancer. *Expert Review of Molecular Diagnostics*, *19*(10), pp. 863–874.

Chugh, P., and Dittmer, D.P., 2012. Potential pitfalls in microRNA profiling. *Wiley Interdisciplinary Reviews: RNA*, *3*(5), pp. 601–616.

Clarke, J., Wu, H.C., Jayasinghe, L., Patel, A., Reid, S., and Bayley, H., 2009. Continuous base identification for single-molecule nanopore DNA sequencing. *Nature Nanotechnology*, *4*(4), pp. 265–270.

Conlon, K.E., Lester, L., and Friedewald, S.M., 2019. Digital mammography. *Advances in Clinical Radiology*, *1*, pp. 19–25.

Conteduca, V., Burgio, S.L., Menna, C., Carretta, E., Rossi, L., Bianchi, E., Masini, C., Amadori, D., and De Giorgi, U., 2014. Chromogranin A is a potential prognostic marker in prostate cancer patients treated with enzalutamide. *The Prostate*, *74*(16), pp. 1691–1696.

Costantini, D., 2019. Understanding diversity in oxidative status and oxidative stress: The opportunities and challenges ahead. *Journal of Experimental Biology*, *222*(13). doi: 10.1242/jeb.194688.

Cote, G.A., Gore, A.J., McElyea, S.D., Heathers, L.E., Xu, H., Sherman, S., & Korc, M. 2014. A pilot study to develop a diagnostic test for pancreatic ductal adenocarcinoma based on differential expression of select miRNA in plasma and bile. *The American Journal of Gastroenterology*, 109(12), p. 1942.

Creighton, C.J., Fountain, M.D., Yu, Z., Nagaraja, A.K., Zhu, H., Khan, M., Olokpa, E., Zariff, A., Gunaratne, P.H., Matzuk, M.M., and Anderson, M.L., 2010. Molecular profiling uncovers a p53-associated role for microRNA-31 in inhibiting the proliferation of serous ovarian carcinomas and other cancers. *Cancer Research*, *70*(5), pp. 1906–1915.

Cuk, K., Zucknick, M., Heil, J., Madhavan, D., Schott, S., Turchinovich, A., Arlt, D., Rath, M., Sohn, C., Benner, A., and Junkermann, H., 2013. Circulating microRNAs in plasma as early detection markers for breast cancer. *International journal of cancer*, *132*(7), pp. 1602–1612.

D'Antona, P., Cattoni, M., Dominioni, L., Poli, A., Moretti, F., Cinquetti, R., Gini, E., Daffrè, E., Noonan, D.M., Imperatori, A., and Rotolo, N., 2019. Serum miR-223: A validated biomarker for detection of early-stage non–small cell lung cancer. *Cancer Epidemiology and Prevention Biomarkers*, *28*(11), pp. 1926–1933.

Dard-Dascot, C., Naquin, D., d'Aubenton-Carafa, Y., Alix, K., Thermes, C., and van Dijk, E., 2018. Systematic comparison of small RNA library preparation protocols for next-generation sequencing. *BMC Genomics*, *19*(1), pp. 1–16.

de Araújo, R.F., Martins, D.B.G., and Borba, M.A.C., 2016. Oxidative stress and disease. In *A Master Regulator of Oxidative Stress-The Transcription Factor Nrf2*. IntechOpen.

De Guire, V., Robitaille, R., Tetreault, N., Guerin, R., Menard, C., Bambace, N., and Sapieha, P., 2013. Circulating miRNAs as sensitive and specific biomarkers for the diagnosis and monitoring of human diseases: promises and challenges. *Clinical Biochemistry*, *46*(10-11), pp. 846–860.

de Kock, R., Borne, B. van den, Soud, M.Y.- El, Belderbos, H., Stege, G., de Saegher, M., van Dongen-Schrover, C., Genet, S., Brunsveld, L., Scharnhorst, V., and Deiman, B., 2021. Circulating biomarkers for monitoring therapy response and detection of disease progression in lung cancer patients. *Cancer Treatment and Research Communications*, *28*, p. 100410.

de Planell-Saguer, M., and Rodicio, M.C., 2013. Detection methods for microRNAs in clinic practice. *Clinical Biochemistry*, *46*(10-11), pp. 869–878.

Del Vescovo, V., Meier, T., Inga, A., Denti, M.A., and Borlak, J., 2013. A cross-platform comparison of affymetrix and Agilent microarrays reveals discordant miRNA expression in lung tumors of c-Raf transgenic mice. *PloS One*, *8*(11), p. e78870.

Detassis, S., Del Vescovo, V., Grasso, M., Masella, S., Cantaloni, C., Cima, L., Cavazza, A., Graziano, P., Rossi, G., Barbareschi, M., and Ricci, L., 2020. miR375-3p distinguishes low-grade neuroendocrine from non-neuroendocrine lung tumors in FFPE samples. *Frontiers in Molecular Biosciences*, *7*, p. 86.

Detassis, S., Grasso, M., Del Vescovo, V., and Denti, M.A., 2017. microRNAs make the call in cancer personalized medicine. *Frontiers in Cell and Developmental Biology*, *5*, p. 86.

d'Herbomez, M., Do Cao, C., Vezzosi, D., Borzon-Chasot, F., and Baudin, E. 2010, September. Chromogranin A assay in clinical practice. In Annales d'endocrinologie (Vol. 71, No. 4, pp. 274–280). Elsevier Masson.

Dietrich, C.S., Gorski, J.W., Davis, C., Erol, L., Dietrich, H., Burgess, B.T., McDowell, A.B., Riggs, M.B., Baldwin, L.A., Miller, R.W., Desimone, C.P., Gallion, H.H., Ueland, F.R., van Nagell, J.R., and Pavlik, E.J., 2020. Cancer risk associated with fluid observed during transvaginal ultrasound. *Gynecologic Oncology*, *159*, pp. 97–98.

Ding, D., Han, S., Zhang, H., He, Y., and Li, Y., 2019. Predictive biomarkers of colorectal cancer. *Computational Biology and Chemistry*, *83*. doi: 10.1016/j.compbiolchem.2019.107106.

Dobbe, E., Gurney, K., Kiekow, S., Lafferty, J.S., and Kolesar, J.M., 2008. Gene-expression assays: New tools to individualize treatment of early-stage breast cancer. *American Journal of Health-System Pharmacy*, *65*(1), pp. 23–28.

Doello, K., Mesas, C., Cabeza, L., Gandara, M.J., Quiñonero, F., and Ortiz, R., 2020. 34P Molecular markers of response to different chemotherapeutic agents in RAS / BRAF mutated colon cancer cell lines. *Annals of Oncology*, *31*, pp. S254–S255.

Doi, H., Takahara, T., Minamoto, T., Matsuhashi, S., Uchii, K., and Yamanaka, H., 2015. Droplet digital polymerase chain reaction (PCR) outperforms real-time PCR in the detection of environmental DNA from an invasive fish species. *Environmental Science & Technology*, *49*(9), pp. 5601–5608.

Donepudi, M.S., Kondapalli, K., Amos, S.J., and Venkanteshan, P., 2014. Breast cancer statistics and markers. *Journal of Cancer Research and Therapeutics*, *10*(3), p. 506.

Dong, H., Lei, J., Ding, L., Wen, Y., Ju, H., and Zhang, X., 2013. MicroRNA: Function, detection, and bioanalysis. *Chemical Reviews*, *113*(8), pp. 6207–6233.

Dong, S.M., Traverso, G., Johnson, C., Geng, L., Favis, R., Boynton, K., Hibi, K., Goodman, S.N., D'Allessio, M., Paty, P., and Hamilton, S.R., 2001. Detecting colorectal cancer in stool with the use of multiple genetic targets. *Journal of the National Cancer Institute*, *93*(11), pp. 858–865.

Du, G., Yu, X., Chen, Y., and Cai, W., 2021. MiR-1-3p suppresses colorectal cancer cell proliferation and metastasis by inhibiting YWHAZ-mediated epithelial-mesenchymal transition. *Frontiers in Oncology*, *11*, p. 264.

Edmondson, R.J., Crosbie, E.J., Nickkho-Amiry, M., Kaufmann, A., Stelloo, E., Nijman, H.W., Leary, A., Auguste, A., Mileshkin, L., Pollock, P., MacKay, H.J., Powell, M.E., Bosse, T., Creutzberg, C.L., and Kitchener, H.C., 2017. Markers of the p53 pathway further refine molecular profiling in high-risk endometrial cancer: A TransPORTEC initiative. *Gynecologic Oncology*, *146*(2), pp. 327–333.

Ell, B., Mercatali, L., Ibrahim, T., Campbell, N., Schwarzenbach, H., Pantel, K., Amadori, D., and Kang, Y., 2013. Tumor-induced osteoclast miRNA changes as regulators and biomarkers of osteolytic bone metastasis. *Cancer Cell*, *24*(4), pp. 542–556.

Esquela-Kerscher, A., and Slack, F.J., 2006. Oncomirs – microRNAs with a role in cancer. *Nature Reviews Cancer*, *6*(4), pp. 259–269.

F, M., V, P., N, M., T, P., F, M., L, Q., Mc, M., G, B., F, C., Ni, T., C, C., F, P., A, C., G, S., and Ac, T., 2021. Transvaginal ultrasound assessment of urinary tract in gynecological oncology patients: A multicenter prospective study. *European Journal of Surgical Oncology*, *47*(5), pp. 1083–1089.

Fang, S., Lee, H.J., Wark, A.W., and Corn, R.M., 2006. Attomole microarray detection of microRNAs by nanoparticle-amplified SPR imaging measurements of surface polyadenylation reactions. *Journal of the American Chemical Society*, *128*(43), pp. 14044–14046.

Fettke, H., Kwan, E.M., Bukczynska, P., Ng, N., Nguyen-Dumont, T., Southey, M.C., Davis, I.D., Mant, A., Parente, P., Pezaro, C., Hauser, C., and Azad, A.A., 2020. Prognostic impact of total plasma cell-free DNA concentration in androgen receptor pathway inhibitor–treated metastatic castration-resistant prostate cancer. *European Urology Focus*. doi: 10.1016/j.euf.2020.07.001.

Fimognari, C., 2015. Role of oxidative RNA damage in chronic-degenerative diseases. *Oxidative medicine and cellular longevity*, *2015*. doi: 10.1155/2015/358713.

Fort, R.S., Mathó, C., Oliveira-Rizzo, C., Garat, B., Sotelo-Silveira, J.R., and Duhagon, M.A., 2018. An integrated view of the role of miR-130b/301b miRNA cluster in prostate cancer. *Experimental Hematology & Oncology*, *7*(1), pp. 1–14.

Funaki, N.O., Tanaka, J., Kasamatsu, T., Ohshio, G., Hosotani, R., Okino, T., and Imamura, M., 1996. Identification of carcinoembryonic antigen mRNA in circulating peripheral blood of pancreatic carcinoma and gastric carcinoma patients. *Life Sciences*, *59*(25-26), pp. 2187–2199.

Garcia-Rico, E., Alvarez-Puebla, R.A., and Guerrini, L., 2018. Direct surface-enhanced Raman scattering (SERS) spectroscopy of nucleic acids: from fundamental studies to real-life applications. *Chemical Society Reviews*, *47*(13), pp. 4909–4923.

Ghai, V., and Wang, K., 2016. Recent progress toward the use of circulating microRNAs as clinical biomarkers. *Archives of toxicology*, *90*(12), pp. 2959–2978.

Giraldez, M.D., Chevillet, J.R., and Tewari, M., 2018. Droplet digital PCR for absolute quantification of extracellular MicroRNAs in plasma and serum: quantification of the cancer biomarker hsa-miR-141. In *Digital PCR* (pp. 459–474). Humana Press, New York, NY.

Giraldez, M.D., Spengler, R.M., Etheridge, A., Godoy, P.M., Barczak, A.J., Srinivasan, S., De Hoff, P.L., Tanriverdi, K., Courtright, A., Lu, S., and Khoory, J., 2018. Comprehensive multi-center assessment of small RNA-seq methods for quantitative miRNA profiling. *Nature Biotechnology*, *36*(8), pp. 746–757.

Glitza, I.C., and Davies, M.A., 2014. Genotyping of cutaneous melanoma. *Chinese Clinical Oncology*, *3*(3), p. 27.

González-Timoneda, M., González-Timoneda, A., Mata, D., Cano, A., and Hidalgo, J.J., 2021. Invasive ovarian seromucinous carcinoma developed on an endometrioma. The determining role of transvaginal ultrasound. *Clinica e Investigacion en Ginecologia y Obstetricia*. doi: 10.1016/j.gine.2021.01.006.

Goodman, C.R., and Speers, C.W., 2021. The role of circulating tumor cells in breast cancer and implications for radiation treatment decisions. *International Journal of Radiation Oncology Biology Physics*. doi: 10.1016/j.ijrobp.2020.08.039.

Grasso, M., Piscopo, P., Confaloni, A., and Denti, M.A., 2014. Circulating miRNAs as biomarkers for neurodegenerative disorders. *Molecules*, *19*(5), pp. 6891–6910.

Grobbee, E.J., van der Vlugt, M., van Vuuren, A.J., Stroobants, A.K., Mallant-Hent, R.C., Lansdorp-Vogelaar, I., Bossuyt, P.M.M., Kuipers, E.J., Dekker, E., and Spaander, M.C.W., 2020. Diagnostic yield of one-time colonoscopy vs one-time flexible sigmoidoscopy vs multiple rounds of mailed fecal immunohistochemical tests in colorectal cancer screening. *Clinical Gastroenterology and Hepatology*, *18*(3), pp. 667–675.e1.

Gu, L.Q., Wanunu, M., Wang, M.X., McReynolds, L., and Wang, Y., 2012. Detection of miRNAs with a nanopore single-molecule counter. *Expert Review of Molecular Diagnostics*, *12*(6), pp. 573–584.

Gu, W., Li, X., and Wang, J., 2014. miR-139 regulates the proliferation and invasion of hepatocellular carcinoma through the WNT/TCF-4 pathway. *Oncology Reports*, *31*(1), pp. 397–404.

Hadjipanteli, A., Elangovan, P., Mackenzie, A., Wells, K., Dance, D.R., and Young, K.C., 2019. The threshold detectable mass diameter for 2D-mammography and digital breast tomosynthesis. *Physica Medica*, *57*, pp. 25–32.

Harbeck, N., Schmitt, M., Paepke, S., Allgayer, H., and Kates, R.E., 2007. Tumor-associated proteolytic factors uPA and PAI-1: critical appraisal of their clinical relevance in breast cancer and their integration into decision-support algorithms. *Critical Reviews in Clinical Laboratory Sciences*, *44*(2), pp. 179–201.

Harbeck, N., Sotlar, K., Wuerstlein, R., and Doisneau-Sixou, S., 2014. Molecular and protein markers for clinical decision making in breast cancer: Today and tomorrow. *Cancer Treatment Reviews*, *40*(3), pp. 434–444.

Hartman, M.R., Ruiz, R.C., Hamada, S., Xu, C., Yancey, K.G., Yu, Y., Han, W., and Luo, D., 2013. Point-of-care nucleic acid detection using nanotechnology. *Nanoscale*, *5*(21), pp. 10141–10154.

He, L., Musick, M.D., Nicewarner, S.R., Salinas, F.G., Benkovic, S.J., Natan, M.J., and Keating, C.D., 2000. Colloidal Au-enhanced surface plasmon resonance for ultrasensitive detection of DNA hybridization. *Journal of the American Chemical Society*, *122*(38), pp. 9071–9077.

Hedegaard, J., Thorsen, K., Lund, M.K., Hein, A.M.K., Hamilton-Dutoit, S.J., Vang, S., Nordentoft, I., Birkenkamp-Demtröder, K., Kruhøffer, M., Hager, H., and Knudsen, B., 2014. Next-generation sequencing of RNA and DNA isolated from paired fresh-frozen and formalin-fixed paraffin-embedded samples of human cancer and normal tissue. *PloS One*, *9*(5), p. e98187.

Henley, R.Y., Carson, S., and Wanunu, M., 2016. Studies of RNA sequence and structure using nanopores. *Progress in Molecular Biology and Translational Science*, *139*, pp. 73–99.

Henning, G.M., Barashi, N.S., and Smith, Z.L., 2021. Advances in biomarkers for detection, surveillance, and prognosis of bladder cancer. *Clinical Genitourinary Cancer*. 10.1016/j.clgc.2020.12.003.

Henry, N.L., and Hayes, D.F., 2012. Cancer biomarkers. *Molecular Oncology*, *6*(2), pp. 140–146.

Herman, J.G., Graff, J.R., Myöhänen, S.B.D.N., Nelkin, B.D., and Baylin, S.B., 1996. Methylation-specific PCR: a novel PCR assay for methylation status of CpG islands. Proceedings of the National Academy of Sciences, *93*(18), pp. 9821–9826.

Hindson, B.J., Ness, K.D., Masquelier, D.A., Belgrader, P., Heredia, N.J., Makarewicz, A.J., Bright, I.J., Lucero, M.Y., Hiddessen, A.L., Legler, T.C., and Kitano, T.K., 2011. High-throughput droplet digital PCR system for absolute quantitation of DNA copy number. *Analytical Chemistry*, *83*(22), pp. 8604–8610.

Homola, J., 2008. Surface plasmon resonance sensors for detection of chemical and biological species. *Chemical Reviews*, *108*(2), pp. 462–493.

Hong, C.Y., Chen, X., Liu, T., Li, J., Yang, H.H., Chen, J.H., and Chen, G.N., 2013. Ultrasensitive electrochemical detection of cancer-associated circulating microRNA in serum samples based on DNA concatamers. *Biosensors and Bioelectronics*, *50*, pp. 132–136.

Hossain, T., Mahmudunnabi, G., Masud, M.K., Islam, M.N., Ooi, L., Konstantinov, K., Al Hossain, M.S., Martinac, B., Alici, G., Nguyen, N.T., and Shiddiky, M.J., 2017. Electrochemical biosensing strategies for DNA methylation analysis. *Biosensors and Bioelectronics*, *94*, pp. 63–73.

Hsu, P., Yang, T., Sheikh-Fayyaz, S., Brody, J., Bandovic, J., Roy, S., Laser, J., Kolitz, J.E., Devoe, C., and Zhang, X., 2014. Mantle cell lymphoma with in situ or mantle zone growth pattern: a study of five cases and review of literature. *International Journal of Clinical and Experimental Pathology*, *7*(3), p. 1042.

Hu, F., Xu, J., and Chen, Y., 2017. Surface plasmon resonance imaging detection of sub-femtomolar microRNA. *Analytical chemistry*, *89*(18), pp. 10071–10077.

Hu, Z., Chen, X., Zhao, Y., Tian, T., Jin, G., Shu, Y., Chen, Y., Xu, L., Zen, K., Zhang, C., and Shen, H., 2010. Serum microRNA signatures identified in a genome-wide serum microRNA expression profiling predict survival of non-small-cell lung cancer. *J Clin Oncol*, *28*(10), pp. 1721–1726.

Huang, X., and Lu, S., 2017. MicroR-545 mediates colorectal cancer cells proliferation through up-regulating epidermal growth factor receptor expression in HOTAIR long non-coding RNA dependent. *Molecular and Cellular Biochemistry*, *431*(1), pp. 45–54.

Huang, Z., Huang, D., Ni, S., Peng, Z., Sheng, W., and Du, X., 2010. Plasma microRNAs are promising novel biomarkers for early detection of colorectal cancer. *International Journal of Cancer*, *127*(1), pp. 118–126.

Huggett, J.F., 2020. The digital MIQE guidelines update: Minimum information for publication of quantitative digital PCR experiments for 2020. *Clinical Chemistry*, *66*(8), pp. 1012–1029.

Hulstaert, E., Morlion, A., Levanon, K., Vandesompele, J., and Mestdagh, P., 2021. Candidate RNA biomarkers in biofluids for early diagnosis of ovarian cancer: A systematic review. *Gynecologic Oncology*.

Inamura, K., 2017. Lung cancer: understanding its molecular pathology and the 2015 WHO classification. *Frontiers in Oncology*, *7*, p. 193.

Iorio, M.V., Visone, R., Di Leva, G., Donati, V., Petrocca, F., Casalini, P., Taccioli, C., Volinia, S., Liu, C.G., Alder, H., and Calin, G.A., 2007. MicroRNA signatures in human ovarian cancer. *Cancer Research*, *67*(18), pp. 8699–8707.

Islam, M.N., Yadav, S., Haque, M.H., Munaz, A., Islam, F., Al Hossain, M.S., Gopalan, V., Lam, A.K., Nguyen, N.T., and Shiddiky, M.J., 2017. Optical biosensing strategies for DNA methylation analysis. *Biosensors and Bioelectronics*, *92*, pp. 668–678.

Islam, M.N., Masud, M.K., Haque, M.H., Hossain, M.S.A., Yamauchi, Y., Nguyen, N.T., and Shiddiky, M.J., 2017. RNA biomarkers: Diagnostic and prognostic potentials and recent developments of electrochemical biosensors. *Small Methods*, *1*(7), p. 1700131.

Iwasaki, H., Shimura, T., and Kataoka, H., 2019. Current status of urinary diagnostic biomarkers for colorectal cancer. *Clinica Chimica Acta*. doi: 10.1016/j.cca.2019.08.011.

Jackson, P.E., Qian, G.S., Friesen, M.D., Zhu, Y.R., Lu, P., Wang, J.B., Wu, Y., Kensler, T.W., Vogelstein, B., and Groopman, J.D., 2001. Specific p53 mutations detected in plasma and tumors of hepatocellular carcinoma patients by electrospray ionization mass spectrometry. *Cancer Research*, *61*(1), pp. 33–35.

Jalil, O., Pandey, C.M., and Kumar, D., 2021. Highly sensitive electrochemical detection of cancer biomarker based on anti-EpCAM conjugated molybdenum disulfide grafted reduced graphene oxide nanohybrid. *Bioelectrochemistry*, *138*. 10.1016/j.bioelechem.2020.107733.

Jiang, M., Zhang, P., Hu, G., Xiao, Z., Xu, F., Zhong, T., Huang, F., Kuang, H., and Zhang, W., 2013. Relative expressions of miR-205-5p, miR-205-3p, and miR-21 in tissues and serum of non-small cell lung cancer patients. *Molecular and Cellular Biochemistry*, *383*(1), pp. 67–75.

Johnson, B.N., and Mutharasan, R., 2014. Biosensor-based microRNA detection: techniques, design, performance, and challenges. *Analyst*, *139*(7), pp. 1576–1588.

Jones, D.H., Caracciolo, J.T., Hodul, P.J., Strosberg, J.R., Coppola, D., and Bui, M.M., 2015. Familial gastrointestinal stromal tumor syndrome: report of 2 cases with KIT exon 11 mutation. *Cancer Control*, *22*(1), pp. 102–108.

Kakimoto, Y., Kamiguchi, H., Ochiai, E., Satoh, F., and Osawa, M., 2015. MicroRNA stability in postmortem FFPE tissues: quantitative analysis using autoptic samples from acute myocardial infarction patients. *PLoS One*, *10*(6), p. e0129338.

Kan, C.W., Hahn, M.A., Gard, G.B., Maidens, J., Huh, J.Y., Marsh, D.J., and Howell, V.M., 2012. Elevated levels of circulating microRNA-200 family members correlate with serous epithelial ovarian cancer. *BMC Cancer*, *12*(1), pp. 1–9.

Khan, N., Cheng, J., Pezacki, J.P., and Berezovski, M.V., 2011. Quantitative analysis of microRNA in blood serum with protein-facilitated affinity capillary electrophoresis. *Analytical Chemistry*, *83*(16), pp. 6196–6201.

Kilic, D., Guler, T., Atigan, A., Avsaroglu, E., Karakaya, Y.A., Kaleli, I., and Kaleli, B., 2020. Predictors of Human papillomavirus (HPV) persistence after treatment of high grade cervical lesions; does cervical cytology have any prognostic value in primary HPV screening? *Annals of Diagnostic Pathology, 49.* doi: 10.1016/j.anndiagpath.2020.151626.

Kilic, T., Topkaya, S.N., and Ozsoz, M., 2013. A new insight into electrochemical microRNA detection: a molecular caliper, p19 protein. *Biosensors and Bioelectronics, 48*, pp. 165–171.

Kim, J., Jun, S.Y., and Maeng, L.S., 2020. The clinical performance of human papillomavirus genotyping using PANArray HPV chip: Comparison to ThinPrep cytology alone and co-testing. *Pathology Research and Practice, 216*(9). doi: 10.1016/j.prp.2020.153121.

Kong, L., Guan, J., and Pumera, M., 2018. Micro-and nanorobots based sensing and biosensing. *Current Opinion in Electrochemistry, 10*, pp. 174–182.

Krismann, M., Todt, B., Schröder, J., Gareis, D., Müller, K.M., Seeber, S., and Schütte, J., 1995. Low specificity of cytokeratin 19 reverse transcriptase-polymerase chain reaction analyses for detection of hematogenous lung cancer dissemination. *Journal of clinical Oncology, 13*(11), pp. 2769–2775.

Kumar, P., Dezso, Z., MacKenzie, C., Oestreicher, J., Agoulnik, S., Byrne, M., Bernier, F., Yanagimachi, M., Aoshima, K., and Oda, Y., 2013. Circulating miRNA biomarkers for Alzheimerd's disease. *PloS One, 8*(7), p. e69807.

Labib, M., Khan, N., Ghobadloo, S.M., Cheng, J., Pezacki, J.P., and Berezovski, M.V., 2013. Three-mode electrochemical sensing of ultralow microRNA levels. *Journal of the American Chemical Society, 135*(8), pp. 3027–3038.

Lawrie, C.H., Gal, S., Dunlop, H.M., Pushkaran, B., Liggins, A.P., Pulford, K., Banham, A.H., Pezzella, F., Boultwood, J., Wainscoat, J.S., and Hatton, C.S., 2008. Detection of elevated levels of tumour-associated microRNAs in serum of patients with diffuse large B-cell lymphoma. *British Journal of Haematology, 141*(5), pp. 672–675.

Lebanony, D., Benjamin, H., Gilad, S., Ezagouri, M., Dov, A., Ashkenazi, K., Gefen, N., Izraeli, S., Rechavi, G., Pass, H., and Nonaka, D., 2009. Diagnostic assay based on hsa-miR-205 expression distinguishes squamous from nonsquamous non–small-cell lung carcinoma. *Journal of Clinical Oncology, 27*(12), pp. 2030–2037.

Lee, T., Wi, J.S., Oh, A., Na, H.K., Lee, J., Lee, K., Lee, T.G., and Haam, S., 2018. Highly robust, uniform and ultra-sensitive surface-enhanced Raman scattering substrates for microRNA detection fabricated by using silver nanostructures grown in gold nanobowls. *Nanoscale, 10*(8), pp. 3680–3687.

Lefeuvre, C., Pivert, A., Guillou-Guillemette, H. Le, Lunel-Fabiani, F., Veillon, P., Le Duc-Banaszuk, A.S., and Ducancelle, A., 2020. Urinary HPV DNA testing as a tool for cervical cancer screening in women who are reluctant to have a Pap smear in France. *Journal of Infection, 81*(2), pp. 248–254.

Leidinger, P., Backes, C., Deutscher, S., Schmitt, K., Mueller, S.C., Frese, K., Haas, J., Ruprecht, K., Paul, F., Stähler, C., and Lang, C.J., 2013. A blood based 12-miRNA signature of Alzheimer disease patients. *Genome Biology, 14*(7), pp. 1–16.

Leshkowitz, D., Horn-Saban, S., Parmet, Y., and Feldmesser, E., 2013. Differences in microRNA detection levels are technology and sequence dependent. *RNA, 19*(4), pp. 527–538.

Li, B., Pu, K., Ge, L., and Wu, X., 2019. Diagnostic significance assessment of the circulating cell-free DNA in ovarian cancer: An updated meta-analysis. *Gene.* doi: 10.1016/j.gene.2019.143993.

Li, W., and Ruan, K., 2009. MicroRNA detection by microarray. *Analytical and Bioanalytical Chemistry, 394*(4), pp. 1117–1124.

Li, Y., Di, C., Li, W., Cai, W., Tan, X., Xu, L., Yang, L., Lou, G., and Yan, Y., 2016. Oncomirs miRNA-221/222 and tumor suppressors miRNA-199a/195 are crucial miRNAs in liver cancer: a systematic analysis. *Digestive Diseases and Sciences, 61*(8), pp. 2315–2327.

Li, Y., and Kowdley, K.V., 2012. MicroRNAs in common human diseases. *Genomics, Proteomics & Bioinformatics, 10*(5), pp. 246–253.

Lin, H.-I., and Chang, Y.-C., 2021. Colorectal cancer detection by immunofluorescence images of circulating tumor cells. *Ain Shams Engineering Journal.* doi: 10.1016/j.asej.2021.01.013.

Lin, S., and Gregory, R.I., 2015. MicroRNA biogenesis pathways in cancer. *Nature Reviews Cancer, 15*(6), pp. 321–333.

Lin, S.R., Huang, M.Y., and Chang, H.J., 2011. Molecular detection of circulating tumor cells with multiple mrna markers by genechip for colorectal cancer early diagnosis and prognosis prediction. *Genomic Medicine, Biomarkers, and Health Sciences.* 10.1016/S2211-4254(11)60003-4.

Lin, Y., Leng, Q., Jiang, Z., Guarnera, M.A., Zhou, Y., Chen, X., Wang, H., Zhou, W., Cai, L., Fang, H., and Li, J., 2017. A classifier integrating plasma biomarkers and radiological characteristics for distinguishing malignant from benign pulmonary nodules. *International Journal of Cancer, 141*(6), pp. 1240–1248.

Liu, C.G., Calin, G.A., Volinia, S., and Croce, C.M., 2008. MicroRNA expression profiling using micro-arrays. *Nature Protocols*, 3(4), p. 563.

Liu, R., Chen, X., Du, Y., Yao, W., Shen, L., Wang, C., Hu, Z., Zhuang, R., Ning, G., Zhang, C., and Yuan, Y., 2012. Serum microRNA expression profile as a biomarker in the diagnosis and prognosis of pancreatic cancer. *Clinical Chemistry*, 58(3), pp. 610–618.

Liu, R., Wang, Q., Li, Q., Yang, X., Wang, K., and Nie, W., 2017. Surface plasmon resonance biosensor for sensitive detection of microRNA and cancer cell using multiple signal amplification strategy. *Biosensors and Bioelectronics*, 87, pp. 433–438.

Liu, S., Wu, J., Xia, Q., Liu, H., Li, W., Xia, X., and Wang, J., 2020. Finding new cancer epigenetic and genetic biomarkers from cell-free DNA by combining SALP-seq and machine learning. *Computational and Structural Biotechnology Journal*, 18, pp. 1891–1903.

Liu, X.G., Zhu, W.Y., Huang, Y.Y., Ma, L.N., Zhou, S.Q., Wang, Y.K., Zeng, F., Zhou, J.H., and Zhang, Y.K., 2012. High expression of serum miR-21 and tumor miR-200c associated with poor prognosis in patients with lung cancer. *Medical Oncology*, 29(2), pp. 618–626.

Lou, W., Liu, J., Ding, B., Chen, D., Xu, L., Ding, J., Jiang, D., Zhou, L., Zheng, S., and Fan, W., 2019. Identification of potential miRNA–mRNA regulatory network contributing to pathogenesis of HBV-related HCC. *Journal of Translational Medicine*, 17(1), pp. 1–14.

Lu, J., Getz, G., Miska, E.A., Alvarez-Saavedra, E., Lamb, J., Peck, D., Sweet-Cordero, A., Ebert, B.L., Mak, R.H., Ferrando, A.A., and Downing, J.R., 2005. MicroRNA expression profiles classify human cancers. *Nature*, 435(7043), pp. 834–838.

Lusi, E.A., Passamano, M., Guarascio, P., Scarpa, A., and Schiavo, L., 2009. Innovative electrochemical approach for an early detection of microRNAs. *Analytical Chemistry*, 81(7), pp. 2819–2822.

Ma, J., Li, N., Guarnera, M., and Jiang, F., 2013. Quantification of plasma miRNAs by digital PCR for cancer diagnosis. *Biomarker insights*, 8, pp. BMI–S13154.

Magario, M.B., Poli-Neto, O.B., Tiezzi, D.G., Angotti Carrara, H.H., Moreira de Andrade, J., and Candido dos Reis, F.J., 2019. Mammography coverage and tumor stage in the opportunistic screening context. *Clinical Breast Cancer*, 19(6), pp. 456–459.

Marcuello, M., Vymetalkova, V., Neves, R.P.L., Duran-Sanchon, S., Vedeld, H.M., Tham, E., van Dalum, G., Flügen, G., Garcia-Barberan, V., Fijneman, R.J., Castells, A., Vodicka, P., Lind, G.E., Stoecklein, N.H., Heitzer, E., and Gironella, M., 2019. Circulating biomarkers for early detection and clinical management of colorectal cancer. *Molecular Aspects of Medicine*. doi: 10.1016/j.mam.2019.06.002.

Matamala, N., Vargas, M.T., González-Cámpora, R., Miñambres, R., Arias, J.I., Menéndez, P., Andrés-León, E., Gómez-López, G., Yanowsky, K., Calvete-Candenas, J., and Inglada-Pérez, L., 2015. Tumor microRNA expression profiling identifies circulating microRNAs for early breast cancer detection. *Clinical Chemistry*, 61(8), pp. 1098–1106.

Meller, A., Nivon, L., and Branton, D., 2001. Voltage-driven DNA translocations through a nanopore. *Physical Review Letters*, 86(15), p. 3435.

Ménard, C., Rezende, F.A., Miloudi, K., Wilson, A., Tétreault, N., Hardy, P., SanGiovanni, J.P., De Guire, V., and Sapieha, P., 2016. MicroRNA signatures in vitreous humour and plasma of patients with exudative AMD. *Oncotarget*, 7(15), p. 19171.

Meng, W., McElroy, J.P., Volinia, S., Palatini, J., Warner, S., Ayers, L.W., Palanichamy, K., Chakravarti, A., and Lautenschlaeger, T., 2013. Comparison of microRNA deep sequencing of matched formalin-fixed paraffin-embedded and fresh frozen cancer tissues. *PloS One*, 8(5), p. e64393.

Meng, X., Müller, V., Milde-Langosch, K., Trillsch, F., Pantel, K., and Schwarzenbach, H., 2016. Diagnostic and prognostic relevance of circulating exosomal miR-373, miR-200a, miR-200b and miR-200c in patients with epithelial ovarian cancer. *Oncotarget*, 7(13), p. 16923.

Mestdagh, P., Hartmann, N., Baeriswyl, L., Andreasen, D., Bernard, N., Chen, C., Cheo, D., D'andrade, P., DeMayo, M., Dennis, L., and Derveaux, S., 2014. Evaluation of quantitative miRNA expression platforms in the microRNA quality control (miRQC) study. *Nature Methods*, 11(8), p. 809.

Mills, N.E., Fishman, C.L., Scholes, J., Anderson, S.E., Rom, W.N., and Jacobson, D.R., 1995. Detection of K-ras oncogene mutations in bronchoalveolar lavage fluid for lung cancer diagnosis. *JNCI: Journal of the National Cancer Institute*, 87(14), pp. 1056–1060.

Mitchell, P.S., Parkin, R.K., Kroh, E.M., Fritz, B.R., Wyman, S.K., Pogosova-Agadjanyan, E.L., Peterson, A., Noteboom, J., O'Briant, K.C., Allen, A., and Lin, D.W., 2008. Circulating microRNAs as stable blood-based markers for cancer detection. *Proceedings of the National Academy of Sciences*, 105(30), pp. 10513–10518.

Mjelle, R., Dima, S.O., Bacalbasa, N., Chawla, K., Sorop, A., Cucu, D., Herlea, V., Sætrom, P., and Popescu,

I., 2019. Comprehensive transcriptomic analyses of tissue, serum, and serum exosomes from hepato-cellular carcinoma patients. *BMC Cancer*, *19*(1), pp. 1–13.

Mo, H., Wang, X., Ma, F., Qian, Z., Sun, X., Yi, Z., Guan, X., Li, L., Liu, B., and Xu, B., 2020. Genome-wide chromosomal instability by cell-free DNA sequencing predicts survival in patients with metastatic breast cancer. *Breast*, *53*, pp. 111–118.

Mohanty, S.S., Sahoo, C.R., and Padhy, R.N., 2021. Role of hormone receptors and HER2 as prospective molecular markers for breast cancer: An update. *Genes and Diseases*. doi: 10.1016/j.gendis. 2020.12.005.

Mollasalehi, H., and Shajari, E., 2021. A colorimetric nano-biosensor for simultaneous detection of pre-valent cancers using unamplified cell-free ribonucleic acid biomarkers. *Bioorganic Chemistry*, *107*. doi: 10.1016/j.bioorg.2020.104605.

Moretti, F., D'Antona, P., Finardi, E., Barbetta, M., Dominioni, L., Poli, A., Gini, E., Noonan, D.M., Imperatori, A., Rotolo, N., and Cattoni, M., 2017. Systematic review and critique of circulating miRNAs as biomarkers of stage I-II non-small cell lung cancer. *Oncotarget*, *8*(55), p. 94980.

Moroney, M.R., Davies, K.D., Wilberger, A.C., Sheeder, J., Post, M.D., Berning, A.A., Fisher, C., Lefkowits, C., Guntupalli, S.R., Behbakht, K., and Corr, B.R., 2019. Molecular markers in recurrent stage I, grade 1 endometrioid endometrial cancers. *Gynecologic Oncology*, *153*(3), pp. 517–520.

Mou, T., Zhu, D., Wei, X., Li, T., Zheng, D., Pu, J., Guo, Z., and Wu, Z., 2017. Identification and interaction analysis of key genes and microRNAs in hepatocellular carcinoma by bioinformatics analysis. *World Journal of Surgical Oncology*, *15*(1), pp. 1–9.

Muinao, T., Deka Boruah, H.P., and Pal, M., 2019. Multi-biomarker panel signature as the key to diagnosis of ovarian cancer. *Heliyon*. doi: 10.1016/j.heliyon.2019.e02826.

Murakami, Y., Aly, H.H., Tajima, A., Inoue, I., and Shimotohno, K., 2009. Regulation of the hepatitis C virus genome replication by miR-199a. *Journal of Hepatology*, *50*(3), pp. 453–460.

Musto, A., Grassetto, G., Marzola, M.C., Rampin, L., Chondrogiannis, S., Maffione, A.M., Colletti, P.M., Perkins, A.C., Fagioli, G., and Rubello, D., 2014. Management of epithelial ovarian cancer from di-agnosis to restaging: An overview of the role of imaging techniques with particular regard to the contribution of 18F-FDG PET/CT. *Nuclear Medicine Communications*, *35*(6), pp. 588–597.

Nagy, A., Lánczky, A., Menyhárt, O., and Győrffy, B., 2018. Validation of miRNA prognostic power in hepatocellular carcinoma using expression data of independent datasets. *Scientific Reports*, *8*(1), pp. 1–9.

Nair, M., Sandhu, S.S., and Sharma, A.K., 2018. Cancer molecular markers: A guide to cancer detection and management. *Seminars in Cancer Biology*. doi: 10.1016/j.semcancer.2018.02.002.

Ng, E.K., Chong, W.W., Jin, H., Lam, E.K., Shin, V.Y., Yu, J., Poon, T.C., Ng, S.S., and Sung, J.J. 2009. Differential expression of microRNAs in plasma of patients with colorectal cancer: a potential marker for colorectal cancer screening. *Gut*, *58*(10), pp. 1375–1381.

Ng, E.K., Li, R., Shin, V.Y., Jin, H.C., Leung, C.P., Ma, E.S., Pang, R., Chua, D., Chu, K.M., Law, W.L., & Law, S.Y. 2013. Circulating microRNAs as specific biomarkers for breast cancer detection. *PloS One*, *8*(1), p. e53141.

Nolen, B.M., and Lokshin, A.E., 2012. Multianalyte assay systems in the differential diagnosis of ovarian cancer. *Expert Opinion on Medical Diagnostics*, *6*(2), pp. 131–138.

Ochoa, A.E., Choi, W., Su, X., Siefker-Radtke, A., Czerniak, B., Dinney, C., and McConkey, D.J., 2016. Specific micro-RNA expression patterns distinguish the basal and luminal subtypes of muscle-invasive bladder cancer. *Oncotarget*, *7*(49), p. 80164.

Ogata-Kawata, H., Izumiya, M., Kurioka, D., Honma, Y., Yamada, Y., Furuta, K., Gunji, T., Ohta, H., Okamoto, H., Sonoda, H., and Watanabe, M., 2014. Circulating exosomal microRNAs as biomarkers of colon cancer. *PloS One*, *9*(4).

Olasagasti, F., Lieberman, K.R., Benner, S., Cherf, G.M., Dahl, J.M., Deamer, D.W., and Akeson, M., 2010. Replication of individual DNA molecules under electronic control using a protein nanopore. *Nature Nanotechnology*, *5*(11), pp. 798–806.

Orazem, M.E., and Tribollet, B., 2008. *Electrochemical Impedance Spectroscopy* (pp. 383–389). Wiley, New Jersey.

Orlicka-Płocka, M., Gurda, D., Fedoruk-Wyszomirska, A., Smolarek, I., and Wyszko, E., 2016. Circulating microRNAs in cardiovascular diseases. *Acta Biochimica Polonica*, *63*(4), pp. 725–729.

Özgür, E., Mayer, Z., Keskin, M., Yörüker, E.E., Holdenrieder, S., and Gezer, U., 2021. Satellite 2 repeat DNA in blood plasma as a candidate biomarker for the detection of cancer. *Clinica Chimica Acta*, *514*, pp. 74–79.

Padhy, R.R., Davidov, A., Madrigal, L., Alcide, G., and Spahiu, A., 2020. Detection of high-risk human papillomavirus RNA in urine for cervical cancer screening with HPV 16 & 18/45 genotyping. *Heliyon*, *6*(4), p. e03745.

Palecek, E., and Bartosik, M., 2012. Electrochemistry of nucleic acids. *Chemical Reviews*, *112*(6), pp. 3427–3481.

Palmela Leitão, T., Miranda, M., Polido, J., Morais, J., Corredeira, P., Alves, P., Oliveira, T., Pereira e Silva, R., Fernandes, R., Ferreira, J., Palma Reis, J., Lopes, T., and Costa, L., 2021. Circulating tumor cell detection methods in renal cell carcinoma: A systematic review. *Critical Reviews in Oncology/Hematology*. doi: 10.1016/j.critrevonc.2021.103331.

Patil, V., Jothimani, D., Narasimhan, G., Danielraj, S., and Rela, M., 2021. Hereditary persistence of alpha-fetoprotein in chronic liver disease-confusing genes! *Journal of Clinical and Experimental Hepatology*. 10.1016/j.jceh.2020.12.008.

Pemmaraju, N., and Cortes, J., 2014. Chronic myeloid leukemia in adolescents and young adults: patient characteristics, outcomes and review of the literature. *Acta Haematologica*, *132*(3-4), pp. 298–306.

Peraza, J.M.T., Pinto, J.I.G., Osuna, L.R.T., and Romero, J.D.J.M., 2014. Primary carcinoma of the peritoneum. Case report and literature review. *Ginecologia y obstetricia de Mexico*, *82*(05), pp. 344–349.

Pérez-Ibave, D.C., Burciaga-Flores, C.H., and Elizondo-Riojas, M.Á., 2018. Prostate-specific antigen (PSA) as a possible biomarker in non-prostatic cancer: A review. *Cancer Epidemiology*. doi: 10.1016/j.canep.2018.03.009.

Petriella, D., De Summa, S., Lacalamita, R., Galetta, D., Catino, A., Logroscino, A.F., Palumbo, O., Carella, M., Zito, F.A., Simone, G., and Tommasi, S., 2016. miRNA profiling in serum and tissue samples to assess noninvasive biomarkers for NSCLC clinical outcome. *Tumor Biology*, *37*(4), pp. 5503–5513.

Pinheiro, L.B., Coleman, V.A., Hindson, C.M., Herrmann, J., Hindson, B.J., Bhat, S., and Emslie, K.R., 2012. Evaluation of a droplet digital polymerase chain reaction format for DNA copy number quantification. *Analytical Chemistry*, *84*(2), pp. 1003–1011.

Portnoy, V., Huang, V., Place, R.F., and Li, L.C., 2011. Small RNA and transcriptional upregulation. *Wiley Interdisciplinary Reviews: RNA*, *2*(5), pp. 748–760.

Precazzini, F., Detassis, S., Imperatori, A.S., Denti, M.A., and Campomenosi, P., 2021. Measurements methods for the development of MicroRNA-based tests for cancer diagnosis. *International Journal of Molecular Sciences*, *22*(3), p. 1176.

Pritchard, C.C., Kroh, E., Wood, B., Arroyo, J.D., Dougherty, K.J., Miyaji, M.M., Tait, J.F., and Tewari, M., 2012. Blood cell origin of circulating microRNAs: A cautionary note for cancer biomarker studies. *Cancer Prevention Research*, *5*(3), pp. 492–497.

Qiu, X., Xu, J., Guo, J., Yahia-Ammar, A., Kapetanakis, N.I., Duroux-Richard, I., Unterluggauer, J.J., Golob-Schwarzl, N., Regeard, C., Uzan, C., and Gouy, S., 2018. Advanced microRNA-based cancer diagnostics using amplified time-gated FRET. *Chemical Science*, *9*(42), pp. 8046–8055.

Rački, N., Dreo, T., Gutierrez-Aguirre, I., Blejec, A., and Ravnikar, M., 2014. Reverse transcriptase droplet digital PCR shows high resilience to PCR inhibitors from plant, soil and water samples. *Plant Methods*, *10*(1), pp. 1–10.

Ramalho-Carvalho, J., Graça, I., Gomez, A., Oliveira, J., Henrique, R., Esteller, M., and Jerónimo, C., 2017. Downregulation of miR-130b~ 301b cluster is mediated by aberrant promoter methylation and impairs cellular senescence in prostate cancer. *Journal of Hematology & Oncology*, *10*(1), pp. 1–13.

Randel, K.R., Schult, A.L., Botteri, E., Hoff, G., Bretthauer, M., Ursin, G., Natvig, E., Berstad, P., Jørgensen, A., Sandvei, P.K., Olsen, M.E., Frigstad, S.O., Darre-Næss, O., Norvard, E.R., Bolstad, N., Kørner, H., Wibe, A., Wensaas, K.A., de Lange, T., and Holme, O., 2021. Colorectal cancer screening with repeated fecal immunochemical test versus sigmoidoscopy: baseline results from a randomized trial. *Gastroenterology*, *160*(4), pp. 1085–1096.e5.

Rao, S.R., Alham, N.K., Upton, E., McIntyre, S., Bryant, R.J., Cerundolo, L., Bowes, E., Jones, S., Browne, M., Mills, I., Lamb, A., Tomlinson, I., Wedge, D., Browning, L., Sirinukunwattana, K., Palles, C., Hamdy, F.C., Rittscher, J., and Verrill, C., 2020. Detailed molecular and immune marker profiling of archival prostate cancer samples reveals an inverse association between TMPRSS2:ERG Fusion Status and Immune Cell Infiltration. *Journal of Molecular Diagnostics*, *22*(5), pp. 652–669.

Rao, Y., Lee, Y., Jarjoura, D., Ruppert, A.S., Liu, C.G., Hsu, J.C., and Hagan, J.P., 2008. A comparison of normalization techniques for microRNA microarray data. *Statistical Applications in Genetics and Molecular Biology*, *7*(1). doi: 10.2202/1544-6115.1287.

Rapisuwon, Suthee, Eveline E. Vietsch, and Anton Wellstein (2016) Circulating biomarkers to monitor cancer progression and treatment. *Computational and Structural Biotechnology Journal 14*, pp. 211–222

Ritter, M., Paradiso, V., Widmer, P., Garofoli, A., Quagliata, L., Eppenberger-Castori, S., Soysal, S.D., Muenst, S., Ng, C.K.Y., Piscuoglio, S., Weber, W., and Weber, W.P., 2020. Identification of somatic mutations in thirty-year-old serum cell-free dna from patients with breast cancer: A feasibility study. *Clinical Breast Cancer*, *20*(5), pp. 413–421.e1.

Romano, G., 2015. Tumor markers currently utilized in cancer care. *Mater Methods*, *5*, p. 1456.

Rotenberg, Z., Weinberger, I., Sagie, A., Fuchs, J., Davidson, E., Sperling, O., and Agmon, J., 1988. Total lactate dehydrogenase and its isoenzymes in serum of patients with non-small-cell lung cancer. *Clinical Chemistry*, *34*(4), pp. 668–670.

Ruiz-Tagle, C., Naves, R., and Balcells, M.E., 2020. Unraveling the role of microRNAs in Mycobacterium tuberculosis infection and disease: Advances and pitfalls. *Infection and Immunity*, *88*(3), pp. e00649-19.

Sanchez-Cespedes, M., Esteller, M., Wu, L., Nawroz-Danish, H., Yoo, G.H., Koch, W.M., Jen, J., Herman, J.G., and Sidransky, D., 2000. Gene promoter hypermethylation in tumors and serum of head and neck cancer patients. *Cancer Research*, *60*(4), pp. 892–895.

Schiffman, Joshua D., Paul G. Fisher, and Peter Gibbs (2015) Early detection of cancer: Past, present, and future. American Society of Clinical Oncology educational book/ASCO. *American Society of Clinical Oncology. Meeting. American Society of Clinical Oncology*.

Schulte, C., and Zeller, T., 2015. microRNA-based diagnostics and therapy in cardiovascular disease – summing up the facts. *Cardiovascular Diagnosis and Therapy*, *5*(1), p. 17.

Shackelford, R.E., Vora, M., Mayhall, K., and Cotelingam, J., 2014. ALK-rearrangements and testing methods in non-small cell lung cancer: a review. *Genes & Cancer*, *5*(1-2), p. 1.

Shi, K.Q., Lin, Z., Chen, X.J., Song, M., Wang, Y.Q., Cai, Y.J., Yang, N.B., Zheng, M.H., Dong, J.Z., Zhang, L., and Chen, Y.P., 2015. Hepatocellular carcinoma associated microRNA expression signature: integrated bioinformatics analysis, experimental validation and clinical significance. *Oncotarget*, *6*(28), p. 25093.

Sidransky, D., 2002. Emerging molecular markers of cancer. *Nature Reviews Cancer*, *2*(3), pp. 210–219.

Silva, A., Bullock, M., and Calin, G., 2015. The clinical relevance of long non-coding RNAs in cancer. *Cancers*, *7*(4), pp. 2169–2182.

Sinha, S., Brown, H., Tabak, J., Fang, Z., Tertre, M.C. du, McNamara, S., Gambaro, K., Batist, G., and Buell, J.F., 2019. Multiplexed Real-Time Polymerase Chain Reaction Cell-free Dna Assay as a Potential Method to Monitor Stage IV Colorectal Cancer. *In: Surgery (United States)* (pp. 534–539). Mosby Inc.

Siomi, M.C., Sato, K., Pezic, D., and Aravin, A.A., 2011. PIWI-interacting small RNAs: The vanguard of genome defence. *Nature Reviews Molecular Cell Biology*, *12*(4), pp. 246–258.

Solanki, A.A., Venkatesulu, B.P., and Efstathiou, J.A., 2021. Will the Use of Biomarkers Improve Bladder Cancer Radiotherapy Delivery? *Clinical Oncology*, *33*(6), pp. e264–e273.

Solin, L.J., Gray, R., Baehner, F.L., Butler, S.M., Hughes, L.L., Yoshizawa, C., Cherbavaz, D.B., Shak, S., Page, D.L., Sledge Jr, G.W., and Davidson, N.E., 2013. A multigene expression assay to predict local recurrence risk for ductal carcinoma in situ of the breast. *Journal of the National Cancer Institute*, *105*(10), pp. 701–710.

Song, T., 2018. Comparison of cervicography and human papillomavirus test as an adjunctive test to pap cytology to detect high-grade cervical neoplasia in country with high prevalence of HPV infection. *Gynecologic Oncology*, *149*, p.238.

Stephan, C., Ralla, B., and Jung, K., 2014. Prostate-specific antigen and other serum and urine markers in prostate cancer. *Biochimica et Biophysica Acta (BBA)-Reviews on Cancer*, *1846*(1), pp. 99–112.

Stone, R., Burnett, A.F., Hitt, W.C., Reynolds, K., O'Brien, K., Beard, J.B., and O'Brien, T.J., 2014. Serine protease matriptase and CA-125 co-testing for ovarian cancer detection. *Gynecologic Oncology*, *133*, p.205.

Suryavanshi, M., Jaipuria, J., Panigrahi, M.K., Goyal, N., Singal, R., Mehta, A., Batra, U., Doval, D.C., and Talwar, V., 2020. CSF cell-free DNA EGFR testing using DdPCR holds promise over conventional modalities for diagnosing leptomeningeal involvement in patients with non-small cell lung cancer. *Lung Cancer*, *148*, pp. 33–39.

Swanton, C., 2012. Intratumor heterogeneity: evolution through space and time. *Cancer Research*, *72*(19), pp. 4875–4882.

Świderska, M. Choromanska, B. , Dabrowska, B., Konarzewska-Duchnowska, E., Choromanska, K., Szczurko, G., Mysliwiec, P., Dadan, J., Ladny, J.R., and Zwierz, K. 2014. The diagnostics of colorectal cancer. *Contemporary Oncology*, *18*(1), p. 1.

Tang, L., and Han, X., 2013. The urokinase plasminogen activator system in breast cancer invasion and metastasis. *Biomedicine & Pharmacotherapy*, *67*(2), pp. 179–182.

Tavallaie, R., De Almeida, S.R., and Gooding, J.J., 2015. Toward biosensors for the detection of circulating microRNA as a cancer biomarker: an overview of the challenges and successes. *Wiley Interdisciplinary Reviews: Nanomedicine and Nanobiotechnology*, 7(4), pp. 580–592.

Tavano, F., Gioffreda, D., Valvano, M.R., Palmieri, O., Tardio, M., Latiano, T.P., Piepoli, A., Maiello, E., Pirozzi, F., and Andriulli, A., 2018. Droplet digital PCR quantification of miR-1290 as a circulating biomarker for pancreatic cancer. *Scientific Reports*, 8(1), pp. 1–11.

Tayanloo-Beik, A., Sarvari, M., Payab, M., Gilany, K., Alavi-Moghadam, S., Gholami, M., Goodarzi, P., Larijani, B., and Arjmand, B., 2020. OMICS insights into cancer histology; metabolomics and proteomics approach. *Clinical Biochemistry*. 10.1016/j.clinbiochem.2020.06.008.

Thakur, S., Tobey, A., Daley, B., Walter, M., Patel, D., Nilubol, N., Kebebew, E., Patel, A., Jensen, K., Vasko, V., and Klubo-Gwiezdzinska, J., 2019. Limited utility of circulating cell-free DNA integrity as a diagnostic tool for differentiating between malignant and benign thyroid nodules with indeterminate cytology (Bethesda Category III). *Frontiers in Oncology*, 9, p. 905.

Thompson, J.L., and Wright, G.P., 2021. The role of breast MRI in newly diagnosed breast cancer: An evidence-based review. *American Journal of Surgery*, 221(3), pp. 525–528.

Traverso, G., Shuber, A., Levin, B., Johnson, C., Olsson, L., Schoetz Jr, D.J., Hamilton, S.R., Boynton, K., Kinzler, K.W., and Vogelstein, B., 2002. Detection of APC mutations in fecal DNA from patients with colorectal tumors. *New England Journal of Medicine*, 346(5), pp. 311–320.

Usadel, H., Brabender, J., Danenberg, K.D., Jerónimo, C., Harden, S., Engles, J., Danenberg, P.V., Yang, S., and Sidransky, D., 2002. Quantitative adenomatous polyposis coli promoter methylation analysis in tumor tissue, serum, and plasma DNA of patients with lung cancer. *Cancer Research*, 62(2), pp. 371–375.

Vaisocherová, H., Šípová, H., Víšová, I., Bocková, M., Špringer, T., Ermini, M.L., Song, X., Krejčík, Z., Chrastinová, L., Pastva, O., and Pimková, K., 2015. Rapid and sensitive detection of multiple microRNAs in cell lysate by low-fouling surface plasmon resonance biosensor. *Biosensors and Bioelectronics*, 70, pp. 226–231.

Valihrach, L., Androvic, P., and Kubista, M., 2020. Circulating miRNA analysis for cancer diagnostics and therapy. *Molecular Aspects of Medicine*. doi: 10.1016/j.mam.2019.10.002.

van der Pol, Y., and Mouliere, F., 2019. Toward the early detection of cancer by decoding the epigenetic and environmental fingerprints of cell-free DNA. *Cancer Cell*. doi: 10.1016/j.ccell.2019.09.003.

Van Houten, B., Santa-Gonzalez, G.A., and Camargo, M., 2018. DNA repair after oxidative stress: Current challenges. *Current Opinion in Toxicology*, 7, pp. 9–16.

van Kruchten, M., Hospers, G.A., Glaudemans, A.W., Hollema, H., Arts, H.J., and Reyners, A.K., 2013. Positron emission tomography imaging of oestrogen receptor-expression in endometrial stromal sarcoma supports oestrogen receptor-targeted therapy: Case report and review of the literature. *European Journal of Cancer*, 49(18), pp. 3850–3855.

Vandghanooni, S., Sanaat, Z., Barar, J., Adibkia, K., Eskandani, M., and Omidi, Y., 2021. Recent advances in aptamer-based nanosystems and microfluidics devices for the detection of ovarian cancer biomarkers. *TrAC – Trends in Analytical Chemistry*, 143, p. 116343.

Vasantharajan, S.S., Eccles, M.R., Rodger, E.J., Pattison, S., McCall, J.L., Gray, E.S., Calapre, L., and Chatterjee, A., 2021. The Epigenetic landscape of Circulating tumour cells. *Biochimica et Biophysica Acta – Reviews on Cancer*. doi: 10.1016/j.bbcan.2021.188514.

Vescovo, V.D., Cantaloni, C., Cucino, A., Girlando, S., Silvestri, M., Bragantini, E., Fasanella, S., Cuorvo, L.V., Palma, P.D., Rossi, G., and Papotti, M.G., 2011. miR-205 expression levels in nonsmall cell lung cancer do not always distinguish adenocarcinomas from squamous cell carcinomas. *American Journal of Surgical Pathology*, 35, pp. 268–275.

Volinia, S., Calin, G.A., Liu, C.G., Ambs, S., Cimmino, A., Petrocca, F., Visone, R., Iorio, M., Roldo, C., Ferracin, M., and Prueitt, R.L., 2006. A microRNA expression signature of human solid tumors defines cancer gene targets. *Proceedings of the National Academy of Sciences*, 103(7), pp. 2257–2261.

Wang, H.N., Crawford, B.M., Fales, A.M., Bowie, M.L., Seewaldt, V.L., and Vo-Dinh, T., 2016. Multiplexed detection of MicroRNA biomarkers using SERS-based inverse molecular sentinel (iMS) Nanoprobes. *The Journal of Physical Chemistry C*, 120(37), pp. 21047–21055.

Wang, J., Chen, J., Chang, P., LeBlanc, A., Li, D., Abbruzzesse, J.L., Frazier, M.L., Killary, A.M., and Sen, S., 2009. MicroRNAs in plasma of pancreatic ductal adenocarcinoma patients as novel blood-based biomarkers of disease. *Cancer Prevention Research*, 2(9), pp. 807–813.

Wang, J., Raimondo, M., Guha, S., Chen, J., Diao, L., Dong, X., Wallace, M.B., Killary, A.M., Frazier, M.L., Woodward, T.A., and Wang, J., 2014. Circulating microRNAs in pancreatic juice as candidate biomarkers of pancreatic cancer. *Journal of Cancer*, 5(8), p. 696.

Wang, J.L., Hu, Y., Kong, X., Wang, Z.H., Chen, H.Y., Xu, J., and Fang, J.Y., 2013. Candidate microRNA biomarkers in human gastric cancer: a systematic review and validation study. *PloS One*, *8*(9), p. e73683.

Wang, L.G., and Gu, J., 2012. Serum microRNA-29a is a promising novel marker for early detection of colorectal liver metastasis. *Cancer Epidemiology*, *36*(1), pp. e61–e67.

Wang, Q., Huang, Z., Ni, S., Xiao, X., Xu, Q., Wang, L., Huang, D., Tan, C., Sheng, W., and Du, X., 2012. Plasma miR-601 and miR-760 are novel biomarkers for the early detection of colorectal cancer. *PloS One*, *7*(9), p. e44398.

Wang, Q., Li, Q., Yang, X., Wang, K., Du, S., Zhang, H., and Nie, Y., 2016. Graphene oxide–gold nanoparticles hybrids-based surface plasmon resonance for sensitive detection of microRNA. *Biosensors and Bioelectronics*, *77*, pp. 1001–1007.

Wang, X., Hou, T., Lin, H., Lv, W., Li, H., and Li, F., 2019. In situ template generation of silver nanoparticles as amplification tags for ultrasensitive surface plasmon resonance biosensing of microRNA. *Biosensors and Bioelectronics*, *137*, pp. 82–87.

Wang, Y., Zheng, D., Tan, Q., Wang, M.X., and Gu, L.Q., 2011. Nanopore-based detection of circulating microRNAs in lung cancer patients. *Nature Nanotechnology*, *6*(10), pp. 668–674.

Wanunu, M., Dadosh, T., Ray, V., Jin, J., McReynolds, L., and Drndić, M., 2010. Rapid electronic detection of probe-specific microRNAs using thin nanopore sensors. *Nature Nanotechnology*, *5*(11), pp. 807–814.

Weng, J.L., Atyah, M., Zhou, C.H., and Ren, N., 2019. Progress in quantitative technique of circulating cell free DNA and its role in cancer diagnosis and prognosis. *Cancer Genetics*. doi: 10.1016/j.cancergen.2019.10.001.

Wenstrom, K., Owen, J., and Boots, L., 1997. Second-trimester maternal serum Ca-125 versus estriol in the multiple-marker screening test for Down syndrome. *Obstetrics & Gynecology*, *89*(3), pp. 359–363.

Westwood, M., Asselt, A.D.I., Ramaekers, B., Whiting, P., Joore, M., Armstrong, N., Noake, C., Ross, J., Severens, H., and Kleijnen, J., 2014. KRAS mutation testing of tumours in adults with metastatic colorectal cancer: a systematic review and cost-effectiveness analysis. *Health Technology Assessment*. doi: 10.3310/hta18620.

Wong, C.C.L., Wong, C.M., Tung, E.K.K., Au, S.L.K., Lee, J.M.F., Poon, R.T.P., Man, K., and Ng, I.O.L., 2011. The microRNA miR-139 suppresses metastasis and progression of hepatocellular carcinoma by down-regulating Rho-kinase 2. *Gastroenterology*, *140*(1), pp. 322–331.

Wu, J.T. 1999. Review of circulating tumor markers: from enzyme, carcinoembryonic protein to oncogene and suppressor gene. *Annals of Clinical & Laboratory Science*, *29*(2), pp. 106–111.

Wu, Y., Tilley, R.D., and Gooding, J.J., 2018. Challenges and solutions in developing ultrasensitive biosensors. *Journal of the American Chemical Society*, *141*(3), pp. 1162–1170.

Xiong, H., Yan, J., Cai, S., He, Q., Peng, D., Liu, Z., and Liu, Y., 2019. Cancer protein biomarker discovery based on nucleic acid aptamers. *International Journal of Biological Macromolecules*. doi: 10.1016/j.ijbiomac.2019.03.165.

Xu, W., Zhang, Z., Zou, K., Cheng, Y., Yang, M., Chen, H., Wang, H., Zhao, J., Chen, P., He, L., and Chen, X., 2017. MiR-1 suppresses tumor cell proliferation in colorectal cancer by inhibition of Smad3-mediated tumor glycolysis. *Cell Death & Disease*, *8*(5), pp. e2761–e2761.

Xue, X., Wang, C., Xue, Z., Wen, J., Han, J., Ma, X., Zang, X., Deng, H., Guo, R., Asuquo, I.P., and Qin, C., 2020. Exosomal miRNA profiling before and after surgery revealed potential diagnostic and prognostic markers for lung adenocarcinoma. *Acta biochimica et biophysica Sinica*, *52*(3), pp. 281–293.

Yanaihara, N., Caplen, N., Bowman, E., Seike, M., Kumamoto, K., Yi, M., Stephens, R.M., Okamoto, A., Yokota, J., Tanaka, T., and Calin, G.A., 2006. Unique microRNA molecular profiles in lung cancer diagnosis and prognosis. *Cancer Cell*, *9*(3), pp. 189–198.

Yanaihara, N., and Harris, C.C., 2013. MicroRNA involvement in human cancers. *Clinical Chemistry*, *59*(12), pp. 1811–1812.

Yang, M., Chen, J., Su, F., Yu, B., Su, F., Lin, L., Liu, Y., Huang, J.D., and Song, E., 2011. Microvesicles secreted by macrophages shuttle invasion-potentiating microRNAs into breast cancer cells. *Molecular Cancer*, *10*(1), pp. 1–13.

Yuan, T., Huang, X., Woodcock, M., Du, M., Dittmar, R., Wang, Y., Tsai, S., Kohli, M., Boardman, L., Patel, T., and Wang, L., 2016. Plasma extracellular RNA profiles in healthy and cancer patients. *Scientific Reports*, *6*, p. 19413.

Zakaria, H.M., Mohamed, A., Omar, H., and Gaballa, N.K., 2020. Alpha-fetoprotein level to total tumor volume as a predictor of hepatocellular carcinoma recurrence after resection. A retrospective cohort study. *Annals of Medicine and Surgery*, *54*, pp. 109–113.

Zhang, Y., Sui, J., Shen, X., Li, C., Yao, W., Hong, W., Peng, H., Pu, Y., Yin, L., and Liang, G., 2017. Differential expression profiles of microRNAs as potential biomarkers for the early diagnosis of lung cancer. *Oncology Reports*, *37*(6), pp. 3543–3553.

Zhao, F., Vesprini, D., Liu, R.S., Olkhov-Mitsel, E., Klotz, L.H., Loblaw, A., Liu, S.K., and Bapat, B., 2019, May. Combining urinary DNA methylation and cell-free microRNA biomarkers for improved monitoring of prostate cancer patients on active surveillance. In *Urologic Oncology: Seminars and Original Investigations* (*Vol. 37*, No. 5, pp. 297–e9). Elsevier.

Zhao, G., Jiang, T., Liu, Y., Huai, G., Lan, C., Li, G., Jia, G., Wang, K., and Yang, M., 2018. Droplet digital PCR-based circulating microRNA detection serve as a promising diagnostic method for gastric cancer. *BMC Cancer*, *18*(1), pp. 1–10.

Zhao, Y.J., Ju, Q., and Li, G.C., 2013. Tumor markers for hepatocellular carcinoma. *Molecular and Clinical Oncology*, *1*(4), pp. 593–598.

Zhao, G., Wang, B., Liu, Y., Zhang, J.G., Deng, S.C., Qin, Q., Tian, K., Li, X., Zhu, S., Niu, Y., and Gong, Q., 2013. miRNA-141, downregulated in pancreatic cancer, inhibits cell proliferation and invasion by directly targeting MAP4K4. *Molecular Cancer Therapeutics*, *12*(11), pp. 2569–2580.

Zhao, S., Lin, H., Chen, S., Yang, M., Yan, Q., Wen, C., Hao, Z., Yan, Y., Sun, Y., Hu, J., and Chen, Z., 2015. Sensitive detection of Porcine circovirus-2 by droplet digital polymerase chain reaction. *Journal of Veterinary Diagnostic Investigation*, *27*(6), pp. 784–788.

Zhou, Q., Zuo, M.Z., He, Z., Li, H.R., and Li, W., 2018. Identification of circulating microRNAs as diagnostic biomarkers for ovarian cancer: A pooled analysis of individual studies. *The International Journal of Biological Markers*, *33*(4), pp. 379–388.

Zvereva, M., Roberti, G., Durand, G., Voegele, C., Nguyen, M.D., Delhomme, T.M., Chopard, P., Fabianova, E., Adamcakova, Z., Holcatova, I., Foretova, L., Janout, V., Brennan, P., Foll, M., Byrnes, G.B., McKay, J.D., Scelo, G., and Le Calvez-Kelm, F., 2020. Circulating tumour-derived KRAS mutations in pancreatic cancer cases are predominantly carried by very short fragments of cell-free DNA. *EBioMedicine*, *55*. doi: 10.1016/j.ebiom.2019.09.042.

4 Biosensor Development: A Way to Achieve a Milestone for Cancer Detection

4.1 INTRODUCTION

Nanoparticles have unique physical and chemical properties for developing the biosensors and chemical sensors with improved sensitivity and lower limits of detection (Khanna, 2008; Palei et al., 2018; Rajkumar and Prabaharan, 2019; Shah, Imran, and Ullah, 2019). Recently, different types of nanoparticles including metallic, non-metallic, semiconductors, and oxides-based biosensors have been used in different sensing systems (Luo et al., 2006; Hou et al., 2017; Li et al., 2017; Parvanian, Mostafavi, and Aghashiri, 2017; Saeed et al., 2017; Shah, Imran, and Ullah, 2019; Mottaghitalab et al., 2019; Peng et al., 2019; Perumal et al., 2019). There are different methods to synthesize nanoparticles such as chemical, physical, and biological. Many synthesis methods have been modified and developed to enhance the specific properties of nanoparticles and to reduce the costs of production (Cho et al., 2013). They are the particles of 1–100 nm size range and are being used for several purposes from industries like cosmetics, solar fuels, and clothing to medical treatments. In this chapter, different synthesis methods of nanoparticles, characterization techniques and methods of sensor fabrication are discussed.

4.2 SYNTHESIS OF NANOPARTICLES

There are different strategies to synthesize nanoparticles including chemical or physical but the toxic effects imposed by these methods can be overcome by an eco-friendly approach, i.e., the biological way of synthesis using plant extracts (Shankar et al., 2004; Ahmad et al., 2011; Balavijayalakshmi and Ramalakshmi, 2017; Joshi et al., 2017; Karimi et al., 2017; Liu et al., 2017; Rajeshkumar and Bharath, 2017; Ramasamy et al., 2017; Rezaie et al., 2017; Sithara et al., 2017), microorganisms (Konishi et al., 2007), fungi (Vigneshwaran et al., 2007), enzymes (Willner et al., 2006), etc.

The synthesis methods are categorized into bottom-up or top-down methods (Figure 4.1).

4.2.1 BOTTOM-UP METHOD

This method involves the building up of nanomaterials from atoms to nanoparticles. Among the bottom-up methods, sol-gel, pyrolysis, chemical vapor deposition (CVD), spinning, and biological methods are the important ones.

   a. **Sol-gel:** It is the most favored method of nanoparticle synthesis. Sol can be defined as the solution of small, solid particles suspended in a liquid medium while the gel is an aggregate of small particles in the solvent. In this process, a chemical solution acts as a precursor for the combined system of distinct particles. The precursor is then disseminated in a solvent by either stirring or shaking, resulting in the solid and liquid phase (Ramesh, 2013). In the final step, nanoparticles are separated by sedimentation, centrifugation, or other filtration methods (Dervin and Pillai, 2017).
   b. **Chemical vapor deposition (CVD):** This method carries out the deposition of gaseous reactants on a substrate in a reaction chamber at a favorable temperature. Due to the

DOI: 10.1201/9781003201946-4

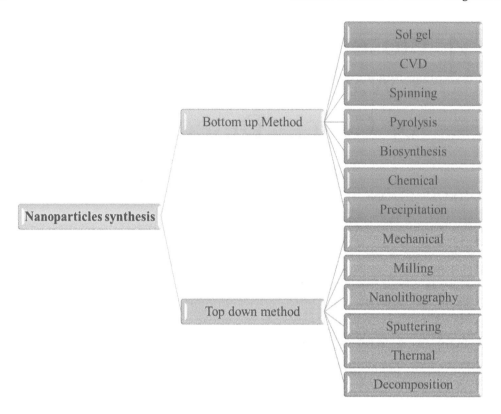

FIGURE 4.1 Common synthesis methods of nanoparticles.

combined interaction of gaseous molecules and heated substrate, chemical vapors develop (Bhaviripudi et al., 2007) which ultimately form a thin film on the substrate
surface that is used after recovering. CVD method is advantageous due to the formation
of highly pure and strong nanoparticles while the specific equipment requirement and
toxic gaseous by-products production are some disadvantages of this method (Adachi
et al., 2003).

c. **Spinning:** Spinning method involves a spinning disc reactor (SDR) having a rotating disc
where the physical factors can be controlled. For the removal of oxygen, the inner part of the
reactor is filled with inert gases to eliminate oxygen inside (Tai et al., 2007). The disc is
swapped at a different speed to pump in the precursor liquid. All the atoms get fused together
after spinning and further precipitated and dried (Mohammadi et al., 2014).

d. **Pyrolysis:** For large-scale production of nanoparticles, pyrolysis is used. Pyrolysis means
burning of a precursor either liquid or vapor into a furnace (Kammler et al., 2001) and further
to recover the nanoparticles, by-product gases are air classified. This is a cost-effective and
high yielding continuous process.

e. **Biosynthesis:** Biosynthesis or green synthesis is a nontoxic and environmental friendly
(Kuppusamy et al., 2016) approach for nanoparticles synthesis where any plant extracts,
microorganisms including fungi, bacteria, actinomycetes, etc. act as precursors instead of
chemicals for the production of nanoparticles (Hasan, 2015; (Baláž et al., 2017; Beyene
et al., 2017; Brobbey et al., 2017; Ho et al., 2017; Tang, Wang and Gomez, 2017; Uddin
et al., 2017; Vijayaraghavan and Ashokkumar, 2017; Parvanian, Mostafavi and Aghashiri,
2017; Saeed et al., 2017; Santhoshkumar et al., 2017; Li et al., 2017; Mamatha et al., 2017;
Menon et al., 2017; Hasanzadeh et al., 2018; Palei et al., 2018; Hong et al., 2019; Htoo et al.,
2019; Mottaghitalab et al., 2019; Peng et al., 2019; Perumal et al., 2019; Rajkumar and

Prabaharan, 2019; Shah et al., 2019). The nanoparticle synthesis can be intracellular or extracellular rendering to nanoparticle locations (Mann, 2001; Hulkoti and Taranath, 2014). Among the microorganisms, both uni- and multicellular microorganisms can reduce the metallic precursors to nanoparticles. These microorganisms like fungi or bacteria produce the intra- or extracellular inorganic materials (Shankar et al., 2004) to accumulate the metals for nanoparticles production. During the intracellular synthesis, ions are transported inside the microbial cells in the presence of enzymes. The nanoparticles are of a smaller size as compared to the extracellular one (Narayanan and Sakthivel, 2010). Mostly extracellular synthesis of nanoparticles is done by fungi due to their large number of secretory components, which are able to reduce and cap the nanoparticles. Due to the presence of effective phytochemicals such as ketones, aldehydes, terpenoids, and other reducing agents, plant extracts are considered as a vast source for metal and metal oxide nanoparticles synthesis (Doble et al., 2010). These phytochemicals are able to reduce the metal salts into metallic nanoparticles. Nanoparticles synthesized by green approach are useful in various fields including optical imaging, catalysis, biomedical diagnostics, etc. (Aguilar, 2012).

f. **Chemical precipitation:** This method is developed for controlling the size of nanoparticles by exploiting precipitation technique. The synthesis process is a reaction between component materials present in a suitable solvent. In this process, a dopant is added before the precipitation of parent solution and by using surfactant agglomeration of particles is avoided. Ultimately, formed nanocrystals can be separated by centrifugation (Konrad et al., 2001; Sharma et al., 2009).

4.2.2 Top-Down Method

The top-down method is also known as a destructive method of nanoparticle synthesis where the bulk materials are reduced to a nanometric scale.

a. **Mechanical milling:** During nanoparticle synthesis, mechanical milling is employed for milling and after that nanoparticles annealing where different components are milled in an inert atmosphere (Chen et al., 2014; Makkar, Agarwala and Agarwala, 2014; Bello, Agunsoye and Hassan, 2015; Panjiar, Gakkhar and Daniel, 2015; Zhang et al., 2015; Ağaoğulları et al., 2017; Bokhonov and Dudina, 2017; Han et al., 2017; Pei et al., 2017; Popov et al., 2017). Among the several top-down methods, it is the most commonly used approach.

b. **Nanolithography:** It is the study of nanometric constructions with a size range of 1 to 100 nm. There are different types of nanolithographic processes such as optical, multiphoton, nanoimprint, and electron-beam lithography (Pimpin and Srituravanich, 2012; Calborean et al., 2015; Hong and Lin, 2015; Bhagoria et al., 2019; Sebastian et al., 2019; Sun et al., 2019; Hu et al., 2020; Liao et al., 2020; Zhang et al., 2020). Usually, lithography is to print shape or structure of - a material on demand that is light sensitive. To produce nanoparticles with a desired shape and structure is the main advantage of nanolithography (Hulteen et al., 1999).

c. **Sputtering:** Sputtering is to deposit the nanoparticles on a surface by the ejection of particles from it by collision with ions (Shah and Gavrin, 2006; Achour et al., 2018; Chung et al., 2018; Verma et al., 2018a, 2018b; Zhang et al., 2018; Cigáň et al., 2018; Dai and Moon, 2018; Jiang et al., 2018; Lee et al., 2018; Liu, 2018; Liu et al., 2018; Mahdhi et al., 2018; Rezaee and Ghobadi, 2018). It is followed by annealing. The shape and size of the nanoparticles are determined by this layer thickness, duration, and temperature of annealing.

d. **Thermal decomposition:** Thermal decomposition is the process of chemical decomposition of a compound at a specific temperature. In this method, nanoparticles are produced by decomposing the metals at their specific decomposition temperature (Salavati-Niasari et al., 2008; Abdullah et al., 2016; Adner et al., 2016; Ahab et al., 2016; Bartůněk et al., 2016; Glasgow et al., 2016; Khayati et al., 2016; Lee et al., 2016; Polteau et al., 2016; Sharifi et al., 2016; Sharma et al., 2016).

4.3 TYPES OF NANOPARTICLES

Different types of nanoparticles (Figure 4.2) have been synthesized and their applications have been reported as explained below

1. **Organic nanoparticles:** Organic nanoparticles are biodegradable and non-toxic. Most commonly known organic nanoparticles are dendrimers, liposomes, micelles, etc. with a sensitivity towards light and heat. These particles are the ideal choice for drug delivery (Alvarez-Trabado et al., 2017; Di Martino et al., 2017; Hashemipour and Ahmad Panahi, 2017; Kumar et al., 2017; Wuttke et al., 2017; Adhikari et al., 2018; Yu et al., 2018; Monarca et al., 2018; Pawar et al., 2018; Poovi and Damodharan, 2018; Sun et al., 2018).

2. **Inorganic nanoparticles:** Nanoparticles made up of non-carbon materials are inorganic nanoparticles. They are of two types:
   i. **Metallic nanoparticles:** Nanoparticles synthesized from metals to nanometric-scale structures are metallic nanoparticles. Any type of metal can be used for synthesis including aluminium, cadmium, cobalt, gold, silver, etc. Nanoparticles have distinct properties such as small dimensions in the nanometer range, high surface-to-volume ratio, specific shapes, and sensitivity to different environmental factors (Salavati-Niasari et al., 2008; Shaganov, Perova and Berwick, 2017; Shojaati et al., 2017; Holm et al., 2017; Khan et al., 2017; Altman, 2019; Bao and Lan, 2019; Wang et al., 2019; Zhou et al., 2020; Kuchur et al., 2020; Liu et al., 2020; Roma et al., 2020).
   ii. **Metal oxide nanoparticles:** Metal oxide nanoparticles are the form of their respective metals with modified properties such as improved reactivity and higher efficiency. Examples of metal oxide nanoparticles include aluminium oxide, iIron oxide, zinc oxide, etc. (Tai et al., 2007; Ashouri et al., 2015; Jiménez-Rojo et al., 2015; Kumar et al., 2015; Mallakpour and Madani, 2015; Akbari et al., 2018;

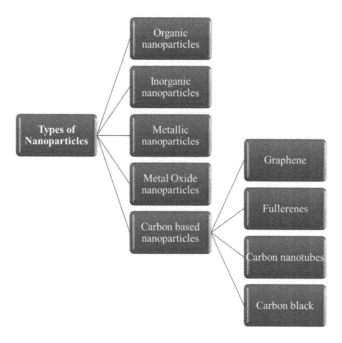

FIGURE 4.2   Different types of nanoparticles.

Zikalala et al., 2018; Lee et al., 2019; Muthuvinothini and Stella, 2019; Sizochenko et al., 2019; Vara and Dave, 2019; Ahangaran and Navarchian, 2020; Gupta et al., 2020; Jabli et al., 2020).

3. **Carbon based:** Carbon-based nanoparticles are classified into graphenes, carbon nanotubes, fullerenes, and carbon black in nanosize (Bhaviripudi et al., 2007; Chen and Liu, 2020; Ghosal et al., 2020; Na et al., 2020; Singh and Khullar, 2020; X. Zhang et al., 2020; Y. Zhang et al., 2020):

   i. **Graphene:** A hexagonal network of carbon atoms creating a honeycomb lattice. Graphene is a carbon allotrope.

   ii. **Fullerenes:** Fullerenes (C60) are a spherical carbon molecule with carbon atoms in sp2 hybridization. About 28 to 1,500 carbon atoms take participation in forming multi-layered fullerenes.

   iii. **Carbon nanotubes (CNTs):** Carbon nanotubes are formed as the graphites are rolled up into cylindrical forms. CNTs are of two types: single and multiwalled. These nanotubes can carry the current density of 1 billion amperes per meter squared, causing it to behave like a superconductor. Besides, it carries a very high mechanical strength and stability that is around 60 times greater than steel (Pal et al., 2011; Jiang et al., 2018; Malik, 2018; Meena and Choudhary, 2018; Tostado-Plascencia et al., 2018; Hoyos-Palacio et al., 2019; Krajewski et al., 2019; Yoosefian et al., 2019).

   iv. **Carbon black:** Carbon black is an amorphous, generally spherical material with 20 to 70 nm diameter. The particles are strongly interconnected with each other, forming an aggregate (Pal et al., 2011).

4.4 CHARACTERIZATION OF NANOPARTICLES

Nanoparticles are usually characterized by their absorption maxima, size, shape, surface charge through UV-Vis spectrophotometer, advanced microscopic techniques such as scanning electron microscope (SEM) or transmission electron microscope (TEM) or atomic force microscopy (AFM) and particle size analyzer (Table 4.1, Figure 4.3) (Jayachandran et al., 2021; Merugu and Gothalwal, 2021; Sivakumar and Nelson Prabu, 2021).

4.4.1 Particle Size

Particle size, its distribution, and shape are the most important factors of nanoparticle characterization. Particle size mostly affects the use of nanoparticles in drug release. Smaller particles are very useful for drug delivery due to their large surface area (Pal et al., 2011; Handol Lee et al., 2019; Kim et al., 2019; Scherer et al., 2019; Singh et al., 2019; Made Joni et al., 2020; Wren et al., 2020; X. Hu et al., 2020). There are several techniques for nanoparticle size distribution, as discussed below.

4.4.2 Dynamic Light Scattering (DLS) Technique

This technique is widely used for the size determination of nano- to micrometer range of particles that follow the Brownian motion in colloidal solution. When the light hits the particles in the solution, due to the Brownian motion of the particles, a Doppler shift occurs that results in a change in wavelength of light. This change in the wavelength is directly related to the size of the particles. Using the DLS technique, it is possible to determine size distribution and motion as well as the diffusion coefficient of the particles (de Assis et al., 2008).

TABLE 4.1

Different Characterization Methods of Nanoparticles and their Working Principle

S.No.	Characterization Method	Characteristic	Principle of Working
1.	Dynamic light scattering (DLS)	Size of the particles	Due to Brownian motion, Doppler shift occurs after passing of light in the solution
2.	Scanning electron microscopy (SEM)	Size and morphology	Electron microscope-based visualization
3.	Transmission electron microscope (TEM)	Size and morphology	Electron microscope-based visualization
4.	Atomic force microscopy (AFM)	Size of the particles	Atomic-scale probe tip based physical scanning
5.	Zeta potential	Surface charge	Measurement of indirect charge

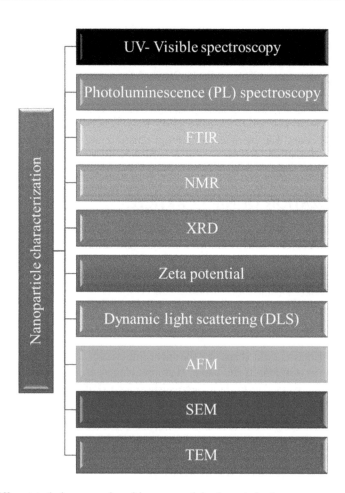

FIGURE 4.3   Different techniques employed in nanoparticle characterization.

4.4.3 SCANNING ELECTRON MICROSCOPY (SEM)

SEM techniques are used for examining the morphology of nanoparticles by direct visualization. It is based on electron microscopy providing various advantages in sizing analysis but limited information about the size distribution analysis. To analyze the sample, it must be prepared in a specific way. First, the sample is converted into a dry powder and then fixed on a holder, followed by coating with any metal (conductive material) through a coater. Then the mounted sample is passed through an electron beam (Jores et al., 2004). The sample surface features are found using the secondary electrons produced by the surface of the sample. This technique is more costly and time consuming than the DLS method.

4.4.4 TRANSMISSION ELECTRON MICROSCOPE (TEM)

The source of energy is also an electron beam in this method but the operation principle differs from SEM. In TEM, the sample preparation in this method is difficult and time consuming due to its ultra-thin requirement so that electrons can be transmitted. Nanoparticles are dispersed on copper grids or films and fixed using a negative staining compound such as uranyl acetate or phosphotungstic acid. Surface characteristics of the nanoparticles are investigated by passing the electron beam through the ultra-thin film (Molpeceres et al., 2000).

4.4.5 ATOMIC FORCE MICROSCOPY (AFM)

AFM proposes ultra-high resolution for the measurement of particle size. It uses the atomic scale probe tip for the physical scanning of sub-micron samples (Zur Mühlen et al., 1996). In this technique, samples can be scanned in contact mode or noncontact mode according to their properties. The topographical map is produced by tapping the surface of the probe through the sample in contact mode while in the non-contact mode, probe drifts over the conducting surface. The ability to image the biological micro- and nanostructures that are non-conducting materials, without any special treatment, is its primary advantage (Shi et al., 2003).

4.4.6 SURFACE CHARGE

To determine the surface charge of the nanoparticles is a very important aspect because it decides the ability to interact with biological compounds and its electrostatic interaction with bioactive components. The colloidal stability is investigated by analyzing the zeta potential of nanoparticles that measures indirect charge. Surface hydrophobicity can also be obtained by the zeta potential (Pangi et al., 2003).

4.4.7 UV-VISIBLE ABSORPTION SPECTROSCOPY

Absorbance spectroscopy gives information about the optical properties. It works on the Beer-Lamberts Law, where the light is passed through the sample and the amount of light absorbed is measured. In a given range of wavelengths, absorbance of the sample can be measured at each wavelength (Subbaiya et al., 2014). The absorbance can also be used to know the sample concentration using the Beer-Lamberts equation.

4.4.8 X-RAY DIFFRACTION (XRD)

XRD is a conventional method of determining the crystalline structure and morphology of the nanoparticles or constituents. The intensity of the diffraction pattern can increase or decrease with the amount of the sample. This method is used to analyze the particle's metallic nature including size and shape, the electron density of the elements, or constituents based on peak positions. Using Bragg's equation, the crystalline size of the particles can be calculated (Yelil Arasi et al., 2012).

Bragg's equation: $n\lambda = 2\ d \sin \theta$
  θ = Angle of incidence
  n = An integer
  d = Crystal planes spaces
  λ = Wavelength

4.4.9 FOURIER TRANSFORM INFRARED (FTIR) SPECTROSCOPY

FTIR is used to analyze the functional groups associated and structural characteristics of biological or plant extracts with nanoparticles. This method is used to determine the infrared spectrum of sample absorption or transmission (Amudha and Shanmugasundaram, 2014) in the frequency range of 600–4,000 cm^{-1}. The specific molecular properties of the sample can be obtained through spectrum data using the spectroscopy software.

4.4.10 NUCLEAR MAGNETIC RESONANCE (NMR) SPECTROSCOPY

Another method to determine the quantitative and structural characteristics of nanomaterials is NMR spectroscopy. It involves the nuclear magnetic resonance properties, which is shown by nuclei possessing a nonzero revolution in a strong magnetic field.

Transitions among spin-up and spin-down states can be explored through electromagnetic radiations in the range of radio waves. NMR can determine the interaction between dia- or antiferromagnetic nanoparticles' surface and ligand. However, the characterization of ferri or ferromagnetic materials is not preferable because of the higher magnetization of these materials, which causes differences in the local magnetic area and ultimately leads to, shifts in signal frequency.

This results in the broadening of the signal peak and inutile measurements (Mourdikoudis et al., 2018). NMR spectroscopy provides the direct molecular-scale information of NPs materialization and morphology in both liquid and solid phases. The capping ligands are also be investigated by NMR to get the electronic structure and atomic composition (Coates, 2006). Additionally, NMR can also be used for investigating the role of capping ligands in particle shape determination. Overall, NMR can examine the chemical changes of nanoparticle precursors in both liquid and solid phases with high 3D chemical resolution. Furthermore, NMR is valuable for the process monitoring of ligands exchange in forming the final product.

4.4.11 PHOTOLUMINESCENCE (PL) SPECTROSCOPY

PL spectroscopy is another method to characterize the nanomaterials. It measures the light that is emitted from the atoms after the absorption. This technique is particularly useful for fluorescent nanomaterials like quantum dots, metal nanoclusters, etc. Recently, the intrinsic photoluminescence of metallic nanoparticles has received significant interest. Furthermore, the PL of metallic NPs is without any photobleaching and photoblinking. So it is regarded as a better optical labeling method for fluorescent molecules. Single and multi photonic PL has been developed via plasmon nanostructures of different shapes (Zhang et al., 2014). In a study of Gong and co-workers, the photoluminescence behavior of gold nanoflower has shown its branched plasmonic nanostructure. This PL characteristic of gold nanoflower was supposed to be dependent on excitation wavelength and polarization properties (Lim et al., 2013).

4.5 PREPARATION OF A SENSOR DEVICE

Biosensors are compact analytical devices resembling natural chemoreception schemes where biological molecules bind to the specific analyte for the generation of biochemical information and then converted into an electric signal by a transducer element. The unique characteristics of biosensors including miniaturized sensor size, fast responses, rapid label-free detection, biocompatibility, easy device tailoring, high reliability of measurements, ultra-low detection limits, and low development costs have helped to attract end users and industries. These biosensors have the potential to engineer new, wearable, implantable biosensors in the clinical diagnosis of different biomarkers (Pantelopoulos and Bourbakis, 2009; Jin et al., 2017).

A typical biosensor comprises of (a) a bioreceptor to which an analyte can bind with high affinity, (b) a biosurface that provides a functioning environment for the bioreceptor, (c) a transducer that converts the biosignal into a readable or electrical signal, and (d) an electronic amplifier that can amplify the signals for proper visualization and evaluation of produced data (Figure 4.4) (Cui et al., 2019).

4.5.1 BIO-TRANSDUCER ELEMENTS

Elements that recognize the interactions between the analyte and the biosensing component and transfer that interaction into readable signals are called transducers. The commonly used biosensors are electrochemical, piezoelectrical, and optical (Jain et al., 2010).

4.5.1.1 Electrochemical Sensors

The working principle of electrochemical sensors is based on the measurement of electrochemical changes occurring on the sensing surface of electrodes after interaction with the analyte. Based on electric signals, it is divided into potentiometric, amperometry, and conductometric. Potentiometric biosensors measure the change in voltage, change in current at a fixed voltage is measured in amperometry, while the measurement of change in charge due to sensing material is conductometry. The advantages of electrochemical biosensors include cost-effectiveness, ease of

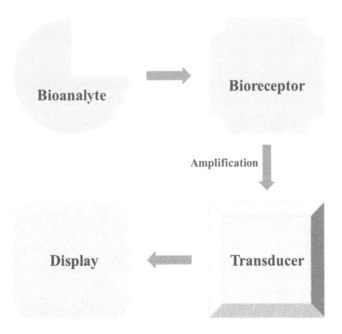

FIGURE 4.4 Design of a typical biosensor.

handling, miniature size, and high sensitivity. Commercially, a sensor based on this approach is used for nucleic acid detection, enzymatic assays such as glucose biosensor, etc. (Gouvêa, 2011; Patel et al., 2016).

In nucleic acid electrochemical biosensors, DNA is immobilized onto the surface of the electrode and allowed to hybridize and electrical conductance variations are measured. Labeling-based indirect methods are also available. The capture and detector agents sandwich the target analyte Heterocyclic dyes, organometals, and ferrocene dyes generally work as capturing agents and are immobilized on electrodes, nanomaterials, or glass chips. The detector agents are coupled to signaling tags such as enzymes, nanoparticles, etc. (Bora et al., 2013; Sin et al., 2014). This technique is used for proteins, antibodies, and nucleic acid detection. Lateral flow immunoassays and immunochromatographic assay strips are the best sandwich methods that are commercially available (e.g., home pregnancy test kits). Qualitative or semiquantitative measurement of signals is possible through photodiode or amperometric sensors (Chin et al., 2012). The advantages of this method include the requirement of one recognition component, less analysis time, and low cost of reagent. It permits the consistent monitoring of data and real-time analysis. The sample components are examined in their native form without any chemical change (Sin et al., 2014; Paleček et al., 2015). Recently, a label-free DNA biosensor was introduced as MFON (metal film on nanosphere). In this system, molecular sentinel is embedded onto a plasmonic "nano wave" chip. It is based on the principle that reduced SERS (surface-enhanced Raman scattering) occurs due to DNA hybridization. This sensor could potentially be used to detect the S-adenosyl methionine domain containing 2 (RSAD2) gene, an inflammatory biomarker (Ngo et al., 2013). Label-free protein assays comprise the detection of aggregated proteins, commonly in neurodegenerative diseases such as Alzheimer's disease (Vargová, 2015), Parkinson's disease, (Breydo et al., 2012), etc.

4.5.1.2 Optical Sensors

In this type of transducer, an output signal is light and obtained due to optical diffraction. In an optical system, light is directed towards the sensor's surface with the help of optical fiber and is reflected. A detector such as a photodiode screens the reflected light, which further helps in calculating the physical alterations that occur on the sensing surface. These types of biosensors are generally applied to detect bacterial pathogens and to study the antigen-antibody kinetics. During the analysis, change in the light properties relates to change in concentration or mass. The measured signals in optical sensors are absorbance, chemiluminescence, fluorescence, and surface plasmon resonance (Watts et al., 1994; Radhika et al., 2012; Syam et al., 2012). Thus, it can simultaneously screen a population of samples. But, the inability to miniaturize and the requirement of a spectrophotometer for signal measurement are the limitations of this method.

4.5.1.3 Piezoelectric Sensors

This biosensor is based on the interactions between the amount of the analyte and the sensing element. It is also called a mass sensor. The sensing element is generally used in vibrating piezoelectric (PZ) quartz crystal. The resonant frequency of this crystal changes after binding an analyte, leading to the generation of an oscillating voltage. The acoustic wave sensor senses this oscillating voltage. This biosensor is mostly used in the case of analysis of gas phases. It has the same limitations as optical sensors, i.e., the requirement of sophisticated instruments and the inability to miniaturize (Patel et al., 2016; Dai et al., 2019).

4.5.1.4 Calorimetric Sensors

Calorimetric sensors work on the principle of adsorption and heat production, the fundamental properties of a reaction. That's why they are also called thermal sensors. During the biological responses, a change in the temperature of the medium is measured and the concentration is evaluated by comparing it with a sensor without any reaction. Therefore, it is best suited for enzymatic reactions. This biosensor is mainly used to detect and quantify pesticides and pathogenic bacteria and serum cholesterol measurement (Vereshchagina et al., 2015; Patel et al., 2016).

4.5.2 NEW GENERATION BIOSENSORS: NANOBIOSENSORS

4.5.2.1 Quantum Dots (QDs)

The sensitivity and specificity of the biosensors can be improved via the coupling of quantum dots. QDs are semiconductors in nature and of nanometer scale with distinct quantum confinement properties.

They possess broad excitation and confined size-tunable emission bandwidth and high stability (Frasco and Chaniotakis, 2009; Nurunnabi et al., 2010). The working principle is based on the fluorescence transduction generated after analyte interaction with quantum dots surface via photoluminescent activation or quenching. Multimodal biosensors based on probes have been developed using the QDs (with surface alterations) for nucleic acids and peptides detection. Quantum dots have broad-spectrum applications ranging from pH detection to quantification of biomolecules (DNA, RNA, proteins). The only limitation of QDs is high toxicity (Wang et al., 2013).

4.5.2.2 Graphene Biosensors

It is a carbon sheet with a honeycomb-like structure. It has a strong electrical conductivity and a larger surface area due to possessing a 2D structure. It helps in the accurate detection of molecules by rapid electron transfer. Graphene is considered more advantageous than carbon nanotubes due to their large surface area, high electrolytic performance, and low cost. It has less interference of transition metal contamination such as iron, nickel, etc.; hence, it exhibits more purity than a carbon nanotube. Due to this property, it also provides an improved platform for studying the electrocatalytic properties of carbon atoms. Due to these properties and its high tensile strength, graphene is considered a better option for biosensor fabrication. Graphene electrodes are generally used for glucose, neurotransmitters, and other small molecule detection. These electrodes work based on oxidation-reduction reactions occurring on their exteriors. The graphene electrodes can be modified (chemically) to increase the electron transfer rate to improve biosensor performance (Kuila et al., 2011; Ruan et al., 2013).

Graphene is also an excellent material for electrodes used in electrochemical biosensors. Graphene-based nanomaterials are used to prepare nanobiosensors, e.g., for detection of glucose and environmental contaminants monitoring, graphene-gold nanoparticles coated with nafion biosensors are reported with very fast response time. These biosensors are non-enzymatic, having high sensitivity and ultra stability. Likewise, graphene-DNA biosensors possess high selectivity and sensitivity to detect specific mutated DNA in human disease.

Biosensors derived from graphene quantum dots (Gdots) are photoluminescent. They also have the unique properties of broad excitation-emission spectral range and quantum confinement. Gdots are more competent than other imaging agents such as Qdots because of their more excellent photostability against photobleaching and less toxicity (Chen et al., 2016). These properties make it to be useful for electrochemical and photosensors. The size-tunable property of Gdots allows evaluation of the ssDNA and enzyme immobilization. Metal ion and amino acid detection are also possible by Gdot based electrochemiluminescence sensor (Peng et al., 2012; Sun et al., 2013).

4.5.2.3 Carbon Nanotubes (CNTs)

CNTs are made up of graphene sheets. They are cylindrical in structure. CNTs have attractive physical and chemical properties, making them suitable for biosensor applications in the biomedical field. Due to having solid atomic bonds, CNTs can sustain a very high temperature and are considered outstanding thermal and electrical conductors. CNT-based immunosensors are prepared by the coating of antibodies on CNTs. It can sense the antigens such as viruses, enzymes, nucleic acids, etc. The working principle of CNT-based sensors involves the electrical conductivity variations that correlate with the distance between the probe and the target analyte. CNTs coupled with electrochemical biosensors provide better sensitivity to electrodes used for immunosensors, enzyme sensors, and nucleic acid biosensors. CNTs can be twisted effortlessly and miniaturized because of their high tensile strength. However, the toxic effects of CNTs do not allow them to be used for bio-medicinal applications. Also, they have limited water solubility and their pharmacokinetics rely on the shape, aggregation, and chemical composition, which is not proven yet. CNTs have been widely explored for their potential application in the field of nucleic acid and enzyme-based biosensors. Its high tensile strength and elasticity help it to be used to substitute bone implants into the bone. Also, it can be used as an artificial joint implant due to its miniature size and defense evasion properties. Due to its nanosize, it is also used for drug peptide and nucleic acid delivery (Rai et al., 2012; Chandra, 2015; Mahapatra et al., 2021).

4.5.2.4 Lab-on-a-Chip

A lab-on-a-chip is a miniature device with the utmost importance in the field of disease diagnosis. This single chip is capable of analyzing multiple parameters, including nucleic acids and proteins. Lab-on-chip technology development is based on microfluidics and molecular biotechnology. Fabrication of these devices employs the implantation of several microchannels with antibodies, antigens, or nucleic acids, making it suitable to analyze the biochemical reactions in a single drop of blood. Polydimethylsiloxane, glass, and thermoplastics are commonly used materials for the fabrication of lab-on-chip devices (Balslev et al., 2006; Patel et al., 2016).

4.5.3 CONSTRUCTION OF BIOSENSOR PLATFORMS

The methods used in the fabrication of biosensors utilize the knowledge gained in semiconductor systems. The bottom-up and top-down are common approaches that are being used. In the bottom-up technique, the self-assembly methods utilize minimization of thermodynamic energy for induction phase separation and produce polymer structures (Prakash et al., 2009; Ma et al., 2016) using the advanced equipment including optical tweezers (Song et al., 2010; Suei et al., 2015) and atomic force microscopy (Ozkan et al., 2016) that help in acquiring higher accuracies for biosensor fabrication. Top-down approaches depend upon the breaking of cutting-edge materials by etching and lithography (Prakash et al., 2012). The methods employed for the immobilization of the biological system on the transducer surface rely specifically on surface-species interactions; hence, it enables the complete control and manipulation of surface properties (a charge, stress, etc.) in biosensor design (Siontorou and Batzias, 2014): The available nanotools present numerous alternatives for engineering different biological moieties as per the requirements of being analytical or regulatory. Their conjugation to transducer elements can be of any type, including electrochemical, mass-based, and optical, reliant on the category of biological response (Bollella et al., 2017).

4.5.4 THE IMMOBILIZATION OF SENSITIVE ELEMENTS ONTO BIOSENSOR FILM

The approaches in immobilization of immunological reagents (i.e., antibodies) onto the sensitive membrane commonly comprise the methods of physical adsorption, crosslinking, covalent binding, self-assembly, and Langumir-Blodgett (L-B) membrane and embedding.

The physical adsorption approach utilizes adsorption of biological constituents onto the inert carrier by electrostatic or hydrophobic interactions. The procedure does not make any alterations in the biological activity of the molecule. The only drawback of the prepared electrode is comparatively lower stability due to weak interactive forces among biological molecules and the carrier. Hence, the use of nanoparticles is beneficial in the immobilization of biological macromolecules due to their large surface area, and strengthen capacities of adsorption.

Recently, the self-assembly method is being utilized for the immobilization of electrode sensitive materials by electrostatic forces. Miao and Bard (2003) prepared an anodic electrogenerated chemiluminescence sensor with tri-n-propylamine (TPrA) as a co-reactant to identify a C-reactive protein (CRP) by utilizing $Ru(bpy)_3^{2+}$ labels.

4.5.5 Types of Electrodes for Biosensor Fabrication

4.5.5.1 Mercury Electrodes

In electrochemical biosensors, the material of the working electrode matters the most. Usually, hanging mercury drop electrodes (HMDE) are used because the new surfaces can be renewed with a recent mercury drop. Furthermore, the electro-inactive potential window of mercury is enormous. Hence the electronegative metals are detected quickly. However, researchers are also using its modified form. E.g., in a study by Adam et al., Cd^{2+} and Zn^{2+} ions were detected using the phytochelatin-modified HMDE via AdSV (Adam et al., 2005). However, it has several drawbacks like inability to measure metallic ions (Hg, Au and Ag, etc.) and exhibiting harmful activities. Mercury thin-film electrodes (MFEs) are an alternative to this problem because less mercury is required for its preparation.

4.5.5.2 Gold and Silver Electrodes

Gold and silver electrodes were used by Kirowa-Eisner et al. to detect cadmium, copper, and lead. Silver was found to be a more competent nanomaterial for lead and cadmium detection due to its high repeatability, stability, and LOD in the nM range (Kirowa-Eisner et al., 1999). Sensitivity can be improved further with advanced techniques of electrode preparation. For example, in a study, arsenic (III) was detected with a LOD value of 1 ppb on gold (Simm et al., 2005b and Sim et al., 2005a) and silver (Simm et al., 2005b) electrodes via ASV supported by ultrasound. Rahman et al. (2010) obtained a LOD value of 0.28 ppb with gold (111)-like polycrystalline electrodes. They have successfully detected As (III) trace levels in tap water. Total inorganic arsenic was detected by differential pulse ASV using gold-coated diamond thin film electrode showing a LOD of 20 ppb (Song and Swain, 2007). In another study by Compton et al., gold electrodes were shown to be highly sensitive for chromium (VI) detection by cyclic voltammetry and a LOD of 228 ppb was achieved. Gold electrodes have shown better performance than the glassy carbon and bore-doped diamond (Welch et al., 2005). At the same time, Kachoosangi and Compton have achieved a LOD of 4.4 ppb for chromium (VI) detection using gold film-modified carbon composite electrodes (Kachoosangi and Compton, 2013).

4.5.5.2.1 Gold Nanoparticle-Modified Electrodes

Limit of detection can be improved by using gold nanoparticle-modified electrodes. In a study, a gold nanoparticle-modified glassy carbon electrode was developed for detection of As (III) and a LOD of 0.0096 ppb was achieved using linear sweep voltammetry (Dai et al., 2004). Jena et al. have developed an extremely sensitive approach. It was constructed on the gold nanoelectrodes ensembles (GNEEs). GNEEs were synthesized by the colloidal synthesis method. It has a thiol-modified 3D silicate network prefabricated on the polycrystalline gold electrode. They have detected arsenic (III) and Hg (II) simultaneously in the presence of Cu (II) using square wave anodic stripping voltammetry (SWASV) with a LOD value of 0.02 ppb. Using this platform, chromium (VI) was also detected with a LOD of 0.1 ppb (Jena and Raj, 2008). Mordegan et al. have developed GNEEs with a polycarbonate membrane to control the nanoelectrodes' density. Using this method, a LOD of 5 ppb was achieved (Mardegan et al., 2012).

4.5.5.2.2 Bismuth Film Electrodes

Bismuth film electrodes (BFEs) were first introduced in 2000 as a substitute for mercury film electrodes (MFEs) (Wang et al., 2000). BFEs are prepared by a coating of thin bismuth films on appropriate electrode materials. The environmentally friendly nature of BFEs is the primary advantage because bismuth and bismuth ions exhibit no toxicity and have analytical properties comparable to MFEs. Furthermore, bismuth can be coated on the same substrate as mercury (Wang, Lu, Hocevar et al., 2001; Wang, Lu, Kirgöz et al., 2001; Oliveira Salles et al., 2009; Serrano et al., 2013). In the case of intermetallic compounds, Cd^{2+}, Pb^{2+}, etc., BFEs were shown to offer better separation between them than the MFEs (Demetriades et al., 2004; Kadara and Tothill, 2004).

There are three methods to deposit bismuth (Economou, 2005): The first method is ex-situ coating or plating, which involves electroplating the bismuth film before the electrode is transferred into the sample solution. An acidic medium is a must for plating because bismuth ions get hydrolyzed at a high pH. The deposition process takes place in 1–8 minutes at the potential range of −0.5 and −1.2 V.

The second method is an in-situ process of coating. Bi (III) ions are added directly into the sample solution in the 400–1,000 mg/L concentration range. In this method, bismuth film is plated onto the electrode surface during the sample analysis. Acidic medium is also required in this method, like the previous method.

The third method is limited to carbon-paste electrodes. In this method, the bulk of electrodes are modified with Bi_2O_3. The ex-situ plating method is more adaptable and handier than in-situ plating because electroplating settings are independent of analysis conditions in the former method. Also, BFEs can be reused and prepared in the ex-situ method. However, BFEs have a lower potential window than MFEs, which is a disadvantage. Bismuth is easily oxidized compared to mercury. That's why it has a more negative anodic limit; though the cathodic limit potential of BFEs is nearly the same as MFEs. The potential window of BFEs is strongly affected by the pH of the sample solution. Predictably, the maximum cathodic possible limit was attained in primary media, while the maximum anodic potential limit was achieved in acidic media.

4.5.5.2.3 Antimony Film Electrodes

In electrochemical stripping analysis, the use of antimony film electrodes (SbFEs) was first reported in the work of Hocevar et al. in 2007 (Hocevar et al., 2007). Antimony has been known as a pH electrode material since 1923 (Uhl and Kestranek, 1923). Like the BFE plating methods, SbFEs also show similar stripping behavior with almost equal sensitivity to the Pb (II). Furthermore, both BFEs and SbFEs offer better sensitivity towards cadmium compared to MFEs, while opposite effects are observed for lead. SbFEs were mainly used for Pb, Cd, and Zn detection in tap and river water samples (Sebez et al., 2013) and Ni detection by making a complex with dimethylglyoxime (Kokkinos et al., 2009; Sopha et al., 2012).

4.5.5.3 Bore-Doped Diamond (BDD)

Conductive diamond films are novel materials that are utilized to a greater extent for electro-analytical applications. These materials possess extraordinary characteristics, including (i) a low and stable background current, (ii) a relatively high electronic transfer rate, (iii) a low molecular adsorption, (iv) resistance to corrosion, and (v) optical transparency. Typically, diamond films are deposited on conductive substrata, including highly doped silicon wafer, molybdenum, tungsten, or titanium using plasma-wave-assisted chemical vapor deposition (PACVD) or hot filament.

In primeval conditions, diamond is one of the best electrical isolators. To have a suitable conductivity to be employed as an electrode for electrochemical measurements (>10 S·cm−1), doping of diamonds is required. The most commonly utilized doping material is boron and these electrodes are known as boron-doped diamonds (BDDs). BDD film can attain resistivity values of <0.05 $\Omega \cdot$cm.

The BDD electrodes acquire much interest for researchers in the stripping analysis of heavy metals due to their utilization over a wide range of potential windows (Weiss et al., 2008). Peilin et al. (1995) reported that BDD possesses better sensitivity for lead (3 nA·mm−2·ppb−1) than the glassy carbon electrode (GCE) (2.4 nA·mm−2·ppb−1), with in-situ plated mercury. However, BDD expresses three to five times reduced sensitivity without mercury plating compared to the mercury-plated GCE (McGaw and Swain, 2006). With ASV, Manivannan et al. have observed that sub-ppb recognition of lead is achievable using −1 V deposition potential for 15 minutes (Manivannan et al., 2005; Dragoe et al., 2006; Chooto et al., 2010). Compton et al. reported that the sonoelectrochemical treatment elevated the sensitivity for Pb (Saterlay et al., 1999), Cd (Banks et al., 2004), Mn (Saterlay et al., 1999), and Ag (Saterlay et al., 2000). The BDD electrodes have been effectively utilized for the simultaneous recognition and quantification of mixtures of HMs: Pb + Cd + Ag (Sonthalia et al., 2004), Zn + Pb + Cd + Cu (El Tall et al., 2007), Pb + Cd + Cu + Hg (Yoon et al., 2010), and Cd + Ni + Pb + Hg (Sbartai et al., 2012). Hutton et al. have observed the factors regulating stripping voltammetry of lead at BDD by utilizing high-resolution microscopy (Hutton et al., 2011). It has been reported that the deposition method was driven to generate a grain-independent homogeneous distribution of Pb nanoparticles on the electrode surface and the significant quantity of Pb remains on the surface after stripping, thus, explaining the non-linear response at high concentrations. Prado et al. analyzed the interactions between Pb and Cu (Prado et al., 2002) at the time of simultaneous detection by ASV. They reported the presence of an extra peak credited to hydrogen production on copper. They recommended that if Cu deposition happens favorably, then Pb deposition takes place on the developed copper deposits that act as active sites for nucleation and growth phenomenon, resulting in coverage of the copper with a solid lead film. Throughout the stripping step, initially, oxidation of lead occurs so that the copper deposits would be rapidly exposed to an acid electrolyte at a potential on which the hydrogen production takes place. Manivannan et al. (2004) analyzed the interactions among Pb and Cd during simultaneous detection. The authors reported that that the peak currents for Cd were 55% lower in the presence of a constant concentration of Pb (5 μM) than those attained without Pb.

In contrast, they also reported that in the presence of a constant concentration of Cd (5 μM), the peak currents for Pb were 40% higher in comparison to those for Pb in the absence of Cd. This variation was elucidated through the model pronounced for Pb and Cu, i.e., metals with higher negative standard potentials tend to deposit on metals with lower negative standard potentials; hence, the difference in peak current is reported for Cd and Pb.

4.5.5.4   Diamond-Like Carbon

Exploration of BDDs as electrode materials has opened the possibility to use amorphous or "diamond-like carbon" (DLC) for electrochemical purposes. It has similar electrochemical properties to the BDD, i.e., broad potential window, high electronic transfer rate, and low background. DLC films can be synthesized at low temperatures, unlike BDDs. Therefore, they can integrate into glass microfluidic devices. Amorphous carbons exist in a mixed microstructure of sp2 and sp3 carbon. They can also comprise hydrogen. There are several microstructures available with varying sp2/sp3 ratios of carbon and hydrogen content.

Films made up of a high percentage of sp3 carbon (>40%) are known as tetrahedral amorphous carbon (Ta-C). Different elaboration methods are there: (a) deposition of ions, (b) filtered cathodic vacuum arc (FCVA), (c) pulsed lased deposition, (d) plasma-enhanced chemical vapor deposition (PECVD), and (e) sputtering (Robertson, 2002). Zeng et al. (2002) have explored the use of sputtered DLC film to detect Pb, Cd, and Cu simultaneously.

Amorphous carbon (Ta-C: N) doped with tetrahedral nitrogen was used in the study of Khun and Liu (2009) and Liu and Liu (2005) to detect single elements such as Zn, Pb, and Hg and detection of Pb, Cu, and Hg simultaneously in a deaerated and agitated KCl solution. Khadro et al. (2011) have deposited undoped a-C and doped a-C: B (8%) film using Femto laser ablation on SiO2 and Si3N4 substrates. Using square wave anodic stripping voltammetry, Pb, Ni, Hg, and Cd were detected. The boron doping effect was shown for Pb detection: The peak height of Pb was increased by 20% using 8% of the bore as a dopant.

4.5.6 STRATEGY FOR IMPROVEMENT OF BIOSENSOR SENSITIVITY

Due to the very low concentration of cancer biomarkers in tissue or blood, the sensitivity of the sensors is a significant indicator to assess the newly developed immunosensors for cancer diagnosis.

4.5.7 ENZYMATIC AMPLIFICATION

Enzymes catalyze specific biological reactions and enzyme-catalyzed reactions can generate exponential amplification effects that are mainly applied in electrochemical sensor detection. During the development of immunosensors, oxidase, i.e., HRP and GOD, is often conjugated on a second antibody. The products of these enzymatic reactions are electrochemically measured. The amplification cycle of an enzymatic reaction helps in the detection of trace amounts of the analyte of antigen (Li et al., 2012).

4.5.8 AMPLIFICATION VIA NANOPARTICLES

Nanoparticles are the nanoscaled materials with a size in the range of 1–100 nm. Nanoparticles employed in the fabrication of immunosensors include graphene, carbon nanotubes, metal nanoparticles, and quantum dots. In this case, the signal amplification in electrochemical immunosensors greatly depends on the enzyme-functionalized nanoparticles used and type of electroactive labels utilized (Shiddiky et al., 2012). There is a provision for the nanoparticle surface to be easily modified and functionalized with enhanced electrical and optical properties, commonly utilized in signal amplification of the recognition by immunosensors.

Gold nanoparticles (AuNPs) have fascinated scientists due to their extensive applications in developing biosensors, including carriers of enzymes and antibodies and improvement of the conductivity (Saha et al., 2012).

4.5.8.1 Optical Transducers

Optical transducers use the absorption of radiations of different wavelengths for simple (Jerónimo et al., 2007) or multiplexed detection (Fan et al., 2008). Optical fibers and waveguide devices are employed to enhance the sensitivity of the sensors by elevating the interactions among the guiding light and the sensor surface. Pu et al. (2010) anticipated a new amplification approach utilizing hybrid nanomaterials (oligomeric silsesquioxane-based fluorescent nanoparticles) as signal amplifiers for biological imaging. These materials possess fluorescent sidechains that can be modified chemically to fine-tune their emission wavelength, diameter, and charge as per the requirements. Similarly, semiconductor nanoparticles possess easily tunable absorbance and fluorescence properties.

4.5.8.2 Mass-Based Transducers

Mass-based transducers sense the changes in mass due to biochemical interaction. They comprise a piezoelectric crystal that oscillates at a specific frequency under the applied electric field. The mass of the crystal and the electrical frequency are applied to affect the oscillation frequency (Saitakis and Gizeli, 2012). Su et al. (2013) prepared a piezoelectric system for the direct recognition of cancer biomarkers based on a lead titanate zirconate ceramic resonator as a transducer. The dual-sensing device possessed two resonators linked in parallel, one acting as the sensing unit and the other as the control unit; thus, minimizing the environmental and physical interference. The device possessed high sensitivity (0.25 ng/mL for prostate-specific antigen and α-fetoprotein) and fast analysis time (<30 minutes) for the 1 μL samples used.

4.5.8.3 Calorimetric Biosensors

Calorimetric biosensors are comparatively less common in cancer diagnostics, but nanotechnology-based modifications have widened the range of their applications. These systems detect heat changes to monitor biochemical interaction, providing information about the substrate concentration indirectly (Bohunicky and Mousa, 2011). Medley et al. (2008) prepared a calorimetric biosensor based on aptamer-linked gold nanoparticles that differentiated among acute leukemia cells and Burkitt's lymphoma cells. Their work established the viability of developing calorimetric platforms with aptamer-based recognition elements for discrimination of normal and cancer cells. Lab-on-chip technology assimilates several steps of diverse analytical events, with a large variety of applications, lower utilization of reagents and samples, and increased portability (Gambari et al., 2003).

Photolithography positions microfluidic channels on cellulose fiber-based paper, whereas screen-printing is used to prepare electrodes on paper (Pires et al., 2014). The screen-printed electrodes can be surface functionalized with enzymes or nucleic acid strands serving as capture probes for different analytes. The utilization of luminescent nanocrystals (quantum dots) as molecular labels pave the way for new possibilities in cellular labeling and visualization (Tothill, 2009). The nanocrystals can be conjugated to molecules for tracing intracellular components or antibody labeling. The narrow emission peaks and spectroscopic properties help in their use in the multiplexed analysis.

REFERENCES

Abdullah, N.H., Zainal, Z., Silong, S., Tahir, M.I.M., Tan, K.B., and Chang, S.K., 2016. Synthesis of zinc sulphide nanoparticles from thermal decomposition of zinc N-ethyl cyclohexyl dithiocarbamate complex. *Materials Chemistry and Physics*, *173*, pp. 33–41.

Achour, A., Islam, M., Solaymani, S., Vizireanu, S., Saeed, K., and Dinescu, G., 2018. Influence of plasma functionalization treatment and gold nanoparticles on surface chemistry and wettability of reactive-sputtered TiO2 thin films. *Applied Surface Science*, *458*, pp. 678–685.

Adachi, M., Tsukui, S., and Okuyama, K., 2003. Nanoparticle synthesis by ionizing source gas in chemical vapor deposition. *Japanese journal of applied physics*, *42*(1A), p. L77.

Adam, V., Zehnalek, J., Petrlova, J., Potesil, D., Sures, B., Trnkova, L., Jelen, F., Vitecek, J., and Kizek, R., 2005. Phytochelatin modified electrode surface as a sensitive heavy-metal ion biosensor. *Sensors*, *5*(1), pp. 70–84.

Adhikari, C., Mishra, A., Nayak, D., and Chakraborty, A., 2018. Metal organic frameworks modified mesoporous silica nanoparticles (MSN): A nano-composite system to inhibit uncontrolled chemotherapeutic drug delivery from Bare-MSN. *Journal of Drug Delivery Science and Technology*, *47*, pp. 1–11.

Adner, D., Noll, J., Schulze, S., Hietschold, M., and Lang, H., 2016. Asperical silver nanoparticles by thermal decomposition of a single-source-precursor. *Inorganica Chimica Acta*, *446*, pp. 19–23.

Ağaoğulları, D., Madsen, S.J., Öğüt, B., Koh, A.L., and Sinclair, R., 2017. Synthesis and characterization of graphite-encapsulated iron nanoparticles from ball milling-assisted low-pressure chemical vapor deposition. *Carbon*, *124*, pp. 170–179.

Aguilar, Z., 2012. *Nanomaterials for Medical Applications*. Newnes.

Ahab, A., Rohman, F., Iskandar, F., Haryanto, F., and Arif, I., 2016. A simple straightforward thermal decomposition synthesis of PEG-covered Gd2O3 (Gd2O3@PEG) nanoparticles. *Advanced Powder Technology*, *27*(4), pp. 1800–1805.

Ahangaran, F., and Navarchian, A.H., 2020. Recent advances in chemical surface modification of metal oxide nanoparticles with silane coupling agents: A review. *Advances in Colloid and Interface Science*. 10.1016/j.cis.2020.102298.

Ahmad, N., Sharma, S., Singh, V.N., Shamsi, S.F., Fatma, A., and Mehta, B.R., 2011. Biosynthesis of silver nanoparticles from Desmodium triflorum: a novel approach towards weed utilization. *Biotechnology Research International*, *2011*. doi: 10.4061/2011/454090.

Akbari, A., Amini, M., Tarassoli, A., Eftekhari-Sis, B., Ghasemian, N., and Jabbari, E., 2018. Transition metal oxide nanoparticles as efficient catalysts in oxidation reactions. *Nano-Structures and Nano-Objects*. doi: 10.1016/j.nanoso.2018.01.006.

Altman, I., 2019. On oxidation kinetics of burning metal nanoparticles. *Chemical Physics Letters*, *735*. doi: 10.1016/j.cplett.2019.136780.

Alvarez-Trabado, J., Diebold, Y., and Sanchez, A., 2017. Designing lipid nanoparticles for topical ocular drug delivery. *International Journal of Pharmaceutics*. doi: 10.1016/j.ijpharm.2017.09.017.

Amudha, M., and Shanmugasundaram, K.K., 2014. Biosynthesis and characterization of silver nanoparticles using the aqueous extract of Vitex negundo Linn. *World Journal of Pharmacy and Pharmaceutical Sciences* (WJPPS), *3*(8), pp. 1385–1393.

Arasi, A.Y., Hemma, M., Tamilselvi, P., and Anbarasan, R., 2012. Synthesis and characterization of SiO2 nanoparticles by sol-gel process. *Indian Journal of Science*, *1*(1), pp. 6–10.

Ashouri, F., Zare, M., and Bagherzade, M., 2015. Manganese and cobalt-terephthalate metal-organic frameworks as a precursor for synthesis of Mn2O3, Mn3O4 and Co3O4 nanoparticles: Active catalysts for olefin heterogeneous oxidation. *Inorganic Chemistry Communications*, *61*, pp. 73–76.

Balavijayalakshmi, J., and Ramalakshmi, V., 2017. Carica papaya peel mediated synthesis of silver nanoparticles and its antibacterial activity against human pathogens. *Journal of Applied Research and Technology*, *15*(5), pp. 413–422.

Baláž, M., Daneu, N., Balážová, L., Dutková, E., Tkáčiková, L., Briančin, J., Vargová, M., Balážová, M., Zorkovská, A., and Baláž, P., 2017. Bio-mechanochemical synthesis of silver nanoparticles with antibacterial activity. *Advanced Powder Technology*, *28*(12), pp. 3307–3312.

Balslev, S., Jorgensen, A.M., Bilenberg, B., Mogensen, K.B., Snakenborg, D., Geschke, O., Kutter, J.P., and Kristensen, A., 2006. Lab-on-a-chip with integrated optical transducers. *Lab on a Chip*, *6*(2), pp. 213–217.

Banks, C.E., Hyde, M.E., Tomčík, P., Jacobs, R., and Compton, R.G., 2004. Cadmium detection via boron-doped diamond electrodes: surfactant inhibited stripping voltammetry. *Talanta*, *62*(2), pp. 279–286.

Bao, Z., and Lan, C.Q., 2019. Advances in biosynthesis of noble metal nanoparticles mediated by photosynthetic organisms – A review. *Colloids and Surfaces B: Biointerfaces*. doi: 10.1016/j.colsurfb.2019.110519.

Bartůněk, V., Průcha, D., Svecová, M., Ulbrich, P., Huber, S., Sedmidubský, D., and Jankovský, O., 2016. Ultrafine ferromagnetic iron oxide nanoparticles: Facile synthesis by low temperature decomposition of iron glycerolate. *Materials Chemistry and Physics*, *180*, pp. 272–278.

Bello, S.A., Agunsoye, J.O., and Hassan, S.B., 2015. Synthesis of coconut shell nanoparticles via a top down approach: Assessment of milling duration on the particle sizes and morphologies of coconut shell nanoparticles. *Materials Letters*, *159*, pp. 514–519.

Beyene, H.D., Werkneh, A.A., Bezabh, H.K., and Ambaye, T.G., 2017. Synthesis paradigm and applications of silver nanoparticles (AgNPs), a review. *Sustainable Materials and Technologies*. doi: 10.1016/j.susmat.2017.08.001.

Bhagoria, P., Sebastian, E.M., Jain, S.K., Purohit, J., and Purohit, R., 2019. Nanolithography and its alternate techniques. *In: Materials Today: Proceedings* (pp. 3048–3053). Elsevier Ltd.

Bhaviripudi, S., Mile, E., Steiner, S.A., Zare, A.T., Dresselhaus, M.S., Belcher, A.M., and Kong, J., 2007. CVD synthesis of single-walled carbon nanotubes from gold nanoparticle catalysts. *Journal of the American Chemical Society*, *129*(6), pp. 1516–1517.

Bohunicky, B., and Mousa, S.A., 2011. Biosensors: the new wave in cancer diagnosis. *Nanotechnology, Science and Applications*, *4*, p. 1.

Bokhonov, B.B., and Dudina, D. V., 2017. Synthesis of ZrC and HfC nanoparticles encapsulated in graphitic shells from mechanically milled Zr-C and Hf-C powder mixtures. *Ceramics International*, *43*(16), pp. 14529–14532.

Bollella, P., Fusco, G., Tortolini, C., Sanzò, G., Favero, G., Gorton, L., and Antiochia, R., 2017. Beyond graphene: electrochemical sensors and biosensors for biomarkers detection. *Biosensors and Bioelectronics*, *89*, pp. 152–166.

Bora, U., Sett, A., and Singh, D., 2013. Nucleic acid-based biosensors for clinical applications. *Biosensors Journal*, *2*(1), pp. 1–8.

Breydo, L., Wu, J.W., and Uversky, V.N., 2012. α-Synuclein misfolding and Parkinson's disease. *Biochimica et Biophysica Acta (BBA)-Molecular Basis of Disease*, *1822*(2), pp. 261–285.

Brobbey, K.J., Haapanen, J., Gunell, M., Mäkelä, J.M., Eerola, E., Toivakka, M., and Saarinen, J.J., 2017. One-step flame synthesis of silver nanoparticles for roll-to-roll production of antibacterial paper. *Applied Surface Science*, *420*, pp. 558–565.

Calborean, A., Martin, F., Marconi, D., Turcu, R., Kacso, I.E., Buimaga-Iarinca, L., Graur, F., and Turcu, I., 2015. Adsorption mechanisms of l-Glutathione on Au and controlled nano-patterning through Dip Pen Nanolithography. *Materials Science and Engineering C, 57*, pp. 171–180.

Chandra, P., 2015. Electrochemical nanobiosensors for cancer diagnosis. *Journal of Analytical & Bioanalytical Techniques, 6*(1), p. e119.

Chen, A., Zhao, C., Yu, Y., and Yang, J., 2016. Graphene quantum dots derived from carbon fibers for oxidation of dopamine. *Journal of Wuhan University of Technology-Mater. Sci. Ed., 31*(6), pp. 1294–1297.

Chen, J., and Liu, B., 2020. Performance evaluation of carbon nanoparticle-based thermal interface materials. *Diamond and Related Materials, 108*. doi: 10.1016/j.diamond.2020.107976

Chen, Z., Zhou, Y., Kang, Z., and Chen, D., 2014. Synthesis of Mn-Zn ferrite nanoparticles by the coupling effect of ultrasonic irradiation and mechanical forces. *Journal of Alloys and Compounds, 609*, pp. 21–24.

Chin, C.D., Linder, V., and Sia, S.K., 2012. Commercialization of microfluidic point-of-care diagnostic devices. *Lab on a Chip, 12*(12), pp. 2118–2134.

Cho, E.J., Holback, H., Liu, K.C., Abouelmagd, S.A., Park, J., and Yeo, Y., 2013. Nanoparticle characterization: State of the art, challenges, and emerging technologies. *Molecular Pharmaceutics, 10*(6), pp. 2093–2110.

Chooto, P., Wararatananurak, P., and Innuphat, C., 2010. Determination of trace levels of Pb (II) in tap water by anodic stripping voltammetry with boron-doped diamond electrode. *Science Asia, 36*(2), pp. 143–157.

Chung, M.W., Cha, I.Y., Ha, M.G., Na, Y., Hwang, J., Ham, H.C., Kim, H.J., Henkensmeier, D., Yoo, S.J., Kim, J.Y., Lee, S.Y., Park, H.S., and Jang, J.H., 2018. Enhanced CO2 reduction activity of polyethylene glycol-modified Au nanoparticles prepared via liquid medium sputtering. *Applied Catalysis B: Environmental, 237*, pp. 673–680.

Cigáň, A., Lobotka, P., Dvurečenskij, A., Škrátek, M., Radnóczi, G., Majerová, M., Czigány, Z., Maňka, J., Vávra, I., and Mičušík, M., 2018. Characterization and magnetic properties of nickel and nickel-iron nanoparticle colloidal suspensions in imidazolium-based ionic liquids prepared by magnetron sputtering. *Journal of Alloys and Compounds, 768*, pp. 625–634.

Coates, J., 2006. Interpretation of infrared spectra, a practical approach. *Encyclopedia of Analytical Chemistry: Applications, Theory and Instrumentation.* doi: 10.1002/9780470027318.a5606

Cui, F., Zhou, Z., and Zhou, H.S., 2019. Measurement and analysis of cancer biomarkers based on electrochemical biosensors. *Journal of The Electrochemical Society, 167*(3), p. 037525.

Dai, M., Wang, Z., Wang, F., Qiu, Y., Zhang, J., Xu, C.Y., Zhai, T., Cao, W., Fu, Y., Jia, D., and Zhou, Y., 2019. Two-Dimensional van der Waals Materials with aligned in-plane polarization and large piezoelectric effect for self-powered piezoelectric sensors. *Nano letters, 19*(8), pp. 5410–5416.

Dai, W., and Moon, M.W., 2018. Carbon-encapsulated metal nanoparticles deposited by plasma enhanced magnetron sputtering. *Vacuum, 150*, pp. 124–128.

Dai, X., Nekrassova, O., Hyde, M.E., and Compton, R.G., 2004. Anodic stripping voltammetry of arsenic (III) using gold nanoparticle-modified electrodes. *Analytical Chemistry, 76*(19), pp. 5924–5929.

de Assis, D.N., Mosqueira, V.C.F., Vilela, J.M.C., Andrade, M.S., and Cardoso, V.N., 2008. Release profiles and morphological characterization by atomic force microscopy and photon correlation spectroscopy of 99mTechnetium-fluconazole nanocapsules. *International Journal of Pharmaceutics, 349*(1–2), pp. 152–160.

Demetriades, D., Economou, A., and Voulgaropoulos, A., 2004. A study of pencil-lead bismuth-film electrodes for the determination of trace metals by anodic stripping voltammetry. *Analytica Chimica Acta, 519*(2), pp. 167–172.

Dervin, S., and Pillai, S.C., 2017. An Introduction to Sol-Gel Processing for Aerogels. *In: Sol-Gel Materials for Energy, Environment and Electronic Applications* (pp. 1–22). Springer, Cham.

Di Martino, A., Guselnikova, O.A., Trusova, M.E., Postnikov, P.S., and Sedlarik, V., 2017. Organic-inorganic hybrid nanoparticles controlled delivery system for anticancer drugs. *International Journal of Pharmaceutics, 526*(1–2), pp. 380–390.

Doble, M., Rollins, K., and Kumar, A., 2010. *Green Chemistry And Engineering.* Academic Press.

Dragoe, D., Spătaru, N., Kawasaki, R., Manivannan, A., Spătaru, T., Tryk, D.A., and Fujishima, A., 2006. Detection of trace levels of Pb2+ in tap water at boron-doped diamond electrodes with anodic stripping voltammetry. *Electrochimica Acta, 51*(12), pp. 2437–2441.

Economou, A., 2005. Bismuth-film electrodes: Recent developments and potentialities for electroanalysis. *TrAC Trends in Analytical Chemistry, 24*(4), pp. 334–340.

El Tall, O., Jaffrezic-Renault, N., Sigaud, M., and Vittori, O., 2007. Anodic stripping voltammetry of heavy metals at nanocrystalline boron-doped diamond electrode. *Electroanalysis: An International Journal Devoted to Fundamental and Practical Aspects of Electroanalysis*, *19*(11), pp. 1152–1159.

Esakkimuthu, T., Sivakumar, D., and Akila, S., 2014. Application of nanoparticles in wastewater treatment. *Pollution Research*, *33*(03), pp. 567–571.

Fan, X., White, I.M., Shopova, S.I., Zhu, H., Suter, J.D., and Sun, Y., 2008. Sensitive optical biosensors for unlabeled targets: A review. *Analytica Chimica Acta*, *620*(1-2), pp. 8–26.

Frasco, M.F., and Chaniotakis, N., 2009. Semiconductor quantum dots in chemical sensors and biosensors. *Sensors*, *9*(9), pp. 7266–7286.

Gambari, R., Borgatti, M., Altomare, L., Manaresi, N., Medoro, G., Romani, A., Tartagni, M., and Guerrieri, R., 2003. Applications to cancer research of "lab-on-a-chip" devices based on dielectrophoresis (DEP). *Technology In Cancer Research & Treatment*, *2*(1), pp. 31–39.

Ghosal, K., Ghosh, S., Ghosh, D., and Sarkar, K., 2020. Natural polysaccharide derived carbon dot based in situ facile green synthesis of silver nanoparticles: Synergistic effect on breast cancer. *International Journal of Biological Macromolecules*, *162*, pp. 1605–1615.

Glasgow, W., Fellows, B., Qi, B., Darroudi, T., Kitchens, C., Ye, L., Crawford, T.M., and Mefford, O.T., 2016. Continuous synthesis of iron oxide (Fe3O4) nanoparticles via thermal decomposition. *Particuology*, *26*, pp. 47–53.

Gouvêa, C.M.C.P., 2011. *Biosensors for Health Applications* (pp. 71–86). InTech.

Gupta, N., Kumar, A., Dhasmana, H., Kumar, V., Kumar, A., Shukla, P., Verma, A., Nutan, G. V., Dhawan, S.K., and Jain, V.K., 2020. Enhanced thermophysical properties of metal oxide nanoparticles embedded magnesium nitrate hexahydrate based nanocomposite for thermal energy storage applications. *Journal of Energy Storage*, *32*. doi: 10.1016/j.est.2020.101773.

Han, G., Chen, D., and Li, X., 2017. Synthesis and catalytic performance of antimony trioxide nanoparticles by ultrasonic-assisted solid-liquid reaction ball milling. *Advanced Powder Technology*, *28*(4), pp. 1136–1140.

Hasan, S., 2015. A review on nanoparticles: Their synthesis and types. *Research Journal of Recent Sciences,* ISSN, *2277*, p. 2502.

Hasanzadeh, M., Feyziazar, M., Solhi, E., Mokhtarzadeh, A., Soleymani, J., Shadjou, N., Jouyban, A., and Mahboob, S., 2019. Ultrasensitive immunoassay of breast cancer type 1 susceptibility protein (BRCA1) using poly (dopamine-beta cyclodextrine-Cetyl trimethylammonium bromide) doped with silver nanoparticles: A new platform in early stage diagnosis of breast cancer and efficient management. *Microchemical Journal*, *145*, pp. 778–783.

Hasanzadeh, M., Sahmani, R., Solhi, E., Mokhtarzadeh, A., Shadjou, N., and Mahboob, S., 2018. Ultrasensitive immunoassay of carcinoma antigen 125 in untreated human plasma samples using gold nanoparticles with flower like morphology: A new platform in early stage diagnosis of ovarian cancer and efficient management. *International Journal of Biological Macromolecules*, *119*, pp. 913–925.

Hashemipour, S., and Ahmad Panahi, H., 2017. Fabrication of magnetite nanoparticles modified with copper based metal organic framework for drug delivery system of letrozole. *Journal of Molecular Liquids*, *243*, pp. 102–107.

Ho, P.Y., Yiu, S.C., Wu, D.Y., Ho, C.L., and Wong, W.Y., 2017. One-step synthesis of platinum nanoparticles by pyrolysis of a polyplatinyne polymer. *Journal of Organometallic Chemistry*, *849–850*, pp. 4–9.

Hocevar, S.B., Švancara, I., Ogorevc, B., and Vytřas, K., 2007. Antimony film electrode for electrochemical stripping analysis. *Analytical Chemistry*, *79*(22), pp. 8639–8643.

Holm, V.R.A., Greve, M.M., and Holst, B., 2017. A theoretical investigation of the optical properties of metal nanoparticles in water for photo thermal conversion enhancement. *Energy Conversion and Management*, *149*, pp. 536–542.

Hong, E., Liu, L., Bai, L., Xia, C., Gao, L., Zhang, L., and Wang, B., 2019. Control synthesis, subtle surface modification of rare-earth-doped upconversion nanoparticles and their applications in cancer diagnosis and treatment. *Materials Science and Engineering C*.

Hong, L.Y., and Lin, H.N., 2015. Fabrication of single titanium oxide nanodot ultraviolet sensors by atomic force microscopy nanolithography. *Sensors and Actuators, A: Physical*, *232*, pp. 94–98.

Hou, W., Xia, F., Alfranca, G., Yan, H., Zhi, X., Liu, Y., Peng, C., Zhang, C., de la Fuente, J.M., and Cui, D., 2017. Nanoparticles for multi-modality cancer diagnosis: Simple protocol for self-assembly of gold nanoclusters mediated by gadolinium ions. *Biomaterials*, *120*, pp. 103–114.

Hoyos-Palacio, L.M., Cuesta Castro, D.P., Ortiz-Trujillo, I.C., Botero Palacio, L.E., Galeano Upegui, B.J., Escobar Mora, N.J., and Carlos Cornelio, J.A., 2019. Compounds of carbon nanotubes decorated with silver nanoparticles via in-situ by chemical vapor deposition (CVD). *Journal of Materials Research and Technology*, 8(6), pp. 5893–5898.

Htoo, K.P.P., Yamkamon, V., Yainoy, S., Suksrichavalit, T., Viseshsindh, W., and Eiamphungporn, W., 2019. Colorimetric detection of PCA3 in urine for prostate cancer diagnosis using thiol-labeled PCR primer and unmodified gold nanoparticles. *Clinica Chimica Acta*, *488*, pp. 40–49.

Hu, X., Yin, D., Chen, X., and Xiang, G., 2020. Experimental investigation and mechanism analysis: Effect of nanoparticle size on viscosity of nanofluids. *Journal of Molecular Liquids*, *314*. doi: 10.1016/j.molliq.2020.113604.

Hu, Y., Li, L., Wang, R., Song, J., Wang, H., Duan, H., Ji, J., and Meng, Y., 2020. High-speed parallel plasmonic direct-writing nanolithography using metasurface-based plasmonic lens. *Engineering*. doi: 10.1016/j.eng.2020.08.019.

Hulkoti, N.I., and Taranath, T.C., 2014. Biosynthesis of nanoparticles using microbes – a review. *Colloids and Surfaces B: Biointerfaces*, *121*, pp. 474–483.

Hulteen, J.C., Treichel, D.A., Smith, M.T., Duval, M.L., Jensen, T.R., and Van Duyne, R.P., 1999. Nanosphere lithography: Size-tunable silver nanoparticle and surface cluster arrays. *The Journal of Physical Chemistry B*, *103*(19), pp. 3854–3863.

Hutton, L.A., Newton, M.E., Unwin, P.R., and Macpherson, J.V., 2011. Factors controlling stripping voltammetry of lead at polycrystalline boron doped diamond electrodes: New insights from high-resolution microscopy. *Analytical Chemistry*, *83*(3), pp. 735–745.

Jabli, M., Al-Ghamdi, Y.O., Sebeia, N., Almalki, S.G., Alturaiki, W., Khaled, J.M., Mubarak, A.S., and Algethami, F.K., 2020. Green synthesis of colloid metal oxide nanoparticles using Cynomorium coccineum: Application for printing cotton and evaluation of the antimicrobial activities. *Materials Chemistry and Physics*, *249*. doi: 10.1016/j.matchemphys.2020.123171.

Jain, Y., Rana, C., Goyal, A., Sharma, N., Verma, M.L., and Jana, A.K., 2010, January. Biosensors, types and applications. *In: BEATS 2010: Proceedings of the 2010 International Conference on Biomedical Engineering and Assistive Technologies, Jalandhar, India* (pp. 1–6). BEATS.

Jayachandran, A., T.R., A., and Nair, A.S., 2021. Green synthesis and characterization of zinc oxide nanoparticles using Cayratia pedata leaf extract. *Biochemistry and Biophysics Reports*, *26*. doi: 10.1016/j.bbrep.2021.100995.

Jena, B.K., and Raj, C.R., 2008. Highly sensitive and selective electrochemical detection of sub-ppb level chromium (VI) using nano-sized gold particle. *Talanta*, *76*(1), pp. 161–165.

Jerónimo, P.C., Araújo, A.N., and Montenegro, M.C.B., 2007. Optical sensors and biosensors based on sol-gel films. *Talanta*, *72*(1), pp. 13–27.

Jiang, L., Li, H., Mu, J., and Ji, Z., 2018. Manipulation of surface plasmon resonance of sputtered gold-nanoparticles on TiO2 nanostructured films for enhanced photoelectrochemical water splitting efficiency. *Thin Solid Films*, *661*, pp. 32–39.

Jiang, Y., Song, H., and Xu, R., 2018. Research on the dispersion of carbon nanotubes by ultrasonic oscillation, surfactant and centrifugation respectively and fiscal policies for its industrial development. *Ultrasonics Sonochemistry*, *48*, pp. 30–38.

Jiménez-Rojo, N., Lete, M.G., Rojas, E., Gil, D., Valle, M., Alonso, A., Moya, S.E., and Goñi, F.M., 2015. Lipidic nanovesicles stabilize suspensions of metal oxide nanoparticles. *Chemistry and Physics of Lipids*, *191*, pp. 84–90.

Jin, H., Gui, R., Yu, J., Lv, W., and Wang, Z., 2017. Fabrication strategies, sensing modes and analytical applications of ratiometric electrochemical biosensors. *Biosensors and Bioelectronics*, *91*, pp. 523–537.

Jores, K., Mehnert, W., Drechsler, M., Bunjes, H., Johann, C., and Mäder, K., 2004. Investigations on the structure of solid lipid nanoparticles (SLN) and oil-loaded solid lipid nanoparticles by photon correlation spectroscopy, field-flow fractionation and transmission electron microscopy. *Journal of Controlled Release*, *95*(2), pp. 217–227.

Joshi, C.G., Danagoudar, A., Poyya, J., Kudva, A.K., and BL, D., 2017. Biogenic synthesis of gold nanoparticles by marine endophytic fungus-Cladosporium cladosporioides isolated from seaweed and evaluation of their antioxidant and antimicrobial properties. *Process Biochemistry*, *63*, pp. 137–144.

Kachoosangi, R.T., and Compton, R.G., 2013. Voltammetric determination of Chromium (VI) using a gold film modified carbon composite electrode. *Sensors and Actuators B: Chemical*, *178*, pp. 555–562.

Kadara, R.O., and Tothill, I.E., 2004. Stripping chronopotentiometric measurements of lead (II) and cadmium (II) in soils extracts and wastewaters using a bismuth film screen-printed electrode assembly. *Analytical and Bioanalytical Chemistry*, *378*(3), pp. 770–775.

Kammler, H.K., Mädler, L., and Pratsinis, S.E., 2001. Flame synthesis of nanoparticles. *Chemical Engineering & Technology: Industrial Chemistry-Plant Equipment-Process Engineering-Biotechnology*, *24*(6), pp. 583–596.

Karimi, M., Jodaei, A., Sadeghinik, A., Ramsheh, M.R., Hafshejani, T.M., Shamsi, M., Orand, F., and Lotfi, F., 2017. Deep eutectic choline chloride-calcium chloride as all-in-one system for sustainable and one-step synthesis of bioactive fluorapatite nanoparticles. *Journal of Fluorine Chemistry*, *204*, pp. 76–83.

Khadro, B., Sikora, A., Loir, A.S., Errachid, A., Garrelie, F., Donnet, C., and Jaffrezic-Renault, N., 2011. Electrochemical performances of B doped and undoped diamond-like carbon (DLC) films deposited by femtosecond pulsed laser ablation for heavy metal detection using square wave anodic stripping voltammetric (SWASV) technique. *Sensors and Actuators B: Chemical*, *155*(1), pp. 120–125.

Khan, Z.U.H., Khan, A., Chen, Y., Shah, N.S., Muhammad, N., Khan, A.U., Tahir, K., Khan, F.U., Murtaza, B., Hassan, S.U., Qaisrani, S.A., and Wan, P., 2017. Biomedical applications of green synthesized Nobel metal nanoparticles. *Journal of Photochemistry and Photobiology B: Biology*. doi: 10.1016/j.jphotobiol.2017.05.034.

Khanna, V.K., 2008. Nanoparticle-based sensors. *Defence Science Journal*, *58*(5), p. 608.

Khayati, G.R., Shahcheraghi, S.H., Lotfi, V., and Darezareshki, E., 2016. Reaction pathway and kinetics of CdO nanoparticles prepared from CdCO3 precursor using thermal decomposition method. *Transactions of Nonferrous Metals Society of China (English Edition)*, *26*(4), pp. 1138–1145.

Khun, N.W., and Liu, E., 2009. Linear sweep anodic stripping voltammetry of heavy metals from nitrogen doped tetrahedral amorphous carbon thin films. *Electrochimica Acta*, *54*(10), pp. 2890–2898.

Kim, B., Song, J., Kim, J.Y., Hwang, J., and Park, D., 2019. The control of particle size distribution for fabricated alumina nanoparticles using a thermophoretic separator. *Advanced Powder Technology*, *30*(10), pp. 2094–2100.

Kirowa-Eisner, E., Brand, M., and Tzur, D., 1999. Determination of sub-nanomolar concentrations of lead by anodic-stripping voltammetry at the silver electrode. *Analytica chimica acta*, *385*(1–3), pp. 325–335.

Kokkinos, C., Economou, A., Raptis, I., and Speliotis, T., 2009. Novel disposable microfabricated antimony-film electrodes for adsorptive stripping analysis of trace Ni (II). *Electrochemistry Communications*, *11*(2), pp. 250–253.

Konishi, Y., Ohno, K., Saitoh, N., Nomura, T., Nagamine, S., Hishida, H., Takahashi, Y., and Uruga, T., 2007. Bioreductive deposition of platinum nanoparticles on the bacterium Shewanella algae. *Journal Of Biotechnology*, *128*(3), pp. 648–653.

Konrad, A., Herr, U., Tidecks, R., Kummer, F., and Samwer, K., 2001. Luminescence of bulk and nano-crystalline cubic yttria. *Journal of Applied Physics*, *90*(7), pp. 3516–3523.

Krajewski, M., Liao, P.Y., Michalska, M., Tokarczyk, M., and Lin, J.Y., 2019. Hybrid electrode composed of multiwall carbon nanotubes decorated with magnetite nanoparticles for aqueous supercapacitors. *Journal of Energy Storage*, *26*. doi: 10.1016/j.est.2019.101020.

Kuchur, O.A., Tsymbal, S.A., Shestovskaya, M. V., Serov, N.S., Dukhinova, M.S., and Shtil, A.A., 2020. Metal-derived nanoparticles in tumor theranostics: Potential and limitations. *Journal of Inorganic Biochemistry*. doi: 10.1016/j.jinorgbio.2020.111117.

Kuila, T., Bose, S., Khanra, P., Mishra, A.K., Kim, N.H., and Lee, J.H., 2011. Recent advances in graphene-based biosensors. *Biosensors and bioelectronics*, *26*(12), pp. 4637–4648.

Kumar, B., Jalodia, K., Kumar, P., and Gautam, H.K., 2017. Recent advances in nanoparticle-mediated drug delivery. *Journal of Drug Delivery Science and Technology*. doi: 10.1016/j.jddst.2017.07.019.

Kumar, J.P., Ramacharyulu, P.V.R.K., Prasad, G.K., and Singh, B., 2015. Montmorillonites supported with metal oxide nanoparticles for decontamination of sulfur mustard. *Applied Clay Science*, *116–117*, pp. 263–272.

Kumar Jena, B., and Retna Raj, C., 2008. Gold nanoelectrode ensembles for the simultaneous electrochemical detection of ultratrace arsenic, mercury, and copper. *Analytical Chemistry*, *80*(13), pp. 4836–4844.

Kuppusamy, P., Yusoff, M.M., Maniam, G.P., and Govindan, N., 2016. Biosynthesis of metallic nanoparticles using plant derivatives and their new avenues in pharmacological applications – an updated report. *Saudi Pharmaceutical Journal*, *24*(4), pp. 473–484.

Lee, C.H., Rai, P., Moon, S.Y., and Yu, Y.T., 2016. Thermal plasma synthesis of Si/SiC nanoparticles from silicon and activated carbon powders. *Ceramics International*, *42*(15), pp. 16469–16473.

Lee, H., Deshmukh, P.R., Kim, J.H., Hyun, H.S., Sohn, Y., and Shin, W.G., 2019. Spray drying formation of metal oxide (TiO2 or SnO2) nanoparticle coated boron particles in the form of microspheres and their physicochemical properties. *Journal of Alloys and Compounds*, *810*.

Lee, H., Kwak, D. Bin, Kim, S.C., and Pui, D.Y.H., 2019. Characterization of colloidal nanoparticles in mixtures with polydisperse and multimodal size distributions using a particle tracking analysis and electrospray-scanning mobility particle sizer. *Powder Technology, 355*, pp. 18–25.

Lee, S.H., Jung, H.K., Kim, T.C., Kim, C.H., Shin, C.H., Yoon, T.S., Hong, A.R., Jang, H.S., and Kim, D.H., 2018. Facile method for the synthesis of gold nanoparticles using an ion coater. *Applied Surface Science, 434*, pp. 1001–1006.

Li, H., Zhao, Y., Zhang, Z., Andaluri, G., and Ren, F., 2019. Electron-beam induced in situ growth of self-supported metal nanoparticles in ion-containing polydopamine. *Materials Letters, 252*, pp. 277–281.

Li, J., Li, S., and Yang, C.F., 2012. Electrochemical biosensors for cancer biomarker detection. *Electroanalysis, 24*(12), pp. 2213–2229.

Li, R., Liu, B., and Gao, J., 2017. The application of nanoparticles in diagnosis and theranostics of gastric cancer. *Cancer Letters*.

Liao, C., Wuethrich, A., and Trau, M., 2020. A material odyssey for 3D nano/microstructures: two photon polymerization based nanolithography in bioapplications. *Applied Materials Today*. doi: 10.1016/j.apmt.2020.100635.

Lim, J., Yeap, S.P., Che, H.X., and Low, S.C., 2013. Characterization Of Magnetic Nanoparticle By Dynamic Light Scattering. *Nanoscale Research Letters, 8*(1), p. 381.

Liu, F., Liu, J., and Cao, X., 2017. Microwave-assisted synthesis silver nanoparticles and their surface enhancement raman scattering. *Xiyou Jinshu Cailiao Yu Gongcheng/Rare Metal Materials and Engineering, 46*(9), pp. 2395–2398.

Liu, L.X., and Liu, E., 2005. Nitrogenated diamond-like carbon films for metal tracing. *Surface and Coatings Technology, 198*(1-3), pp. 189–193.

Liu, N., Mamat, X., Jiang, R., Tong, W., Huang, Y., Jia, D., Li, Y., Wang, L., Wågberg, T., and Hu, G., 2018. Facile high-voltage sputtering synthesis of three-dimensional hierarchical porous nitrogen-doped carbon coated Si composite for high performance lithium-ion batteries. *Chemical Engineering Journal, 343*, pp. 78–85.

Liu, X., Zhang, Y., Wang, S., Liu, C., Wang, T., Qiu, Z., Wang, X., Waterhouse, G.I.N., Xu, C., and Yin, H., 2020. Performance comparison of surface plasmon resonance biosensors based on ultrasmall noble metal nanoparticles templated using bovine serum albumin. *Microchemical Journal, 155*. doi: 10.1016/j.microc.2020.104737.

Liu, Y., 2018. Synthesis of AuPd alloy nanoparticles on ZnO nanorod arrays by sputtering for surface enhanced Raman scattering. *Materials Letters, 224*, pp. 26–28.

Luo, X., Morrin, A., Killard, A.J., and Smyth, M.R., 2006. Application of nanoparticles in electrochemical sensors and biosensors. *Electroanalysis: An International Journal Devoted to Fundamental and Practical Aspects of Electroanalysis, 18*(4), pp. 319–326.

Ma, W., Xu, L., Wang, L., Kuang, H., and Xu, C., 2016. Orientational nanoparticle assemblies and bio-sensors. *Biosensors and Bioelectronics, 79*, pp. 220–236.

Made Joni, I., Vanitha, M., Panatarani, C., and Faizal, F., 2020. Dispersion of amorphous silica nanoparticles via beads milling process and their particle size analysis, hydrophobicity and anti-bacterial activity. *Advanced Powder Technology, 31*(1), pp. 370–380.

Mahapatra, S., Srivastava, V.R., and Chandra, P., 2021. Nanobioengineered sensing technologies based on cellulose matrices for detection of small molecules, macromolecules, and cells. *Biosensors, 11*(6), p. 168.

Mahdhi, H., Djessas, K., and Ben Ayadi, Z., 2018. Synthesis and characteristics of Ca-doped ZnO thin films by rf magnetron sputtering at low temperature. *Materials Letters, 214*, pp. 10–14.

Makkar, P., Agarwala, R.C., and Agarwala, V., 2014. Wear characteristics of mechanically milled TiO2 nanoparticles incorporated in electroless Ni-P coatings. *Advanced Powder Technology, 25*(5), pp. 1653–1660.

Malik, S., 2018. Nanotubes from Atlantis: Magnetite in pumice as a catalyst for the growth of carbon nanotubes. *Polyhedron, 152*, pp. 90–93.

Mallakpour, S., and Madani, M., 2015. A review of current coupling agents for modification of metal oxide nanoparticles. *Progress in Organic Coatings*. doi: 10.1016/j.porgcoat.2015.05.023.

Mamatha, R., Khan, S., Salunkhe, P., Satpute, S., Kendurkar, S., Prabhune, A., Deval, A., and Chaudhari, B.P., 2017. Rapid synthesis of highly monodispersed silver nanoparticles from the leaves of Salvadora persica. *Materials Letters, 205*, pp. 226–229.

Manivannan, A., Kawasaki, R., Tryk, D.A., and Fujishima, A., 2004. Interaction of Pb and Cd during anodic stripping voltammetric analysis at boron-doped diamond electrodes. *Electrochimica Acta, 49*(20), pp. 3313–3318.

Manivannan, A., Ramakrishnan, L., Seehra, M.S., Granite, E., Butler, J.E., Tryk, D.A., and Fujishima, A., 2005. Mercury detection at boron doped diamond electrodes using a rotating disk technique. *Journal of Electroanalytical Chemistry*, *577*(2), pp. 287–293.

Mann, S., 2001. *Biomineralization: Principles and Concepts in Bioinorganic Materials Chemistry* (Vol. 5). *Oxford University Press* on Demand.

Mardegan, A., Scopece, P., Lamberti, F., Meneghetti, M., Moretto, L.M., and Ugo, P., 2012. Electroanalysis of trace inorganic arsenic with gold nanoelectrode ensembles. *Electroanalysis*, *24*(4), pp. 798–806.

McGaw, E.A., and Swain, G.M., 2006. A comparison of boron-doped diamond thin-film and Hg-coated glassy carbon electrodes for anodic stripping voltammetric determination of heavy metal ions in aqueous media. *Analytica Chimica Acta*, *575*(2), pp. 180–189.

Medley, C.D., Smith, J.E., Tang, Z., Wu, Y., Bamrungsap, S., and Tan, W., 2008. Gold nanoparticle-based colorimetric assay for the direct detection of cancerous cells. *Analytical Chemistry*, *80*(4), pp. 1067–1072.

Meena, S., and Choudhary, S., 2018. Effects of functionalization of carbon nanotubes on its spin transport properties. *Materials Chemistry and Physics*, *217*, pp. 175–181.

Menon, S., S., R., and S., V.K., 2017. A review on biogenic synthesis of gold nanoparticles, characterization, and its applications. *Resource-Efficient Technologies*, *3*(4), pp. 516–527.

Merugu, R., and Gothalwal, R., 2021. Green synthesis and characterisation of Indium Tin oxide nanoparticles using toddy palm from Borassus flabellifer. *Materials Today: Proceedings*. doi: 10.1016/j.matpr.2021 .04.447

Miao, W., and Bard, A.J., 2003. Electrogenerated Chemiluminescence. 72. Determination of Immobilized DNA and C-Reactive Protein on Au (111) Electrodes Using Tris (2, 2 '-bipyridyl) ruthenium (II) Labels. *Analytical Chemistry*, *75*(21), pp. 5825–5834.

Mohammadi, S., Harvey, A., and Boodhoo, K.V., 2014. Synthesis of TiO2 nanoparticles in a spinning disc reactor. *Chemical Engineering Journal*, *258*, pp. 171–184.

Molpeceres, J., Aberturas, M.R., and Guzman, M., 2000. Biodegradable nanoparticles as a delivery system for cyclosporine: preparation and characterization. *Journal of Microencapsulation*, *17*(5), pp. 599–614.

Monarca, L., Ragonese, F., Mancinelli, L., Mecca, C., Sportoletti, P., Arcuri, C., Fioretti, B., and Costantino, F., 2018. Synthesis and characterization of UiO-66 metal-organic frameworks nanoparticles and their evaluation as drug delivery carriers in U251 glioblastoma cells. *Journal of Biotechnology*, *280*, p. S79.

Mottaghitalab, F., Farokhi, M., Fatahi, Y., Atyabi, F., and Dinarvand, R., 2019. New insights into designing hybrid nanoparticles for lung cancer: Diagnosis and treatment. *Journal of Controlled Release*. doi: 10.1016/j.jconrel.2019.01.009.

Mourdikoudis, S., Pallares, R.M., and Thanh, N.T., 2018. Characterization techniques for nanoparticles: Comparison and complementarity upon studying nanoparticle properties. *Nanoscale*, *10*(27), pp. 12871–12934.

Muthuraman, A., Rishitha, N., and Mehdi, S., 2018. Role of Nanoparticles in Bioimaging, Diagnosis and Treatment of Cancer Disorder. In: *Design of Nanostructures for Theranostics Applications* (pp. 529–562). Elsevier.

Muthuvinothini, A., and Stella, S., 2019. Green synthesis of metal oxide nanoparticles and their catalytic activity for the reduction of aldehydes. *Process Biochemistry*, *77*, pp. 48–56.

Na, H., Choi, H., Oh, J.W., Kim, Y.E., Choi, S.R., Park, J.Y., and Cho, Y.S., 2020. Graphitic carbon-based core-shell platinum electrocatalysts processed using nickel nanoparticle template for oxygen reduction reaction. *Applied Surface Science*, *533*. doi: 10.1016/j.apsusc.2020.147519.

Narayanan, K.B., and Sakthivel, N., 2010. Biological synthesis of metal nanoparticles by microbes. *Advances In Colloid and Interface Science*, *156*(1-2), pp. 1–13.

Ngo, H.T., Wang, H.N., Fales, A.M., and Vo-Dinh, T., 2013. Label-free DNA biosensor based on SERS molecular sentinel on nanowave chip. *Analytical Chemistry*, *85*(13), pp. 6378–6383.

Nurunnabi, M., Cho, K.J., Choi, J.S., Huh, K.M., and Lee, Y.K., 2010. Targeted near-IR QDs-loaded micelles for cancer therapy and imaging. *Biomaterials*, *31*(20), pp. 5436–5444.

Oliveira Salles, M., Ruas de Souza, A.P., Naozuka, J., de Oliveira, P.V., and Bertotti, M., 2009. Bismuth modified gold microelectrode for Pb (II) determination in wine using alkaline medium. *Electroanalysis: An International Journal Devoted to Fundamental and Practical Aspects of Electroanalysis*, *21*(12), pp. 1439–1442.

Ozkan, A.D., Topal, A.E., Dana, A., Guler, M.O., and Tekinay, A.B., 2016. Atomic force microscopy for the investigation of molecular and cellular behavior. *Micron*, *89*, pp. 60–76.

Pal, S.L., Jana, U., Manna, P.K., Mohanta, G.P., and Manavalan, R., 2011. Nanoparticle: An overview of preparation and characterization. *Journal of Applied Pharmaceutical Science*, *1*(6), pp. 228–234.

Paleček, E., Tkáč, J., Bartosik, M., Bertók, T., Ostatná, V., and Paleček, J., 2015. Electrochemistry of nonconjugated proteins and glycoproteins. Toward sensors for biomedicine and glycomics. *Chemical reviews*, *115*(5), pp. 2045–2108.

Palei, N.N., Mohanta, B.C., Sabapathi, M.L., and Das, M.K., 2018. Lipid-based nanoparticles for cancer diagnosis and therapy. *In*: *Organic Materials as Smart Nanocarriers for Drug Delivery* (pp. 415–470). Elsevier.

Pangi Z., Beletsi A., and Evangelatos K., 2003. PEG-ylated nanoparticles for biological and pharmaceutical application. *Advanced Drug Delivery Review*, *24*, pp. 403–419.

Panjiar, H., Gakkhar, R.P., and Daniel, B.S.S., 2015. Strain-free graphite nanoparticle synthesis by mechanical milling. *Powder Technology*, *275*, pp. 25–29.

Pantelopoulos, A., and Bourbakis, N.G., 2009. A survey on wearable sensor-based systems for health monitoring and prognosis. *IEEE Transactions on Systems, Man, and Cybernetics, Part C (Applications and Reviews)*, *40*(1), pp. 1–12.

Parvanian, S., Mostafavi, S.M., and Aghashiri, M., 2017. Multifunctional nanoparticle developments in cancer diagnosis and treatment. *Sensing and Bio-Sensing Research*. doi: 10.1016/j.sbsr.2016.08.002

Patel, S., Nanda, R., Sahoo, S., and Mohapatra, E., 2016. Biosensors in health care: the milestones achieved in their development towards lab-on-chip-analysis. *Biochemistry Research International*, *2016*. doi: 10.1155/2016/3130469.

Pawar, A., Thakkar, S., and Misra, M., 2018. A bird's eye view of nanoparticles prepared by electrospraying: Advancements in drug delivery field. *Journal of Controlled Release*. doi: 10.1016/j.jconrel.2018.07.036.

Pei, L., Tsuzuki, T., Dodd, A., and Saunders, M., 2017. Synthesis of calcium chlorapatite nanoparticles and nanorods via a mechanically-induced solid-state displacement reaction and subsequent heat treatment. *Ceramics International*, *43*(14), pp. 11410–11414.

Peilin, Z., Jianzhong, Z., Shenzhong, Y., Xikang, Z., and Guoxiong, Z., 1995. Electrochemical characterization of boron-doped polycrystalline diamond thin-film electrodes. *Fresenius' Journal of Analytical Chemistry*, *353*(2), pp. 171–173.

Peng, J., Gao, W., Gupta, B.K., Liu, Z., Romero-Aburto, R., Ge, L., Song, L., Alemany, L.B., Zhan, X., Gao, G., and Vithayathil, S.A., 2012. Graphene quantum dots derived from carbon fibers. *Nano letters*, *12*(2), pp. 844–849.

Peng, J., Yang, Q., Shi, K., Xiao, Y., Wei, X., and Qian, Z., 2019. Intratumoral fate of functional nanoparticles in response to microenvironment factor: Implications on cancer diagnosis and therapy. *Advanced Drug Delivery Reviews*.

Perumal, V., Sivakumar, P.M., Zarrabi, A., Muthupandian, S., Vijayaraghavalu, S., Sahoo, K., Das, A., Das, S., Payyappilly, S.S., and Das, S., 2019. Near infra-red polymeric nanoparticle based optical imaging in Cancer diagnosis. *Journal of Photochemistry and Photobiology B: Biology*, *199*. doi: 10.1016/j.jphotobiol.2019.111630.

Pimpin, A., and Srituravanich, W., 2012. Review on micro-and nanolithography techniques and their applications. *Engineering Journal*, *16*(1), pp. 37–56.

Pires, N.M.M., Dong, T., Hanke, U., and Hoivik, N., 2014. Recent developments in optical detection technologies in lab-on-a-chip devices for biosensing applications. *Sensors*, *14*(8), pp. 15458–15479.

Polteau, B., Tessier, F., Cheviré, F., Cario, L., Odobel, F., and Jobic, S., 2016. Synthesis of Ni-poor NiO nanoparticles for p-DSSC applications. *In*: *Solid State Sciences* (pp.37–42). Elsevier Masson SAS.

Poovi, G., and Damodharan, N., 2018. Lipid nanoparticles: A challenging approach for oral delivery of BCS Class-II drugs. *Future Journal of Pharmaceutical Sciences*, *4*(2), pp. 191–205.

Popov, V.A., Shelekhov, E. V., Prosviryakov, A.S., Presniakov, M.Y., Senatulin, B.R., Kotov, A.D., Khomutov, M.G., and Khodos, I.I., 2017. Application of nanodiamonds for in situ synthesis of TiC reinforcing nanoparticles inside aluminium matrix during mechanical alloying. *Diamond and Related Materials*, *75*, pp. 6–11.

Prado, C., Wilkins, S.J., Marken, F., and Compton, R.G., 2002. Simultaneous Electrochemical Detection and Determination of Lead and Copper at Boron-Doped Diamond Film Electrodes. *Electroanalysis: An International Journal Devoted to Fundamental and Practical Aspects of Electroanalysis*, *14*(4), pp. 262–272.

Prakash, S., Karacor, M.B., and Banerjee, S., 2009. Surface modification in microsystems and nanosystems. *Surface Science Reports*, *64*(7), pp. 233–254.

Prakash, S., Pinti, M., and Bhushan, B., 2012. Theory, fabrication and applications of microfluidic and nanofluidic biosensors. *Philosophical Transactions of the Royal Society A: Mathematical, Physical and Engineering Sciences*, *370*(1967), pp. 2269–2303.

Pu, K.Y., Li, K., and Liu, B., 2010. Cationic oligofluorene-substituted polyhedral oligomeric silsesquioxane as light-harvesting unimolecular nanoparticle for fluorescence amplification in cellular imaging. *Advanced Materials*, *22*(5), pp. 643–646.

Radhika, S., Davis, K.J., Pratheesh, M.D., Anoopraj, R., and Joseph, B.S., 2012. Biosensors: A novel approach for pathogen detection. *Vet Scan*, *7*(1), pp. 14–18.

Rahman, M.R., Okajima, T., and Ohsaka, T., 2010. Selective detection of As (III) at the Au (111)-like polycrystalline gold electrode. *Analytical Chemistry*, *82*(22), pp. 9169–9176.

Rai, M., Gade, A., Gaikwad, S., Marcato, P.D., and Durán, N., 2012. Biomedical applications of nanobio-sensors: the state-of-the-art. *Journal of the Brazilian Chemical Society*, *23*(1), pp. 14–24.

Rajeshkumar, S., and Bharath, L. V., 2017. Mechanism of plant-mediated synthesis of silver nanoparticles – a review on biomolecules involved, characterisation and antibacterial activity. *Chemico-Biological Interactions*. doi: 10.1016/j.cbi.2017.06.019.

Rajkumar, S., and Prabaharan, M., 2019. Multi-functional core-shell Fe3O4 @Au nanoparticles for cancer diagnosis and therapy. *Colloids and Surfaces B: Biointerfaces*, *174*, pp. 252–259.

Ramasamy, M., Lee, J.H., and Lee, J., 2017. Direct one-pot synthesis of cinnamaldehyde immobilized on gold nanoparticles and their antibiofilm properties. *Colloids and Surfaces B: Biointerfaces*, *160*, pp. 639–648.

Ramesh, S., 2013. Sol-gel synthesis and characterization of nanoparticles. *Journal of Nanoscience*, *2013*. 10.1155/2013/929321.

Rezaee, S., and Ghobadi, N., 2018. Synthesis of Ag-Cu-Pd alloy thin films by DC-magnetron sputtering: Case study on microstructures and optical properties. *Results in Physics*, *9*, pp. 1148–1154.

Rezaie, A.B., Montazer, M., and Rad, M.M., 2017. Photo and biocatalytic activities along with UV protection properties on polyester fabric through green in-situ synthesis of cauliflower-like CuO nanoparticles. *Journal of Photochemistry and Photobiology B: Biology*, *176*, pp. 100–111.

Robertson, J., 2002. Diamond-like amorphous carbon. *Materials Science and Engineering: R: Reports*, *37*(4-6), pp. 129–281.

Roma, J., Matos, A.R., Vinagre, C., and Duarte, B., 2020. Engineered metal nanoparticles in the marine environment: A review of the effects on marine fauna. *Marine Environmental Research*. doi: 10.1016/j.marenvres.2020.105110.

Ruan, C., Shi, W., Jiang, H., Sun, Y., Liu, X., Zhang, X., Sun, Z., Dai, L., and Ge, D., 2013. One-pot preparation of glucose biosensor based on polydopamine–graphene composite film modified enzyme electrode. *Sensors and Actuators B: Chemical*, *177*, pp. 826–832.

Saeed, A.A., Sánchez, J.L.A., O'Sullivan, C.K., and Abbas, M.N., 2017. DNA biosensors based on gold nanoparticles-modified graphene oxide for the detection of breast cancer biomarkers for early diagnosis. *Bioelectrochemistry*, *118*, pp. 91–99.

Saha, K., Agasti, S.S., Kim, C., Li, X., and Rotello, V.M., 2012. Gold nanoparticles in chemical and biological sensing. *Chemical Reviews*, *112*(5), pp. 2739–2779.

Saitakis, M., and Gizeli, E., 2012. Acoustic sensors as a biophysical tool for probing cell attachment and cell/surface interactions. *Cellular and Molecular Life Sciences*, *69*(3), pp. 357–371.

Salavati-Niasari, M., Davar, F., and Mir, N., 2008. Synthesis and characterization of metallic copper nano-particles via thermal decomposition. *Polyhedron*, *27*(17), pp. 3514–3518.

Santhoshkumar, J., Rajeshkumar, S., and Venkat Kumar, S., 2017. Phyto-assisted synthesis, characterization and applications of gold nanoparticles – a review. *Biochemistry and Biophysics Reports*. doi: 10.1016/j.bbrep.2017.06.004.

Saterlay, A.J., Agra-Gutiérrez, C., Taylor, M.P., Marken, F., and Compton, R.G., 1999. Sono-Cathodic Stripping Voltammetry of Lead at a Polished Boron-Doped Diamond Electrode: Application to the Determination of Lead in River Sediment. *Electroanalysis: An International Journal Devoted to Fundamental and Practical Aspects of Electroanalysis*, *11*(15), pp. 1083–1088.

Saterlay, A.J., Foord, J.S., and Compton, R.G., 1999. Sono-cathodic stripping voltammetry of manganese at a polished boron-doped diamond electrode: application to the determination of manganese in instant tea. *Analyst*, *124*(12), pp. 1791–1796.

Saterlay, A.J., Marken, F., Foord, J.S., and Compton, R.G., 2000. Sonoelectrochemical investigation of silver analysis at a highly boron-doped diamond electrode. *Talanta*, *53*(2), pp. 403–415.

Sbartai, A., Namour, P., Errachid, A., Krejči, J., Šejnohová, R., Renaud, L., Larbi Hamlaoui, M., Loir, A.S., Garrelie, F., Donnet, C., and Soder, H., 2012. Electrochemical boron-doped diamond film microcells micromachined with femtosecond laser: application to the determination of water framework directive metals. *Analytical chemistry*, *84*(11), pp. 4805–4811.

Scherer, M.D., Sposito, J.C.V., Falco, W.F., Grisolia, A.B., Andrade, L.H.C., Lima, S.M., Machado, G., Nascimento, V.A., Gonçalves, D.A., Wender, H., Oliveira, S.L., and Caires, A.R.L., 2019. Cytotoxic and genotoxic effects of silver nanoparticles on meristematic cells of Allium cepa roots: A close analysis of particle size dependence. *Science of the Total Environment*, *660*, pp. 459–467.

Scremin, J., and Sartori, E.R., 2018. Simultaneous determination of nifedipine and atenolol in combined dosage forms using a boron-doped diamond electrode with differential pulse voltammetry. *Canadian Journal of Chemistry*, *96*(1), pp. 1–7.

Sebastian, E.M., Jain, S.K., Purohit, R., Dhakad, S.K., and Rana, R.S., 2019. Nanolithography and its current advancements. *In: Materials Today: Proceedings* (pp. 2351–2356). Elsevier Ltd.

Sebez, B., Ogorevc, B., Hocevar, S.B., and Veber, M., 2013. Functioning of antimony film electrode in acid media under cyclic and anodic stripping voltammetry conditions. *Analytica Chimica Acta*, *785*, pp. 43–49.

Serrano, N., Alberich, A., Díaz-Cruz, J.M., Ariño, C., and Esteban, M., 2013. Coating methods, modifiers and applications of bismuth screen-printed electrodes. *TrAC Trends in Analytical Chemistry*, *46*, pp. 15–29.

Shaganov, I.I., Perova, T.S., and Berwick, K., 2017. The effect of the local field and dipole-dipole interactions on the absorption spectra of noble metals and the plasmon resonance of their nanoparticles. *Photonics and Nanostructures – Fundamentals and Applications*, *27*, pp. 24–31.

Shah, M.R., Imran, M., and Ullah, S., 2019. Potential role of gold nanoparticles in cancer diagnosis and targeted drug delivery. *In: Nanocarriers for Cancer Diagnosis and Targeted Chemotherapy* (pp. 267–286). Elsevier.

Shah, M.R., Imran, M., and Ullah, S., 2019. Surface-Functionalized Magnetic Nanoparticles in Cancer-Drug Delivery and Diagnosis. *In: Nanocarriers for Cancer Diagnosis and Targeted Chemotherapy*. Elsevier, 107–128.

Shah, P., and Gavrin, A., 2006. Synthesis of nanoparticles using high-pressure sputtering for magnetic domain imaging. *Journal of Magnetism and Magnetic Materials*, *301*(1), pp. 118–123.

Shankar, S.S., Rai, A., Ahmad, A., and Sastry, M., 2004. Rapid synthesis of Au, Ag, and bimetallic Au core–Ag shell nanoparticles using Neem (Azadirachta indica) leaf broth. *Journal of Colloid and Interface Science*, *275*(2), pp. 496–502.

Shankar, S.S., Rai, A., Ankamwar, B., Singh, A., Ahmad, A., and Sastry, M., 2004. Biological synthesis of triangular gold nanoprisms. *Nature Materials*, *3*(7), pp. 482–488.

Sharifi, I., Zamanian, A., and Behnamghader, A., 2016. Synthesis and characterization of $FeO.6ZnO.4Fe2O4$ ferrite magnetic nanoclusters using simple thermal decomposition method. *Journal of Magnetism and Magnetic Materials*, *412*, pp. 107–113.

Sharma, A.B., Sharma, M., and Pandey, R.K., 2009. Synthesis, Properties and Potential Applications of Semiconductor Quantum Particles". *Asian Journal of Chemistry*, *21*(10), pp. S033–S038.

Sharma, J.K., Srivastava, P., Ameen, S., Akhtar, M.S., Singh, G., and Yadava, S., 2016. Azadirachta indica plant-assisted green synthesis of Mn3O4 nanoparticles: Excellent thermal catalytic performance and chemical sensing behavior. *Journal of Colloid and Interface Science*, *472*, pp. 220–228.

Shi, H.G., Farber, L., Michaels, J.N., Dickey, A., Thompson, K.C., Shelukar, S.D., Hurter, P.N., Reynolds, S.D., and Kaufman, M.J., 2003. Characterization of crystalline drug nanoparticles using atomic force microscopy and complementary techniques. *Pharmaceutical Research 20*, pp. 479–484.

Shiddiky, M.J., Rauf, S., Kithva, P.H., and Trau, M., 2012. Graphene/quantum dot bionanoconjugates as signal amplifiers in stripping voltammetric detection of EpCAM biomarkers. *Biosensors and Bioelectronics*, *35*(1), pp. 251–257.

Shojaati, F., Riazi, M., Mousavi, S.H., and Derikvand, Z., 2017. Experimental investigation of the inhibitory behavior of metal oxides nanoparticles on asphaltene precipitation. *Colloids and Surfaces A: Physicochemical and Engineering Aspects*, *531*, pp. 99–110.

Simm, A.O., Banks, C.E., and Compton, R.G., 2005a. The electrochemical detection of arsenic (III) at a silver electrode. *Electroanalysis: An International Journal Devoted to Fundamental and Practical Aspects of Electroanalysis*, *17*(19), pp. 1727–1733.

Simm, A.O., Banks, C.E., and Compton, R.G., 2005b. Sonoelectroanalytical detection of ultra-trace arsenic. *Electroanalysis: An International Journal Devoted to Fundamental and Practical Aspects of Electroanalysis*, *17*(4), pp. 335–342.

Sin, M.L., Mach, K.E., Wong, P.K., and Liao, J.C., 2014. Advances and challenges in biosensor-based diagnosis of infectious diseases. *Expert Review of Molecular Diagnostics*, *14*(2), pp. 225–244.

Singh, N., and Khullar, V., 2020. On-sun testing of volumetric absorption based concentrating solar collector employing carbon soot nanoparticles laden fluid. *Sustainable Energy Technologies and Assessments*, *42*. doi: 10.1016/j.seta.2020.100868.

Singh, P., Bodycomb, J., Travers, B., Tatarkiewicz, K., Travers, S., Matyas, G.R., and Beck, Z., 2019. Particle size analyses of polydisperse liposome formulations with a novel multispectral advanced nanoparticle tracking technology. *International Journal of Pharmaceutics*, *566*, pp. 680–686.

Siontorou, C.G., and Batzias, F.A., 2014. A methodological combined framework for roadmapping biosensor research: A fault tree analysis approach within a strategic technology evaluation frame. *Critical Reviews in Biotechnology*, *34*(1), pp. 31–55.

Sithara, R., Selvakumar, P., Arun, C., Anandan, S., and Sivashanmugam, P., 2017. Economical synthesis of silver nanoparticles using leaf extract of Acalypha hispida and its application in the detection of Mn(II) ions. *Journal of Advanced Research*, *8*(6), pp. 561–568.

Sivakumar, S., and Nelson Prabu, L., 2021. Synthesis and characterization of α-MnO2 nanoparticles for supercapacitor application. *Materials Today: Proceedings*. doi: 10.1016/j.matpr.2021.03.528.

Sizochenko, N., Syzochenko, M., Fjodorova, N., Rasulev, B., and Leszczynski, J., 2019. Evaluating genotoxicity of metal oxide nanoparticles: Application of advanced supervised and unsupervised machine learning techniques. *Ecotoxicology and Environmental Safety*, *185*. doi: 10.1016/j.ecoenv. 2019.109733.

Song, G., Chen, M., Chen, C., Wang, C., Hu, D., Ren, J., and Qu, X., 2010. Design of proton-fueled tweezers for controlled, multi-function DNA-based molecular device. *Biochimie*, *92*(2), pp. 121–127.

Song, Y., and Swain, G.M., 2007. Total inorganic arsenic detection in real water samples using anodic stripping voltammetry and a gold-coated diamond thin-film electrode. *Analytica Chimica Acta*, *593*(1), pp. 7–12.

Sonthalia, P., McGaw, E., Show, Y., and Swain, G.M., 2004. Metal ion analysis in contaminated water samples using anodic stripping voltammetry and a nanocrystalline diamond thin-film electrode. *Analytica Chimica Acta*, *522*(1), pp. 35–44.

Sopha, H., Jovanovski, V., Hocevar, S.B., and Ogorevc, B., 2012. In-situ plated antimony film electrode for adsorptive cathodic stripping voltammetric measurement of trace nickel. *Electrochemistry Communications*, *20*, pp. 23–25.

Su, L., Zou, L., Fong, C.C., Wong, W.L., Wei, F., Wong, K.Y., Wu, R.S., and Yang, M., 2013. Detection of cancer biomarkers by piezoelectric biosensor using PZT ceramic resonator as the transducer. *Biosensors and Bioelectronics*, *46*, pp. 155–161.

Subbaiya, R., Shiyamala, M., Revathi, K., Pushpalatha, R., Masilamani Selvam, M. 2014. Biological synthesis of silver nanoparticles from *Nerium oleander* and its antibacterial and antioxidant property. *International Journal of Current Microbiology and Applied Sciences*, *3*(1), pp. 83–87.

Suei, S., Raudsepp, A., Kent, L.M., Keen, S.A., Filichev, V.V., and Williams, M.A., 2015. DNA visualization in single molecule studies carried out with optical tweezers: Covalent versus non-covalent attachment of fluorophores. *Biochemical and Biophysical Research Communications*, *466*(2), pp. 226–231.

Sun, H., Wu, L., Wei, W., and Qu, X., 2013. Recent advances in graphene quantum dots for sensing. *Materials Today*, *16*(11), pp. 433–442.

Sun, S., Xiao, Q.R., Wang, Y., and Jiang, Y., 2018. Roles of alcohol desolvating agents on the size control of bovine serum albumin nanoparticles in drug delivery system. *Journal of Drug Delivery Science and Technology*, *47*, pp. 193–199.

Sun, X., Hourwitz, M.J., Das, S., Fourkas, J., and Losert, W., 2019. Cell motility and nanolithography. *In: Three-Dimensional Microfabrication Using Two-Photon Polymerization* (527–540). Elsevier.

Syam, R., Davis, K.J., Pratheesh, M.D., Anoopraj, R., and Joseph, B.S., 2012. Biosensors: a novel approach for pathogen detection. *Vet Scan| Online Veterinary Medical Journal*, *7*(1), pp.101–102.

Tai, C.Y., Tai, C.T., Chang, M.H., and Liu, H.S., 2007. Synthesis of magnesium hydroxide and oxide nanoparticles using a spinning disk reactor. *Industrial & Engineering Chemistry Research*, *46*(17), pp. 5536–5541.

Tang, J., Wang, H., and Gomez, A., 2017. Controlled nanoparticle synthesis via opposite-polarity electrospray pyrolysis. *Journal of Aerosol Science*, *113*, pp. 201–211.

Tostado-Plascencia, M.M., Sanchez-Tizapa, M., and Zamudio-Ojeda, A., 2018. Synthesis and characterization of multiwalled carbon nanotubes functionalized with chlorophyll-derivatives compounds extracted from Hibiscus tiliaceus. *Diamond and Related Materials*, *89*, pp. 151–162.

Tothill, I.E., 2009, February. Biosensors for cancer markers diagnosis. In *Seminars In Cell & Developmental Biology* (Vol. 20, No. 1, pp. 55–62). Academic Press.

Uddin, I., Ahmad, K., Khan, A.A., and Kazmi, M.A., 2017. Synthesis of silver nanoparticles using Matricaria recutita (Babunah) plant extract and its study as mercury ions sensor. *Sensing and Bio-Sensing Research, 16*, pp. 62–67.

Uhl, A., and Kestranek, W., 1923. The electrometric titration of acids and bases with the antimony indicator electrodes. *Monatsh. Chem, 44*, pp. 29–34.

Vara, J.A., and Dave, P.N., 2019. Metal oxide nanoparticles as catalyst for thermal behavior of AN based composite solid propellant. *Chemical Physics Letters, 730*, pp. 600–607.

Vargová, V., 2015. *Oxidace a redukce proteinů na elektricky nabitých površích* (Doctoral dissertation, Masarykova univerzita, Přírodovědecká fakulta).

Vereshchagina, E., Tiggelaar, R.M., Sanders, R.G., Wolters, R.A., and Gardeniers, J.G., 2015. Low power micro-calorimetric sensors for analysis of gaseous samples. *Sensors and Actuators B: Chemical, 206*, pp. 772–787.

Verma, M., Kumar, V., and Katoch, A., 2018a. Sputtering based synthesis of CuO nanoparticles and their structural, thermal and optical studies. *Materials Science in Semiconductor Processing, 76*, 55–60.

Verma, M., Kumar, V., and Katoch, A., 2018b. Synthesis of ZrO2 nanoparticles using reactive magnetron sputtering and their structural, morphological and thermal studies. *Materials Chemistry and Physics, 212*, pp. 268–273.

Vigneshwaran, N., Ashtaputre, N.M., Varadarajan, P.V., Nachane, R.P., Paralikar, K.M., and Balasubramanya, R.H., 2007. Biological synthesis of silver nanoparticles using the fungus Aspergillus flavus. *Materials Letters, 61*(6), pp. 1413–1418.

Vijayaraghavan, K., and Ashokkumar, T., 2017. Plant-mediated biosynthesis of metallic nanoparticles: A review of literature, factors affecting synthesis, characterization techniques and applications. *Journal of Environmental Chemical Engineering*. doi: 10.1016/j.jece.2017.09.026.

Wang, H., Rao, H., Luo, M., Xue, X., Xue, Z., and Lu, X., 2019. Noble metal nanoparticles growth-based colorimetric strategies: From monocolorimetric to multicolorimetric sensors. *Coordination Chemistry Reviews*. doi: 10.1016/j.ccr.2019.06.020.

Wang, J., Lu, J., Hocevar, S.B., Farias, P.A., and Ogorevc, B., 2000. Bismuth-coated carbon electrodes for anodic stripping voltammetry. *Analytical Chemistry, 72*(14), pp. 3218–3222.

Wang, J., Lu, J., Hocevar, S.B., and Ogorevc, B., 2001. Bismuth-coated screen-printed electrodes for stripping voltammetric measurements of trace lead. *Electroanalysis: An International Journal Devoted to Fundamental and Practical Aspects of Electroanalysis, 13*(1), pp. 13–16.

Wang, J., Lu, J., Kirgöz, Ü.A., Hocevar, S.B., and Ogorevc, B., 2001. Insights into the anodic stripping voltammetric behavior of bismuth film electrodes. *Analytica Chimica Acta, 434*(1), pp. 29–34.

Wang, Y., Hu, R., Lin, G., Roy, I., and Yong, K.T., 2013. Functionalized quantum dots for biosensing and bioimaging and concerns on toxicity. *ACS Applied Materials & Interfaces, 5*(8), pp. 2786–2799.

Watts, H.J., Lowe, C.R., and Pollard-Knight, D.V., 1994. Optical biosensor for monitoring microbial cells. *Analytical Chemistry, 66*(15), pp. 2465–2470.

Weiss, E., Groenen-Serrano, K., and Savall, A., 2008. A comparison of electrochemical degradation of phenol on boron doped diamond and lead dioxide anodes. *Journal of Applied Electrochemistry, 38*(3), pp. 329–337.

Welch, C.M., Nekrassova, O., and Compton, R.G., 2005. Reduction of hexavalent chromium at solid electrodes in acidic media: Reaction mechanism and analytical applications. *Talanta, 65*(1), pp. 74–80.

Willner, I., Baron, R., and Willner, B., 2006. Growing metal nanoparticles by enzymes. *Advanced Materials, 18*(9), pp. 1109–1120.

Wren, S., Minelli, C., Pei, Y., and Akhtar, N., 2020. Evaluation of particle size techniques to support the development of manufacturing scale nanoparticles for application in pharmaceuticals. *Journal of Pharmaceutical Sciences, 109*(7), pp. 2284–2293.

Wuttke, S., Lismont, M., Escudero, A., Rungtaweevoranit, B., and Parak, W.J., 2017. Positioning metal-organic framework nanoparticles within the context of drug delivery – a comparison with mesoporous silica nanoparticles and dendrimers. *Biomaterials, 123*, pp. 172–183.

Yadav, T.P., Yadav, R.M., and Singh, D.P., 2012. Mechanical milling: A top down approach for the synthesis of nanomaterials and nanocomposites. *Nanoscience and Nanotechnology, 2*(3), pp. 22–48.

Yoon, J.H., Yang, J.E., Kim, J.P., Bae, J.S., Shim, Y.B., and Won, M.S., 2010. Simultaneous detection of Cd (II), Pb (II), Cu (II), and Hg (II) ions in dye waste water using a boron doped diamond electrode with DPASV. *Bulletin of the Korean Chemical Society, 31*(1), pp. 140–145.

Yoosefian, M., Sabaei, S., and Etminan, N., 2019. Encapsulation efficiency of single-walled carbon nanotube for Ifosfamide anti-cancer drug. *Computers in Biology and Medicine, 114.* doi: 10.1016/j.compbiomed.2019.103433.

Yu, L., Chen, Y., Lin, H., Du, W., Chen, H., and Shi, J., 2018. Ultrasmall mesoporous organosilica nanoparticles: Morphology modulations and redox-responsive biodegradability for tumor-specific drug delivery. *Biomaterials, 161,* pp. 292–305.

Zeng, A., Liu, E., Tan, S.N., Zhang, S., and Gao, J., 2002. Stripping voltammetric analysis of heavy metals at nitrogen doped diamond-like carbon film electrodes. *Electroanalysis: An International Journal Devoted to Fundamental and Practical Aspects of Electroanalysis, 14*(18), pp. 1294–1298.

Zhang, K., Wang, Z., Chen, G., Wang, Y., and Wei, J., 2020. GeTe photoresist films for both positive and negative heat-mode nanolithography. *Materials Letters, 261.* doi: 10.1016/j.matlet.2019.127019.

Zhang, K.X., Wen, X., Yao, C.B., Li, J., Zhang, M., Li, Q.H., Sun, W.J., and Wu, J. Da, 2018. Synthesis, structural and optical properties of silver nanoparticles uniformly decorated ZnO nanowires. *Chemical Physics Letters, 698,* pp. 147–151.

Zhang, T., Lu, G., Shen, H., Shi, K., Jiang, Y., Xu, D., and Gong, Q., 2014. Photoluminescence of a single complex plasmonic nanoparticle. *Scientific Reports, 4*(1), pp. 1–7.

Zhang, X., Xiang, D., Zhu, W., Zheng, Y., Harkin-Jones, E., Wang, P., Zhao, C., Li, H., Wang, B., and Li, Y., 2020. Flexible and high-performance piezoresistive strain sensors based on carbon nanoparticles@polyurethane sponges. *Composites Science and Technology, 200.* doi: 10.1016/j.compscitech.2020.108437.

Zhang, Y., Jiang, R., Wang, Z., Xue, Y., Sun, J., and Guo, Y., 2020. (Fe,N-codoped carbon nanotube)/(Fe-based nanoparticle) nanohybrid derived from Fe-doped g-C3N4: A superior catalyst for oxygen reduction reaction. *Journal of Colloid and Interface Science, 579,* pp. 391–400.

Zhang, Z., Yao, G., Zhang, X., Ma, J., and Lin, H., 2015. Synthesis and characterization of nickel ferrite nanoparticles via planetary ball milling assisted solid-state reaction. *Ceramics International, 41*(3), pp. 4523–4530.

Zhou, Q., Liu, L., Liu, N., He, B., Hu, L., and Wang, L., 2020. Determination and characterization of metal nanoparticles in clams and oysters. *Ecotoxicology and Environmental Safety, 198.* doi: 10.1016/j.ecoenv.2020.110670.

Zikalala, N., Matshetshe, K., Parani, S., and Oluwafemi, O.S., 2018. Biosynthesis protocols for colloidal metal oxide nanoparticles. *Nano-Structures and Nano-Objects.* doi: 10.1016/j.nanoso.2018.07.010.

Zur Mühlen, A., Zur Mühlen, E., Niehus, H., and Mehnert, W., 1996. Atomic force microscopy studies of solid lipid nanoparticles. *Pharmaceutical Research, 13*(9), pp. 1411–1416.

5 Recent Advances in Diagnosis: Nano-Based Approach

5.1 INTRODUCTION

The electrochemical analysis technique is commonly employed in analyte testing to provide a low detection limit, high sensitivity, and a wide linear detection range (Thévenot et al., 2001; Pacheco et al., 2018; Ruiyi et al., 2018; Y. Yang et al., 2018; Aparicio-Martínez et al., 2019; Dhara and Debiprosad, 2019; Rajaji et al., 2019; Wen et al., 2019; Zhang et al., 2019; Dai et al., 2020; Xu et al., 2020). Electrochemical immunosensors offer the benefits of high sensitivity of the electrochemical analysis. The explicit recognition of immunochemical reaction is appropriate in the analysis of biochemical substrates in trace concentrations (Deepa et al., 2020; Dumore and Mukhopadhyay 2020; Glasscott et al., 2020; Gliga et al., 2020; Incebay et al., 2020; Yazdi et al., 2020; Cetinkaya et al., 2021; Cho et al., 2021; Ibrahim 2021; Qian et al., 2021; Tabish et al., 2021; Verma et al., 2021; Xia et al., 2021; Xing et al., 2021; Zhang et al., 2021). The prepared electrochemical biosensors are prepared by immobilization of the biomolecule with the capability to improve the sensitivity during sensor design and construction, thus enabling different types of signals to be recorded (Sadik et al., 2009). The signals recorded by the electrochemical immunoassays can be differentiated into different classes (Tang et al., 2011; Ben Messaoud et al., 2017; Bollella et al., 2017; Forster et al., 2017; Kongsuphol et al., 2017; Li et al., 2017; Lim et al., 2017; Lotfi Zadeh Zhad et al., 2017; Pursey et al., 2017; Shahzad et al., 2017; S. Xu et al., 2017; Song et al., 2021; Raziq et al., 2021), such as the potentiometric method (Wang et al., 2010), electrochemical luminescence (ECL) method (Sardesai et al., 2011), amperometric method (He et al., 2008), gravimetric method (Joo et al., 2012), AC impedance and capacitance method (Carrara et al., 2009), surface plasmon resonance (Uludag and Tothill, 2012), piezoelectric quartz crystal method, etc. Among these approaches, the electrochemical luminescence has a comparatively higher sensitivity. As discussed in the previous chapter, among various sensing techniques for the detection of cancer, the most efficient technique is an electrochemical biosensor. The different types of electrochemical biosensors can be classified as follows.

5.2 AMPEROMETRIC BIOSENSORS

Amperometric biosensors are found to be more sensitive and suitable for mass production in comparison to potentiometric biosensors (Cui et al., 2005; Akhtar et al., 2017; Amouzadeh Tabrizi et al., 2017; Amouzadeh Tabrizi et al., 2017; Giannetto et al., 2017; Jolly et al., 2017; Tang et al., 2017; Xue et al., 2017, 2021; Yuan et al., 2017; Jalalvand, Goicoechea and Gu, 2019; Liu et al., 2019; Wang and Ma, 2019; X. Liang et al., 2019(a) and (c); Eryiğit et al., 2021; Fan et al., 2021; Huang et al., 2021; Takeda et al., 2021; Xiao et al., 2021). In this type of biosensor, a noble metal or a screen-printed coating shielded by the biorecognition element works as a working electrode. Another cost-effective alternative is carbon paste with a fixed enzyme (Mehrvar and Abdi, 2004; Guerrieri et al., 2019; Lee et al., 2019; Maity and Kumar, 2019). An electroactive species is converted at the electrode surface and the current is measured after applying the potential (Magner, 1998). Two or three working electrode conformations are possible for amperometric biosensors. The two-electrode system consists of the only reference and working (including immobilized biorecognition element) electrodes. One primary disadvantage of the two-electrode configuration is limited potential control on the working electrode in the case of higher current and due to this, the short linear range is achieved. This difficulty can be solved through a third electrode and that is an auxiliary electrode. By the employment of auxiliary electrode, voltage is applied in between the working and reference electrodes while the current flow occurs between the working and counter (auxiliary) electrodes. On a large scale, the use of these biosensors was reported for different types of analytes such as glucose, sialic acid (Yemini et al., 2007), and lactate (Marzouk et al., 2007). Amperometric biosensors have also been described for biological agents analysis such as *Bacillus cereus* as well as nerve agents (Pohanka et al., 2007) and the serological analysis of *Francisella tularensis* (Pohanka et al., 2007). Organophosphates and carbamates can also be detected at a faster rate by this sensing method based on enzyme inhibition of AChE (acetylcholinesterase) and butyrylcholinesterase (BChE) (Krejcova et al., 2005; Sotiropoulou and Chaniotakis, 2005). Amperometric biosensors were assessed also for executing tests of nucleic acid acting as a biorecognition element or biomarker (Rossetti, et al., 2001). There are several amperometric biosensors commercially available. The most commonly available biosensors are glucose biosensors such as SIRE P201 (Chemel AB, Lund, Sweden), FreeStyle Freedom Blood Glucose Monitoring System, and Precision Xtra (Abbot Diabetes Care, Alameda, CA, USA), etc.

5.3   POTENTIOMETRIC BIOSENSORS

This method is based on ion-sensitive field-effect transistors (ISFETs) and consists of ion-selective electrodes (ISEs). Ions accumulated at the ion-selective membrane interface are majorly responsible for the primary output signal. The current flow over the electrode is nearly zero. The electrode (Newman and Setford, 2006; Aydin et al., 2019; Negi et al., 2019; Singh et al., 2019; Solovieva et al., 2019; Yan et al., 2019; Altunkök et al., 2021; Dejous and Krishnan, 2021; Musa et al., 2021; Nemčeková and Labuda, 2021; Tabish et al., 2021; Walker et al., 2021) tracks the ions generated from the reaction of enzyme catalysis. For instance, immobilization of glucose oxidase is carried out on a pH electrode surface. Glucose does not disturb the pH of the working media; nevertheless, the product formed from the enzymatic reaction, i.e., gluconate, causes acidification. A biorecognition component is fixed on the outer layer or enclosed inside the membrane. Previously, the pH glass electrode worked as a physicochemical transducer (Buerk, 2014). Physicochemical transducers based on semiconductors are currently more common. LAPS (light addressable potentiometric sensor) and ISFETs-based methods are more suitable for biosensor assembly. The ISFET principle (Yoshinobu et al., 2005) works by a generation of local potential through the surface ions of a solution. This potential controls the flow of current through a silicon semiconductor. The LAPS (D'Agostino et al., 2006) method is based on the activation of semiconductor via a light-emitting diode (LED). Detection of herbicide atrazine using a potentiometric biosensor based on molecularly imprinted polymers was done in the range of 3×10^{-5} to 1×10^{-3} M (Kitade et al., 2004); molecularly imprinted polymers were also employed for the detection of the neurotransmitter serotonin (Karakuş et al., 2006). For creatine examination, a potentiometric biosensor with a poly (vinyl chloride) ammonium membrane, co-immobilized with enzymes, i.e., urease and creatinase, was used (Korpan, et al., 2006). For glycoalkaloids detection, ISFET with fixed butyryl-cholinesterase was used (Timur and Telefoncu, 2004). Similarly, for organophosphate pesticide assay, a simple pH electrode fabricated with acetylcholinesterase (AChE) was utilized (Ercole et al., 2002). This biosensor was used for the *Escherichia coli* test with a detection limit of 10 cells/ml with the specific primary antibody immobilized on the LAPS and for sandwich formation, a secondary antibody was used (Ghindilis et al., 1998).

5.4   IMPEDIMETRIC BIOSENSORS

In this type of biosensor, properties like impedance (Z), capacitance (C), and resistance of components (R) are used (Azzouzi et al., 2016; Benvidi et al., 2016; Ding et al., 2016; J. Liu et al., 2016;

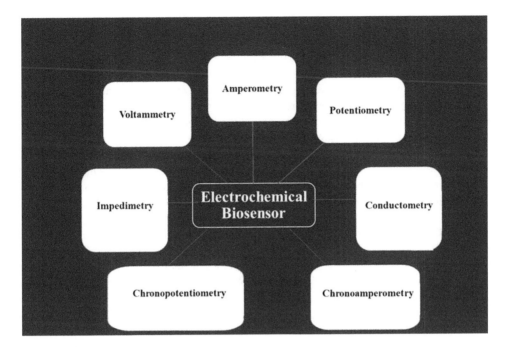

FIGURE 5.1 Different measurement methods of electrochemical biosensors.

Kara et al., 2016; Silva et al., 2016; Top et al., 2016; Yeh et al., 2016; Thangsunan et al., 2021; Bolat, Akbal Vural et al., 2021; Bolat, Vural et al., 2021; Díaz-Fernández et al., 2021; Eksin et al., 2021). Conductance is simply the inverse value of the resistance. This is the reason why it is also called a conductometric sensor. It consists of simply two electrodes with the alternating voltage applied with amplitudes in the range of a few to 100 mV. This biosensor is used for the urea assay using urease as a bio-recognition element.

Alternatively, these biosensors have been effectively employed for monitoring the growth of microorganisms based on conductive metabolite production (Davis et al., 2007). Impedimetric biosensors are less common in comparison to potentiometric and amperometric biosensors due to the production of false-positive results; however, it has some promising methods. Impedance tests have been used to screen DNA fragment hybridization, amplified by PCR (polymerase chain reaction) (Ouerghi et al., 2002). For the ethanol level testing in alcoholic beverages (*Saccharomyces cerevisiae),* impedance biosensors consisting of immobilized yeasts have been used (Spanggaard et al., 1994).

The different measuring techniques employed by electrochemical biosensors are explained below (Figure 5.1).

5.5 DIFFERENT TYPES OF ELECTROCHEMICAL SENSORS

5.5.1 AMPEROMETRY

This method is most commonly employed, in which current is continuously measured when the electrochemical species are oxidized or reduced at a fixed potential. Amperometry is widely used for cancer or cancer biomarker detection (Carvajal et al., 2018; Chung et al., 2018; Dong et al., 2018; Kumar et al., 2018; Narwal et al., 2018; Wang and Ma, 2018; Y. Yang et al., 2018; Arévalo et al., 2021; Bhatt et al., 2021; Shandilya et al., 2021; Hassanpour and Hasanzadeh, 2021; Medici et al., 2021; Promphet et al., 2021). For example, Kim and co-workers have detected lung cancer biomarkers Annexin II and MUC5AC through the immunosensing method. Similarly, gold nanoparticle-modified graphene nanocomposite was used for the biosensing of human cervical cancer biomarker miR-21 (Zhang et al., 2014).

Usually, amperometric biosensors' working flow is (i) labeling of an antibody with an enzyme or nanoparticles; (ii) this assembly is permitted to bind with analyte by using a primary antibody; (iii) by measuring the current at an applied potential, the analyte concentration is quantified (Chikkaveeraiah et al., 2012). The amperometric biosensors are mainly reliant on the properties of the electrode because a signal response occurs at the sensor-electrode surface. Fu has designed a novel amperometric immunosensor to detect CA125. Anti-CA125 antibodies were immobilized onto gold hollow microspheres and glassy carbon electrodes modified with polythionine (Fu, 2007). Due to the larger surface area, gold hollow microspheres have increased the anti-CA125 molecule numbers. Additionally, modified electrodes have also created a biocompatible milieu for proteins. The linear range was found to be between 4.5 to 36.5 U/mL with a LOD value of 1.3 U/mL.

For label-free detection of CA-15-3, an electrochemical immunosensor was developed by Li et al., based on N-doped graphene-modified electrodes (Li et al., 2013). These N-doped graphene sheets are highly conductive, significantly increasing the electron transfer rate from the electrode surface and facilitates the sensitivity of CA15-3 detection. This immunosensor exhibited a LOD value of 0.012 U/mL in a linear range of 0.1–20 U/mL. Wan et al. have developed another novel biosensor for cancer detection based on nanomaterials (Wan et al., 2011). A carbon nanotube-based sandwich immunosensor was developed for sensitive estimation of PSA and interleukin 8 (IL-8). They have used disposable screen-printed electrodes. The capture antibodies (Abs) have adhered to the electrode via covalent linkage and then the capture Abs were attached via amine residues. The detection limit was 5 pg/mL and 8 pg/mL for PSA and IL-8, respectively.

Currently, scientists have developed a biosensor for long non-coding RNA (lncRNA) detection, upregulated in the case of liver cancer (HULC) (Liu et al., 2015). This biosensor platform consisted of graphene oxide/Au/horseradish peroxidase coated with PtPd bimetallic nanodendrites/ nanoflowers (PtPd BND/BNF@GO/Au/HRP) to increase the sensitivity and catalytic efficiency. Gold particles are capped with thionine or probe individually to improve the detection probe‡s binding and reduce the electronic background. The biosensor has provided a LOD value of 0.247 fM/mL for lncRNA HULC detection. The biosensor showed high sensitivity, stability, and better reproducibility for lncRNA HULC detection in hepatocellular carcinoma diagnosis inspite of being a complex and laborious technique.

5.5.1.1 Electrochemical Sensors for Cancer Biomarkers Detection Based on Amperometry

Different biomarkers have been reported for their use in biosensor-based cancer detection (Figure 5.2). These biomarkers include the following.

5.5.1.1.1 Embryonic Antigen Biomarkers

During embryonic development, embryonic antigens are excreted from the embryonic tissue. In the late embryonic stage, they were reduced and disappear gradually. But in the state of cancerous cells, these embryonic antigens are resynthesized sometimes and mostly expressed. Normally, it includes alpha-fetoprotein (AFP) and human chorionic gonadotropin (hCG), which are mostly utilized biomarkers in the diagnosis of cancer (Kojima et al., 2003; Wilson, 2005; Lin and Ju, 2005; Iyama et al., 2016; McGrath et al., 2016; Moreira et al., 2016; F. Liu et al., 2016; Riegler et al., 2016; Wang et al., 2016, 2018; Huang et al., 2017; Yadav et al., 2019; He et al., 2019; Cai et al., 2020; Carmicheal et al., 2020; Fattahi, Khosroushahi and Hasanzadeh, 2020; Negahdary, 2020; Solhi and Hasanzadeh, 2020; Zhang et al., 2020; Zhao et al., 2020, 2021; Zhou et al., 2021).

In a study (Ho et al., 2009), a sensitive electrochemical platform was developed for the detection of carcinoembryonic antigen (CEA), a protein tumor biomarker. This assay was based on detection with a screen-printed graphite electrode shielded by anti-CEA antibodies modified with

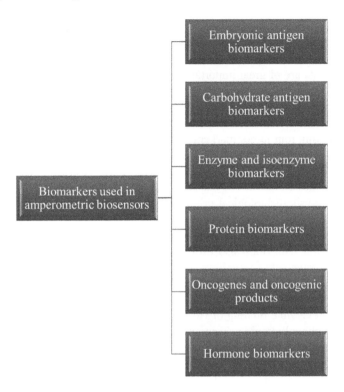

FIGURE 5.2   Different types of cancer biomarkers used in amperometric biosensors.

carbon nanoparticle/poly (ethylene imine). The signal was amplified with cadmium sulfide (CdS) nanocrystals and square wave anodic stripping voltammetry was done for the recording of current signals received from the CEA-CdS. The detection limit was found to be 32 pg/mL.

Other nanomaterials such as graphene oxide, quantum dots, etc. due to their conductive properties, have been also used as electrochemical luminescence (ECL) sensors for detection of embryonic antigens (Tang et al., 2011; Su et al., 2010; Wei et al., 2010; Fang et al., 2012; Deng et al., 2011; Wang and Hu, 2009). In the same way, the ECL sensor was developed by using immobilizing support of CdS quantum dots-alginate and cadmium selenide/zinc sulfide quantum dots as a label for AFP detection. The detection limit of this fabricated immunosensor was 20 ag/mL (Guo et al., 2012).

5.5.1.1.2 Carbohydrate Antigen Biomarkers
Carbohydrate antigens are another type of cancer antigens, where the differences in the sugar chain develop. They are produced in cancerous cells but rarely found in normal tissues. The identified carbohydrate antigens are CA125, CA19-9, CA15-3, etc.; CA125 antigens act as biomarkers for the detection of endometrial cancer, epithelial ovarian cancer, lung cancer, etc. (Kuerer, 2006; Devillers et al., 2017; Hasanzadeh et al., 2018; Honda et al., 2018; Abbas et al., 2019; Saadati et al., 2019; Abdel Ghafar et al., 2020; Sadasivam et al., 2020; Xu et al., 2020; Bertokova et al., 2021; Ge et al., 2021; Luo et al., 2021; McDowell et al., 2021; Park et al., 2021; Shekari et al., 2021; Subramani et al., 2021). CA15-3 is associated with breast and ovarian cancer diagnosis (Dai et al., 2003; Wu et al., 2008; Fu et al., 2009; Fu, 2010). In a study, an electrochemical immunosensor was developed for CA-125 detection. It was done by using a glassy carbon electrode modified with porous polythionine and gold microspheres with immobilized CA-125 (Fu, 2007). The gold microspheres help in creating the biocompatible environment for higher amplification of anti-CA125 molecules. Detection was based on the

amperometric current change signals received from antigen-antibody interactions and the detection limit was found to be 1.3 U/mL when the CA-125 in the concentration range of 4.5 to 36.5 U/mL was used.

Multibiomarker tests are of great importance in clinical diagnosis. In a study (Jia et al., 2012), label-free multiple tumor markers were developed based on a potentiometric sensor. The sensor was fabricated by combining arrayed chips of silicon nitride-silicon dioxide (Si_3N_4-SiO_2-Si) organized on silicon wafers. An l-3,4-dihydroxyphenylalanine (l-dopa) originated chip was biofunctionalized by linking four tumor markers AFP, CEA, CA19-9, and Ferritin antigens-antibodies with it for detection of unlabeled antibodies and antigens.

5.5.1.1.3 Enzyme and Isozyme Biomarkers

Enzymes are the catalysts of biological origin, which regulates biological metabolism. They are also used as tumor markers for cancer diagnosis (Ludwig and Weinstein, 2005). Among the different classes of enzymes, isozymes, despite having a different amino acid sequence, catalyze the same chemical reaction. Isozymes also differ in their immunogenicity, kinetic parameters, and other regulatory properties, etc.

The prostate-specific antigen (PSA) has been recognized as a very important and reliable tool for prostate cancer detection. In a study (Liu et al., 2008), an enzyme-based immunosensor was developed in which reversible binding of Horseradish-peroxidase (HRP)-anti-PSA was done with gold, modified with phenylboronic acid monolayers. A decrease in the electrocatalytic signals of the electrode modified with HRP-anti-PSA was observed by the reduced H_2O_2. The amperometric immunosensor has shown a linear increase in the relative intensity for two PSA concentrations ranging from 2 to 15 ng/mL and 15 to 120 ng/mL.

5.5.1.1.4 Protein Biomarkers

Protein markers associated with cancer are closely linked with the occurrence and development of cancer as well as treatment response. Therefore, for early diagnosis and targeted drug delivery for tumors, an assay of these protein markers is crucial (Maurya et al., 2007). Protein markers usually comprise Bence-Jones protein (Dhodapkar et al., 1997), epidermal growth factor (EGF) (Sharma et al., 2007), beta2 microglobulin, insulin-like growth factor (IGF), etc. (LeRoith and Roberts, 2003). Despite different studies on detection methods, there are very few electrochemical sensor-based applications reported for protein cancer marker detection. In a study (Lai et al., 2007), an aptamer-based electrochemical sensor was developed for platelet-derived growth factor (PDGF) detection in blood serum directly. The voltammetric current fluctuations resulting from PDGF-aptamer modified with methylene blue were observed. PDGF at the concentration of 1 nM was detected directly into the undiluted blood serum and 1.25 ng/mL concentration was detected after a twofold dilution of serum with a buffer. This label-free sensor is reusable.

5.5.1.1.5 Oncogene and Oncogenic Production

Any change in the functional genome sequence results in gene expression alteration. These genetic changes lead to specific epigenetic alterations, which can be used to facilitate disease treatment. For instance, after getting a particular stimulus, few cells undergoing cell death can produce specific DNA methylation patterns (Levenson and Melnikov, 2012). DNA is excreted due to circulating cancer cells lysis. Metastatic cancerous cells have the ability to circulate freely or in bound form in the blood and it is possible to detect circulating DNA in patient plasma (Board et al., 2007).

In a study, an electrochemical sensor was developed for oral cancer based on mRNA markers through the amplification of signals by the hairpin probe (HP) (Wei et al., 2008). The HP specificity without a linker was investigated through the cross-detection of two targets. In the absence of the target, an effective complex was not formed due to the closing of HP. HRP complex was formed only in the presence of a target. 3,3',5,5'-tetramethylbenzidine (TMB) was used as a substrate and causes signal amplification by regeneration of reactive form of HRP. The *p53* gene is a known tumor suppressor. Mutation in a *p53* gene commonly occurs in malignant cancers. In a study, an electrochemical sensor based on a biochip for the *p53* DNA sequence detection was developed (Marquette et al., 2006). This bioship was fabricated using the screen-printing method. Single-stranded DNA containing a 5'-C_6-NH_2- linker, covalently bonded to the surface, acted as a probe for complementary strand detection and enabled the target sequence detection from 1 to 200 nM.

5.5.1.1.6 Hormone Biomarkers

Hormones are secreted by specific cells inside the body. Due to malignancy, normal tissues start releasing peptide hormones and lead to the corresponding disease. Hence, the upsurging concentrations of these hormones can be used as tumor markers. Electrochemical detection of these hormones is mainly based on amperometry and ECL sensors (Ahmed et al., 2008; Sánchez et al., 2008; Mani et al., 2011; Moreno-Guzmán et al., 2012). In a study (Chen et al., 2005), an amperometric immunosensor was fabricated for human chorionic gonadotropin (hCG) detection in serum, by immobilizing the hCG on a glassy carbon electrode with titania gel and using electrochemistry of labeled HRP with hCG antibody. There was a conjugation of HRP anti-hCG and immobilized hCG, and it was observed that direct transfer of electron occurred through the HRP at the prepared electrode. HRP current was decreased when 2.5 to 12.5 mIU/mL of hCG was used as a result of competition between the analyte hCG and immobilized HRP-anti-hCG-hCG.

5.5.2 CHRONOAMPEROMETRY

Chronoamperometry is another type of amperometric technique, where the steady current is measured as a time function after applying a square-wave potential to the working electrode. Chronoamperometric immunosensor was used in a study for tumor necrosis factor-α (TNF-α) detection (Barhoumi et al., 2019). This technique has shown its good clinical performance in a significant concentration range with 8% precision and LOD of 0.3 pg mL−1. Similarly, in another study, the chronoamperometric method was used for MDA-MB-231 (breast cancer cell line) viability assessment (Tsai et al., 2013).

5.5.3 POTENTIOMETRY

Potentiometry measures the difference of electrical potential between the pair of electrodes under zero cell current. This approach is also utilized for cancer biomarker monitoring (Guo et al., 2011; Farzin and Shamsipur, 2018; Qian et al., 2019; Anzar et al., 2020; Ding and Qin, 2020; Ranjan et al., 2020; Karimi-Maleh et al., 2021; Rajapaksha et al., 2021; S. Zhang et al., 2021; Tabish et al., 2021; Walker and Dick, 2021). In the potentiometric measurements, the relationship between concentration and potential is directed via the Nernst equation. It proposes low LOD values, which is very crucial for the detection of cancer because, at the early stage of cancer, biomarkers are present at a very low concentration (Grieshaber et al., 2008). In a study, Light-Addressable Potentiometric Sensor (LAPS) was developed for a liver cancer biomarker, human phosphatase regenerating liver-3 (hPRL-3), detection. In this study, a linear range (LR) of 0.04−400 nM and 0−105 cells/mL was achieved for hPRL-3 and mammary adenocarcinoma cells (MDA-MB-231) detection, respectively (Jia et al., 2007). In another study, carcinoembryonic antigen (CEA) detection was performed in human colon cancer cells. In this assay, hydroxyl-terminated alkanethiol, which was self-assembled and surface molecular imprinted, was used with template biomolecules on a silicon chip coated with gold, for detection. In this work, an LR of 2.5–250 ng/mL was observed. The potentiometric biosensor was also developed for the detection of hyaluronan-linked protein 1 (HAPLN1), which is associated with malignant pleural mesothelioma. This method was label-free and the LOD in the picomolar range was observed (Mathur et al., 2013).

Molecular imprinting by utilizing artificial materials offers a substitute to the identification of a wide range of substances (Pauling, 1940; Cram, 1988; Mosbach, 1994; Piletsky et al., 2001; Haupt, 2003; Mahony et al., 2005; Zhou et al., 2005; Rao et al., 2006). The application of surface molecular imprinting by using self-assembled monolayers to design sensing elements resulted in the detection of cancer biomarkers and various other proteins. These elements comprise of a gold-coated silicon chip onto which hydroxyl-terminated alkanethiol molecules and template biomolecule are co-adsorbed and the thiol molecules are chemically conjugated to the metal substrate and self-assembled into highly ordered monolayers; the biomolecules can be detached, forming the footprint cavities in the monolayer matrix for this kind of template molecules. Re-adsorption of the biomolecules to the sensing chip alters its potential, which can be estimated potentiometrically. This method was employed to detect carcinoembryonic antigen (CEA) in the case of purified CEA and the culture medium of a CEA-producing human colon cancer cell line. The CEA assay, authenticated against a standard immunoassay, sensitive (detection range 2.5–250 ng/mL), and specific (no cross-reactivity with hemoglobin; no response by a non-imprinted sensor). Similar observations were attained for human amylase. Additionally, the virions of poliovirus in a specific manner (no cross-reactivity to adenovirus, no response by a non-imprinted sensor) were detected using the developed sensor (Wang et al., 2010). The results determine the utilization of molecular imprinting principles to create a new sensor for the identification and quantification of protein cancer biomarkers and other protein-based macro-molecular structures such as the capsid of a virion.

In another study (Solovieva et al., 2019), the developed sensors have been used to detect different cationic and anionic species in urine with high sensitivity. The system‡s response signifies a complex chemical fingerprint of a urine sample relatable to the patient status (Pascual et al., 2016) through multivariate modeling. Eighty-nine urine samples were studied (43 from cancer patients confirmed by prostatic puncture biopsy and 46 from the healthy control group). A variety of multivariate cataloging methods was applied to the potentiometric data. The accurate results were achieved with a logistic regression model yielding 100% sensitivity and 93% specificity in the independent test set of samples. The study affirms the efficiency of a potentiometric multisensor system based on cross-sensitive anion-, cation-, and RedOx-sensitive sensors in distinguishing urine samples from PCa patients (prostate cancer) and healthy control group. Utilization of logistic regression to the multisensor data enables developing a classification model with 100% sensitivity and 93% specificity as evaluated with independent test sets. The benefits of the process include overall simplicity, comparatively low price, short measuring time, and the absence of sample pretreatment stages in the developed system. Despite these positive results, a lot is needed to be accomplished for thorough authentication of the approach, specifically from the point of view of experimental design.

In a different work, a potentiometric microarray based on hybridization was developed for exosomal miRNA detection (Goda et al., 2012). Additionally, tumorous cells were also targeted in consuming potentiometric techniques for the study of neighboring micro-environment electrochemistry.

5.5.4 CHRONOPOTENTIOMETRY

Chronopotentiometry is the measurement of potential as a function of time (t) under constant current. In a study, the chronopotentiometric stripping (CPS) technique was used to examine the PSA and its relations with lectins that can identify PSA glycans present in healthy ones or prostate cancer patients (Belicky et al., 2017).

5.5.5 VOLTAMMETRY

Voltammetry is a type of electro-analytical system (Figure 5.3). In this method, analyte analysis is performed by quantifying the resultant current at a varying potential. Potential can vary in many

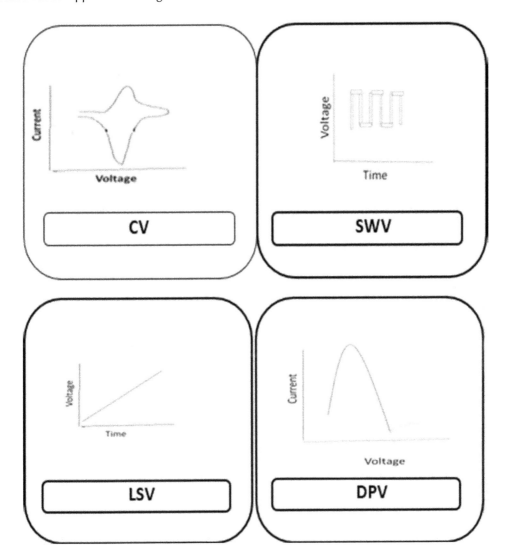

FIGURE 5.3 Voltammogram patterns of different voltammetry methods.

ways, describing various forms of voltammetry methods such as cyclic voltammetry (CV), differential pulse voltammetry (DPV), linear sweep voltammetry (LSV), square wave voltammetry (SWV), and stripping voltammetry (SV). SWV and DPV are mostly used because of imposing high sensitivity (Chandra and Segal, 2016). Voltammetry methods have also been extensively used for the detection of cancer biomarkers in various matrices including biomarkers, IL-10, HER2, CA153, PSA, Osteopontin (OPN), etc. (Hasanzadeh et al., 2017; Raji et al., 2015).

For early, fast, and more sensitive detection of miRNAs, direct hybridization and label-free voltammetric-based methods (Islam, Masud et al., 2018; Islam, Gorgannezhad et al., 2018; Kangkamano et al., 2018; Salahandish et al., 2018; Cui et al., 2019; Lu et al., 2019; Zouari et al., 2020a) are preferred. In a study (Zouari et al., 2020b), a disposable carbon electrode was used as a substrate having the nanostructures reduced graphene oxide (rGO) modified with pyrene carboxylic acid (PCA) and gold nanoparticles modified with 6-ferrocenylhexanethiol (Fc-SH). On the CO_2H fractions of PCA/rGO, an amino-terminal DNA capture probe was attached covalently as a signal molecule. Oxidation peak current reduction of Fc was recorded via differential pulse voltammetry after the hybridization process. This happened due to electron transfer hindrance upon

DNA-RNA duplex construction on the surface of the electrode. The oncogene miRNA-21 determination has permitted quantification of the target with LOD values in the femtomolar range, i.e., 5 fM, with the capability to specify single-base mismatched sequences in the incubation step of 30 minutes. This platform permitted the tiny amount of target miRNA determination from breast cancer (BC) cell lines or serum samples of breast cancer patients without the prior requirement of extraction, purification, or amplification of the genetic material.

An electrochemical DNA biosensor (Wang, 2006; Rasheed and Sandhyarani, 2014) was used for specific gene detection associated with breast cancer. It was based on the poly(amidoamine) dendrimers having self-assembled ferrocene. DNA hybridization on the gold electrode was investigated by using cyclic and differential pulse voltammetry. This biosensor has a high sensitivity of 0.13 μA/(ng/ml) for the detection range of 3–20 nM with a LOD value of 0.38 nM (Senel et al., 2019). The biosensor has high selectivity and specificity for target DNA, demonstrating the vital signal difference from non-complementary sequences and the mismatches and employed as BRCA1 gene mutation model.

CA15-3 (Cancer antigen) and HER2-ECD (extracellular domain of the human epidermal growth receptor) are very well-known breast cancer biomarkers (Bhatt et al., 2010; Duffy et al., 2010; Lam et al., 2012; Tsé et al., 2012; Marques et al., 2014; Ravelli et al., 2015; Rama and Costa-García, 2016). For follow-up and diagnosis of this disease, it can be elucidated by the concentration levels of these two biomarkers. Hence, a disposable electrochemical-based immunosensor was developed for CA 15-3 and HER2-ECD detection simultaneously in a study. This immunosensor was fabricated using a screen-printed carbon electrode. The surface of the carbon electrode was first modified with electrodeposited gold nanoparticles and then, separately, coating was done with either anti-human CA 15-3 or anti-human HER2-ECD antibody. After incubating the biomarkers and monoclonal biotin-labeled antibodies, the antigen-antibody interaction was sensed by linear sweep voltammetry of metallic silver generated by enzyme alkaline phosphatase. The LOD value of this immunosensor was observed to be a simultaneous detection of CA 15-3 and HER2-ECD. These values obtained from the sensor can help in the non-invasive regulation of these biomarkers in breast cancer patients (Marques et al., 2018).

5.5.6 IMPEDIMETRIC SENSING

The impedimetric technique is another useful method for the detection of cancer biomarkers. This method possesses high sensitivity and fast speed (An et al., 2018; Zandi et al., 2019). Most importantly, impedimetric can be utilized for real-time, on-site, and long-term detection (Parekh et al., 2018). Among the different impedance methods, electrochemical impedance spectroscopy (EIS) is the most useful method. EIS requires an excitation voltage of 5 or 10 mV, which is much less in comparison to voltammetry methods requiring ~200–600 mV for excitation (Benvidi et al., 2015). This property makes EIS a more safe detection technology. Moreover, the EIS offers multiple factors of the biosensing surface. Change of the charge transfer resistance (Ret) can be determined by using any redox couple such as ferri-ferrocyanide. Generally, Ret is inversely related to the electron transfer rate. The capacitance of the electrical double layer and Ret define dielectric feature and isolation characters of interfacial electrode-electrolyte. The broad frequency detection range (10^{-4}–10^6 Hz) makes this strategy valuable for the analysis of diffusion and kinetics characteristics.

Carcinoembryonic antigen (CEA) is a cell surface protein (Zhou et al., 2014; Kumar et al., 2015; Li et al., 2016; Li et al., 2017). For CEA detection, Liu et al. have developed an unique approach by combining the EIS and UV-Vis spectroscopy (Zeng et al., 2015). This optoelectronic immunosensor was designed using gold-ITO and thiol-functionalized-nanogold (TDN) labeled with an anti-CEA antibody. The linear range reported for EIS was 0.05–80 ng mL-1 and for UV-VIS method 0.5–80 ng mL-1. Lower LOD was observed in the case of EIS (1 pg mL-1) than the UV method (2 pg mL-1) (Zeng et al., 2015).

The sensitivity of the EIS immunosensor was improved by incorporating the combination of gold nanoparticles and biologically well-matched polymeric nanoparticles γ-PGA-DA@CS, synthesized by the self-assembly of chitosan (CS) and dopamine-modified poly (γ-glutamic acid) (γ-PGA-DA) (T. Xu et al., 2017). In the electrophoretic deposition process, γ-PGADA@ CS has covered the GCE. Afterward, AuNPs were attached on γ-PGA-DA@CS film, making an extremely stable immobilization technique for monoclonal anti-CEA antibody attachment. Using this immunosensor, EIS signal response was changed from 20 femtogram/mL to 20 nanogram/mL with LOD value of 10 fg/mL.

Zoroub et al. have prepared an EIS immunosensor for EGFR determination (Elshafey et al., 2013). The central part of the biosensor is made up of a polycrystalline gold electrode shielded with gold nanoparticles, a self-assembled amine-terminated monolayer, a linker (1,4-phenylene diisothiocyanate), and protein G. The biosensor showed a linear response range from 1 picogram/mL to 1 microgram/mL with a limit of the detection value of 0.34 picogram/mL (Elshafey et al., 2013).

HER2 (ErbB2), a cancer biomarker, is one of the most broadly used biomarkers. Its average level in healthy people is 4 to 14 ng/mL, while cancer patients rise to 14–75 ng/m. It is overexpressed in about 20–30% of breast cancer patients and is also connected with the destructive form of the disease and poor diagnosis (Sinibaldi et al., 2017; Arya et al., 2018). Its overexpression is also found in other cancers such as gastric, ovarian, and oesophageal cancers. Presently, there are two systems for analysis of HER2, i.e., immunochemistry and in-situ fluorescence hybridization (Iqbal and Iqbal, 2014).

Pradhan and colleagues (2021) have developed a new method for impedimetric biosensors involving four electrodes fabricated by photolithography. It was used for cytotoxicity evaluation of tamoxifen, a non-steroidal anti-estrogenic drug (Grenman et al., 1988) in cervical cancer cell lines. The impedance of cells was calculated by the electric cell-substrate impedance sensing (ECIS) (Xiao and Luong, 2003; Ceriotti et al., 2007; Daza et al., 2013; Gu et al., 2018) technique in the frequency range of 100 Hz to 1 MHz. The results have shown a substantial reduction in the number of HeLa cells due to tamoxifen on the surfaces of the electrode of impedimetric biosensors in a dose-dependent manner.

The impedance results obtained by this novel method were compared with the results of different predictable techniques, such as 3-(4,5-dimethylthiazol-2-yl)-2,5-diphenyl tetrazolium bromide (MTT), live-dead cell assay, and flow cytometry for estimation of cytotoxicity caused by the tamoxifen. The impedimetric cytotoxicity of tamoxifen on the HeLa cells progression and proliferation perfectly correlates with conventional methods. The IC50 value acquired from this approach was comparable to the IC50 value of the MTT assay. Furthermore, the observation obtained from the impedimetric biosensors that substantial cell death and HeLa cells detachment occur in a dose-dependent manner is also manifested through morphological analysis.

In another study, label-free and rapid impedimetric biosensor was developed to estimate cancer cells quantitatively with high glycoprotein expression. Graphene quantum dots (GQDs) of the 2–9 nm size have been distributed homogeneously on the surface of Fe_3O_4 via covalent bonding. Concovalin A (ConA), a carbohydrate-binding protein (Inbar and Sachs, 1969; Khan et al., 2016; Arai et al., 2013; Liu et al., 2020), was then embedded with GQDs by physical mixing for the fabrication of a ConA-GQD@Fe3O4 nanosensing probe. A LOD of 246 and 367 cells/mL for HeLa and MCF-7 is achieved, respectively. Using impedimetric sensing, 16.7 and 13.1 times higher signals were obtained in HeLa and MCF-7 cells than their primary sensor electrodes (Chowdhury et al., 2018).

Lee et al. (2015) developed an impedimetric biosensor for the detection the Engrailed-2 (EN2) protein (a biomarker for prostate cancer) (Bose et al., 2008; Morgan et al., 2011; Killick et al., 2013; Do Carmo Silva et al., 2020). The EN2 sturdily binds to a precise DNA sequence (5'-TAATTA-3'). The electrochemical detection of EN2 was accomplished by dipping the functionalized gold electrode in different concentrations of the EN2 protein. The working electrode was altered by the electrodeposition of gold nanoparticles, leading to an elevation in the sensitivity of the sensing device. Then, the interaction between EN2 proteins and DNA probes on the electrode was analyzed by EIS. The establishment of the complex on the electrode surface intrudes electron transfer among the solution and the electrode, leading to a higher resistance at the electrode surface. The high-resistance signal was acquired by EN2 protein due to interactions among the specific DNA sequence and EN2 protein. The enumerated detection limit was observed to be 5.62 fM. Substantially, the biosensor‡s specificity and applicability were affirmed using various proteins and an artificial urine medium. The impedance signals elevated in the cases of EN2, thus implying that the system displayed high selectivity to only EN2 (Lee et al., 2015).

5.5.7 CONDUCTOMETRIC SENSING

This technique is considered a prominent method for bioanalytical applications due to its miniaturized nature, the low voltage required for driving, and large-scale production without requiring a reference electrode. To detect the alpha-fetoprotein (AFP), a liver cancer biomarker (Mizejewski, 1994; Wong et al., 2015; Komaki et al., 2017; Nguyen et al., 2017; Attia et al., 2018; Moazeni et al., 2018; S. Yang et al., 2018) conductometric immunoassay method was developed. It was based on enzyme-doped silica beads of nanometer size. At first, urease was doped onto the silica nanoparticles by the reverse micelle method. After that, anti-AFP antibodies labeled with arginase were conjugated covalently onto the surface of the synthesized nanoparticle. A sandwich-type immunoassay was executed in an antibody-coated microplate captured with a monoclonal anti-AFP. In this assay, the recognition element used was bienzymatic-functionalized silica nanobeads. Upon introducing the L-arginine, the substrate got divided into urea and L-ornithine by the enzymatic reaction of arginase (Hermanson, 2008). Urea then broke down to ammonia (NH_3) and bicarbonate (HCO) ions via urease activity, leading to a change in the internal conductivity of the detection solution at the interdigitated conductometric transducer. The designed immune sensing method has shown excellent conductometric response towards the AFP under optimum conditions in the linear range 0.01–100 ng per mL with LOD of 4.8 pg/mL (Liang et al., 2019(b)). Significantly better reproducibility, greater specificity, and higher accuracy were developed for liver cancer patient serum samples. Liang 2019

In a study for prostate cancer marker detection, for the first time, gold screen printed electrodes coated with tetracyanoquinodimethane (TCNQ)-doped thin films of copper-MOF and $Cu_3(BTC)_2$ were used. It was a proven, highly sensitive method. The synthesized material and the prepared electrode were characterized by different instrumental techniques for analyzing their morphological and spectroscopic characters. Doping of $Cu_3(BTC)_2$ with TCNQ enhanced the conductance of base material by the magnitude of order 9 (from 10^{-12} to 10^{-3} S). This sensing platform was further improved by using antibodies to develop an immunosensor to detect prostate cancer antigen (PSA). The anti-PSA coupled with TCNQ- $Cu_3(BTC)_2$ has detected a PSA in a linear range of 0.1–100 ng/mL with a LOD value of 0.06 ng/mL. This sensor can detect the PSA specifically inspite of the presence of other proteins (Belicky et al., 2017).

REFERENCES

Abbas, M., Ahmed, A., Khan, G.J., Baig, M.M.F.A., Naveed, M., Mikrani, R., Cao, T., Naeem, S., Shi, M., and Dingding, C., 2019. Clinical evaluation of carcinoembryonic and carbohydrate antigens as cancer biomarkers to monitor palliative chemotherapy in advanced stage gastric cancer. *Current Problems in Cancer*, *43*(1), pp. 5–17.

Abdel Ghafar, M.T., Gharib, F., Al-Ashmawy, G.M., and Mariah, R.A., 2020. Serum high-temperature-required protein A2: A potential biomarker for the diagnosis of breast cancer. *Gene Reports*, 20.

Ahmed, M.U., Hossain, M.M., and Tamiya, E., 2008. Electrochemical biosensors for medical and food applications. *Electroanalysis: An International Journal Devoted to Fundamental and Practical Aspects of Electroanalysis*, *20*(6), pp. 616–626.

Akhtar, M.H., Hussain, K.K., Gurudatt, N.G., and Shim, Y.B., 2017. Detection of Ca2+-induced acetylcholine released from leukemic T-cells using an amperometric microfluidic sensor. *Biosensors and Bioelectronics*, *98*, pp. 364–370.

Altunkök, N., Biçen Ünlüer, O., Birlik Ozkütük, E., and Ersoz, A., 2021. Development of potentiometric biosensor for diagnosis of prostate cancer. *Materials Science and Engineering B: Solid-State Materials for Advanced Technology*, *263*. doi: 10.1016/j.mseb.2020.114789.

Amouzadeh Tabrizi, M., Shamsipur, M., Saber, R., and Sarkar, S., 2017. Flow injection amperometric sandwich-type aptasensor for the determination of human leukemic lymphoblast cancer cells using MWCNTs-Pdnano/PTCA/aptamer as labeled aptamer for the signal amplification. *Analytica Chimica Acta*, *985*, pp. 61–68.

Amouzadeh Tabrizi, M., Shamsipur, M., Saber, R., Sarkar, S., and Sherkatkhameneh, N., 2017. Flow injection amperometric sandwich-type electrochemical aptasensor for the determination of adenocarcinoma gastric cancer cell using aptamer-Au@Ag nanoparticles as labeled aptamer. *Electrochimica Acta*, *246*, pp. 1147–1154.

An, L., Wang, G., Han, Y., Li, T., Jin, P., and Liu, S., 2018. Electrochemical biosensor for cancer cell detection based on a surface 3D micro-array. *Lab on a Chip*, *18*(2), pp. 335–342.

Anzar, N., Rahil Hasan, M., Akram, M., Yadav, N., and Narang, J., 2020. Systematic and validated techniques for the detection of ovarian cancer emphasizing the electro-analytical approach. *Process Biochemistry*. doi: 10.1016/j.procbio.2020.04.006.

Aparicio-Martínez, E., Ibarra, A., Estrada-Moreno, I.A., Osuna, V., and Dominguez, R.B., 2019. Flexible electrochemical sensor based on laser scribed graphene/Ag nanoparticles for non-enzymatic hydrogen peroxide detection. *Sensors and Actuators, B: Chemical*, *301*. doi: 10.1016/j.snb.2019.127101.

Arai, K., Tsutsumi, H., and Mihara, H., 2013. A monosaccharide-modified peptide phage library for screening of ligands to carbohydrate-binding proteins. *Bioorganic & Medicinal Chemistry Letters*, *23*(17), pp. 4940–4943.

Arévalo, B., ben Hassine, A., Valverde, A., Serafín, V., Montero-Calle, A., Raouafi, N., Camps, J., Arenas, M., Barderas, R., Yáñez-Sedeño, P., Campuzano, S., and Pingarrón, J.M., 2021. Electrochemical immunoplatform to assist in the diagnosis and classification of breast cancer through the determination of matrix-metalloproteinase-9. *Talanta*, *225*. doi: 10.1016/j.talanta.2020.122054.

Arya, S.K., Zhurauski, P., Jolly, P., Batistuti, M.R., Mulato, M., and Estrela, P., 2018. Capacitive aptasensor based on interdigitated electrode for breast cancer detection in undiluted human serum. *Biosensors and Bioelectronics*, *102*, pp. 106–112.

Attia, M.S., Youssef, A.O., Khan, Z.A., and Abou-Omar, M.N., 2018. Alpha fetoprotein assessment by using a nano optical sensor thin film binuclear Pt-2-aminobenzimidazole-Bipyridine for early diagnosis of liver cancer. *Talanta*, *186*, pp. 36–43.

Aydin, R., Asan, A., Attar, A., and Isildak, I., 2019. Trace analysis of amines in cheese serum with liquid chromatographic potentiometric detection by using amine-selective electrode. *Arabian Journal of Chemistry*, *12*(8), pp. 4533–4540.

Azzouzi, S., Patra, H.K., Ben Ali, M., Abbas, M.N., Dridi, C., Errachid, A., and Turner, A.P.F., 2016. Citrate-selective electrochemical μ-sensor for early stage detection of prostate cancer. *Sensors and Actuators, B: Chemical*, *228*, pp. 335–346.

Barhoumi, L., Bellagambi, F.G., Vivaldi, F.M., Baraket, A., Clément, Y., Zine, N., Ben Ali, M., Elaissari, A., and Errachid, A., 2019. Ultrasensitive immunosensor array for TNF-α detection in artificial saliva using polymer-coated magnetic microparticles onto screen-printed gold electrode. *Sensors*, *19*(3), p. 692.

Belicky, S., Černocká, H., Bertok, T., Holazova, A., Réblová, K., Paleček, E., Tkac, J., and Ostatná, V., 2017. Label-free chronopotentiometric glycoprofiling of prostate specific antigen using sialic acid recognizing lectins. *Bioelectrochemistry*, *117*, pp. 89–94.

Ben Messaoud, N., Ghica, M.E., Dridi, C., Ben Ali, M., and Brett, C.M.A., 2017. Electrochemical sensor based on multiwalled carbon nanotube and gold nanoparticle modified electrode for the sensitive detection of bisphenol A. *Sensors and Actuators, B: Chemical*, *253*, pp. 513–522.

Benvidi, A., Firouzabadi, A.D., Tezerjani, M.D., Moshtaghiun, S.M., Mazloum-Ardakani, M., and Ansarin, A., 2015. A highly sensitive and selective electrochemical DNA biosensor to diagnose breast cancer. *Journal of Electroanalytical Chemistry*, *750*, pp. 57–64.

Benvidi, A., Tezerjani, M.D., Jahanbani, S., Mazloum Ardakani, M., and Moshtaghioun, S.M., 2016. Comparison of impedimetric detection of DNA hybridization on the various biosensors based on modified glassy carbon electrodes with PANHS and nanomaterials of RGO and MWCNTs. *Talanta*, *147*, pp. 621–627.

Bertokova, A., Bertok, T., Jane, E., Hires, M., Ďubjaková, P., Novotná, O., Belan, V., Fillo, J., and Tkac, J., 2021. Detection of N,N-diacetyllactosamine (LacdiNAc) containing free prostate-specific antigen for early stage prostate cancer diagnostics and for identification of castration-resistant prostate cancer patients. *Bioorganic and Medicinal Chemistry*, *39*. doi: 10.1016/j.bmc.2021.116156.

Bhardwaj, S.K., Sharma, A.L., Bhardwaj, N., Kukkar, M., Gill, A.A., Kim, K.H., and Deep, A., 2017. TCNQ-doped Cu-metal organic framework as a novel conductometric immunosensing platform for the quantification of prostate cancer antigen. *Sensors and Actuators B: Chemical*, *240*, pp. 10–17.

Bhatt, A.N., Mathur, R., Farooque, A., Verma, A., and Dwarakanath, B.S., 2010. Cancer biomarkers-current perspectives. *Indian Journal of Medical Research*, *132*(2), pp. 129–149.

Bhatt, R., Mishra, A., and Bajpai, A.K., 2021. Role of diaminonaphthalene based polymers as sensors in detection of biomolecules: A review. *Results in Materials*, *9*, pp. 100174.

Board, R.E., Knight, L., Greystoke, A., Blackhall, F.H., Hughes, A., Dive, C., and Ranson, M., 2007. DNA methylation in circulating tumour DNA as a biomarker for cancer. *Biomarker Insights*, *2*, p.117727190700200003.

Bolat, G., Akbal Vural, O., Tugce Yaman, Y., and Abaci, S., 2021. Label-free impedimetric miRNA-192 genosensor platform using graphene oxide decorated peptide nanotubes composite. *Microchemical Journal*, *166*, p. 106218.

Bolat, G., Vural, O.A., Yaman, Y.T., and Abaci, S., 2021. Polydopamine nanoparticles-assisted impedimetric sensor towards label-free lung cancer cell detection. *Materials Science and Engineering C*, *119*. doi: 10.1016/j.msec.2020.111549.

Bollella, P., Fusco, G., Tortolini, C., Sanzò, G., Favero, G., Gorton, L., and Antiochia, R., 2017. Beyond graphene: Electrochemical sensors and biosensors for biomarkers detection. *Biosensors and Bioelectronics*. doi: 10.1016/j.bios.2016.03.068.

Bose, S.K., Bullard, R.S., and Donald, C.D., 2008. Oncogenic role of engrailed-2 (en-2) in prostate cancer cell growth and survival. *Translational Oncogenomics*, *3*, p. 37.

Buerk, D.G., 2014. *Biosensors: Theory and Applications*. CRC Press.

Cai, Q., Zhang, P., He, B., Zhao, Z., Zhang, Y., Peng, X., Xie, H., and Wang, X., 2020. Identification of diagnostic DNA methylation biomarkers specific for early-stage lung adenocarcinoma. *Cancer Genetics*, *246–247*, pp. 1–11.

Carmicheal, J., Patel, A., Dalal, V., Atri, P., Dhaliwal, A.S., Wittel, U.A., Malafa, M.P., Talmon, G., Swanson, B.J., Singh, S., Jain, M., Kaur, S., and Batra, S.K., 2020. Elevating pancreatic cystic lesion stratification: Current and future pancreatic cancer biomarker(s). *Biochimica et Biophysica Acta – Reviews on Cancer*. doi: 10.1016/j.bbcan.2019.188318.

Carrara, S., Bhalla, V., Stagni, C., Benini, L., Ferretti, A., Valle, F., Gallotta, A., Riccò, B., and Samorì, B., 2009. Label-free cancer markers detection by capacitance biochip. *Sensors and Actuators B: Chemical*, *136*(1), pp. 163–172.

Carvajal, S., Fera, S.N., Jones, A.L., Baldo, T.A., Mosa, I.M., Rusling, J.F., and Krause, C.E., 2018. Disposable inkjet-printed electrochemical platform for detection of clinically relevant HER-2 breast cancer biomarker. *Biosensors and Bioelectronics*, *104*, pp. 158–162.

Ceriotti, L., Ponti, J., Broggi, F., Kob, A., Drechsler, S., Thedinga, E., Colpo, P., Sabbioni, E., Ehret, R., and Rossi, F., 2007. Real-time assessment of cytotoxicity by impedance measurement on a 96-well plate. *Sensors and Actuators B: Chemical*, *123*(2), pp. 769–778.

Cetinkaya, A., Kaya, S.I., Ozcelikay, G., Atici, E.B., and Ozkan, S.A., 2021. A molecularly imprinted electrochemical sensor based on highly selective and an ultra-trace assay of anti-cancer drug axitinib in its dosage form and biological samples. *Talanta*, *233*, p. 122569.

Chandra, P., and Segal, E., 2016. *Nanobiosensors for Personalized and Onsite Biomedical Diagnosis*. The Institution of Engineering and Technology.

Chen, J., Yan, F., Dai, Z., and Ju, H., 2005. Reagentless amperometric immunosensor for human chorionic gonadotrophin based on direct electrochemistry of horseradish peroxidase. *Biosensors and Bioelectronics*, *21*(2), pp. 330–336.

Chikkaveeraiah, B.V., Bhirde, A.A., Morgan, N.Y., Eden, H.S., and Chen, X., 2012. Electrochemical immunosensors for detection of cancer protein biomarkers. *ACS Nano*, *6*(8), pp. 6546–6561.

Cho, C.H., Kim, J.H., Kim, J., Yun, J.W., Park, T.J., and Park, J.P., 2021. Re-engineering of peptides with high binding affinity to develop an advanced electrochemical sensor for colon cancer diagnosis. *Analytica Chimica Acta*, *1146*, pp. 131–139.

Chowdhury, A.D., Ganganboina, A.B., Park, E.Y., and Doong, R.A., 2018. Impedimetric biosensor for detection of cancer cells employing carbohydrate targeting ability of Concanavalin A. *Biosensors and Bioelectronics*, *122*, pp. 95–103.

Chung, S., Chandra, P., Koo, J.P., and Shim, Y.B., 2018. Development of a bifunctional nanobiosensor for screening and detection of chemokine ligand in colorectal cancer cell line. *Biosensors and Bioelectronics*, 100, pp. 396–403.

Cram, D.J., 1988. The design of molecular hosts, guests, and their complexes (Nobel lecture). *Angewandte Chemie International Edition in English*, 27(8), pp. 1009–1020.

Cui, F., Zhou, Z., and Zhou, H.S., 2019. Measurement and analysis of cancer biomarkers based on electrochemical biosensors. *Journal of The Electrochemical Society*, 167(3), p. 037525.

Cui, X., Liu, G., and Lin, Y., 2005. Amperometric biosensors based on carbon paste electrodes modified with nanostructured mixed-valence manganese oxides and glucose oxidase. *Nanomedicine: Nanotechnology, Biology and Medicine*, 1(2), pp. 130–135.

D'Agostino, G., Alberti, G., Biesuz, R., and Pesavento, M., 2006. Potentiometric sensor for atrazine based on a molecular imprinted membrane. *Biosensors and Bioelectronics*, 22(1), pp. 145–152.

Dai, S., Zhou, Y., Cheng, G., He, P., and Fang, Y., 2020. Dual-signal electrochemical sensor for detection of cancer cells by the split primer ligation-triggered catalyzed hairpin assembly. *Talanta*, 217. doi: 10.1016/j.talanta.2020.121079.

Dai, Z., Yan, F., Chen, J., and Ju, H., 2003. Reagentless amperometric immunosensors based on direct electrochemistry of horseradish peroxidase for determination of carcinoma antigen-125. *Analytical Chemistry*, 75(20), pp. 5429–5434.

Davis, F., Hughes, M.A., Cossins, A.R., and Higson, S.P., 2007. Single gene differentiation by DNA-modified carbon electrodes using an AC impedimetric approach. *Analytical Chemistry*, 79(3), pp. 1153–1157.

Daza, P., Olmo, A., Canete, D., and Yufera, A., 2013. Monitoring living cell assays with bio-impedance sensors. *Sensors and Actuators B: Chemical*, 176, pp. 605–610.

Deepa, Nohwal, B., and Pundir, C.S., 2020. An electrochemical CD59 targeted noninvasive immunosensor based on graphene oxide nanoparticles embodied pencil graphite for detection of lung cancer. *Microchemical Journal*, 156. doi: 10.1016/j.microc.2020.104957.

Dejous, C., and Krishnan, U.M., 2021. Sensors for diagnosis of prostate cancer: Looking beyond the prostate specific antigen. *Biosensors and Bioelectronics*. doi: 10.1016/j.bios.2020.112790.

Deng, S., Hou, Z., Lei, J., Lin, D., Hu, Z., Yan, F., and Ju, H., 2011. Signal amplification by adsorption-induced catalytic reduction of dissolved oxygen on nitrogen-doped carbon nanotubes for electro-chemiluminescent immunoassay. *Chemical Communications*, 47(44), pp. 12107–12109.

Devillers, M., Ahmad, L., Korri-Youssoufi, H., and Salmon, L., 2017. Carbohydrate-based electrochemical biosensor for detection of a cancer biomarker in human plasma. *Biosensors and Bioelectronics*, 96, pp. 178–185.

Dhara, K., and Debiprosad, R.M., 2019. Review on nanomaterials-enabled electrochemical sensors for ascorbic acid detection. *Analytical Biochemistry*. doi: 10.1016/j.ab.2019.113415.

Dhodapkar, M.V., Merlini, G., and Solomon, A., 1997. Biology and therapy of immunoglobulin deposition diseases. *Hematology/Oncology Clinics of North America*, 11(1), pp. 89–110.

Díaz-Fernández, A., Miranda-Castro, R., de-los-Santos-Álvarez, N., Lobo-Castañón, M.J., and Estrela, P., 2021. Impedimetric aptamer-based glycan PSA score for discrimination of prostate cancer from other prostate diseases. *Biosensors and Bioelectronics*, 175. doi: 10.1016/j.bios.2020.112872.

Ding, J., and Qin, W., 2020. Recent advances in potentiometric biosensors. *TrAC – Trends in Analytical Chemistry*. doi: 10.1016/j.trac.2019.115803.

Ding, L.L., Ge, J.P., Zhou, W.Q., Gao, J.P., Zhang, Z.Y., and Xiong, Y., 2016. Nanogold-functionalized g-C3N4 nanohybrids for sensitive impedimetric immunoassay of prostate-specific antigen using enzymatic biocatalytic precipitation. *Biosensors and Bioelectronics*, 85, pp. 212–219.

Do Carmo Silva, J., Vesely, S., Novak, V., Luksanova, H., Prusa, R., and Babjuk, M., 2020. Is Engrailed-2 (EN2) a truly promising biomarker in prostate cancer detection?. *Biomarkers*, 25(1), pp. 34–39.

Dong, W., Ren, Y., Bai, Z., Yang, Y., Wang, Z., Zhang, C., and Chen, Q., 2018. Trimetallic AuPtPd nanocomposites platform on graphene: Applied to electrochemical detection and breast cancer diagnosis. *Talanta*, 189, pp. 79–85.

Dowling, C.M., Herranz Ors, C., and Kiely, P.A., 2014. Using real-time impedance-based assays to monitor the effects of fibroblast-derived media on the adhesion, proliferation, migration and invasion of colon cancer cells. *Bioscience Reports*, 34(4). doi: 10.1042/BSR20140031.

Duffy, M.J., Evoy, D., and McDermott, E.W., 2010. CA 15-3: Uses and limitation as a biomarker for breast cancer. *Clinica Chimica Acta*, 411(23-24), pp. 1869–1874.

Dumore, N.S., and Mukhopadhyay, M., 2020. Sensitivity enhanced SeNPs-FTO electrochemical sensor for hydrogen peroxide detection. *Journal of Electroanalytical Chemistry*, 878. doi: 10.1016/j.jelechem.2020.114544.

Eksin, E., Torul, H., Yarali, E., Tamer, U., Papakonstantinou, P., and Erdem, A., 2021. Paper-based electrode assemble for impedimetric detection of miRNA. *Talanta*, 225. doi: 10.1016/j.talanta.2020.122043.

Elshafey, R., Tavares, A.C., Siaj, M., and Zourob, M., 2013. Electrochemical impedance immunosensor based on gold nanoparticles–protein G for the detection of cancer marker epidermal growth factor receptor in human plasma and brain tissue. *Biosensors and Bioelectronics*, 50, pp. 143–149.

Ercole, C., Del Gallo, M., Pantalone, M., Santucci, S., Mosiello, L., Laconi, C., and Lepidi, A., 2002. A biosensor for Escherichia coli based on a potentiometric alternating biosensing (PAB) transducer. *Sensors and Actuators B: Chemical*, 83(1-3), pp. 48–52.

Eryiğit, M., Gür, E.P., Hosseinpour, M., Özer, T.Ö., and Doğan, H.Ö., 2021. Amperometric detection of dopamine on Prussian blue nanocube-decorated electrochemically reduced graphene oxide hybrid electrode. *Materials Today: Proceedings*. doi: 10.1016/j.matpr.2021.03.277.

Fan, X., Deng, D., Chen, Z., Qi, J., Li, Y., Han, B., Huan, K., and Luo, L., 2021. A sensitive amperometric immunosensor for the detection of carcinoembryonic antigen using ZnMn2O4@reduced graphene oxide composites as signal amplifier. *Sensors and Actuators, B: Chemical*, 339. doi: 10.1016/j.snb.2021.129852.

Fang, X., Han, M., Lu, G., Tu, W., and Dai, Z., 2012. Electrochemiluminescence of CdSe quantum dots for highly sensitive competitive immunosensing. *Sensors and Actuators B: Chemical*, 168, pp. 271–276.

Farzin, L., and Shamsipur, M., 2018. Recent advances in design of electrochemical affinity biosensors for low level detection of cancer protein biomarkers using nanomaterial-assisted signal enhancement strategies. *Journal of Pharmaceutical and Biomedical Analysis*. doi: 10.1016/j.jpba.2017.07.042.

Fattahi, Z., Khosroushahi, A.Y., and Hasanzadeh, M., 2020. Recent progress on developing of plasmon biosensing of tumor biomarkers: Efficient method towards early stage recognition of cancer. *Biomedicine and Pharmacotherapy*. doi: 10.1016/j.biopha.2020.110850.

Forster, R.J., Spain, E., and Adamson, K., 2017. Electrochemical sensing of cancer cells. *Current Opinion in Electrochemistry*, 3(1), pp. 63–67.

Fu, X., Wang, J., Li, N., Wang, L., and Pu, L., 2009. Label-free electrochemical immunoassay of carci-noembryonic antigen in human serum using magnetic nanorods as sensing probes. *Microchimica Acta*, 165(3-4), pp. 437–442.

Fu, X.H., 2007. Electrochemical immunoassay for carbohydrate antigen-125 based on polythionine and gold hollow microspheres modified glassy carbon electrodes. *Electroanalysis: An International Journal Devoted to Fundamental and Practical Aspects of Electroanalysis*, 19(17), pp. 1831–1839.

Fu, X.H., 2007. Electrochemical immunoassay for carbohydrate antigen-125 based on polythionine and gold hollow microspheres modified glassy carbon electrodes. *Electroanalysis: An International Journal Devoted to Fundamental and Practical Aspects of Electroanalysis*, 19(17), pp. 1831–1839.

Fu, X.H., 2010. Poly (amidoamine) dendrimer-functionalized magnetic beads as an immunosensing probe for electrochemical immunoassay for carbohydrate antigen-125 in human serum. *Analytical Letters*, 43(3), pp. 455–465.

Ge, X.-Y., Feng, Y.-G., Cen, S.-Y., Wang, A.-J., Mei, L.-P., Luo, X., and Feng, J.-J., 2021. A label-free electrochemical immnnunosensor based on signal magnification of oxygen reduction reaction catalyzed by uniform PtCo nanodendrites for highly sensitive detection of carbohydrate antigen 15-3. *Analytica Chimica Acta*, 338750. doi: 10.1016/j.aca.2021.338750.

Ghindilis, A.L., Atanasov, P., Wilkins, M., and Wilkins, E., 1998. Immunosensors: electrochemical sensing and other engineering approaches. *Biosensors and Bioelectronics*, 13(1), pp. 113–131.

Giannetto, M., Bianchi, M.V., Mattarozzi, M., and Careri, M., 2017. Competitive amperometric im-munosensor for determination of p53 protein in urine with carbon nanotubes/gold nanoparticles screen-printed electrodes: A potential rapid and noninvasive screening tool for early diagnosis of urinary tract carcinoma. *Analytica Chimica Acta*, 991, pp. 133–141.

Glasscott, M.W., Vannoy, K.J., Iresh Fernando, P.U.A., Kosgei, G.K., Moores, L.C., and Dick, J.E., 2020. Electrochemical sensors for the detection of fentanyl and its analogs: Foundations and recent advances. *TrAC – Trends in Analytical Chemistry*, 132. doi: 10.1016/j.trac.2020.116037.

Gliga, L.E., Iacob, B.C., Cheșcheș, B., Florea, A., Barbu-Tudoran, L., Bodoki, E., and Oprean, R., 2020. Electrochemical platform for the detection of adenosine using a sandwich-structured molecularly imprinted polymer-based sensor. *Electrochimica Acta*, 354. doi: 10.1016/j.electacta.2020.136656.

Goda, T., Masuno, K., Nishida, J., Kosaka, N., Ochiya, T., Matsumoto, A., and Miyahara, Y., 2012. A label-free electrical detection of exosomal microRNAs using microelectrode array. *Chemical Communications*, 48(98), pp. 11942–11944.

Grenman, S., Shapira, A., and Carey, T.E., 1988. In vitro response of cervical cancer cell lines CaSki, HeLa, and ME-180 to the antiestrogen tamoxifen. *Gynecologic Oncology, 30*(2), pp. 228–238.

Grieshaber, D., MacKenzie, R., Vörös, J., and Reimhult, E., 2008. Electrochemical biosensors-sensor principles and architectures. *Sensors, 8*(3), pp. 1400–1458.

Gu, A.Y., Kho, D.T., Johnson, R.H., Graham, E.S., and O'Carroll, S.J., 2018. In vitro wounding models using the electric cell-substrate impedance sensing (ECIS)-Zθ technology. *Biosensors, 8*(4), p. 90.

Guerrieri, A., Ciriello, R., Crispo, F., and Bianco, G., 2019. Detection of choline in biological fluids from patients on haemodialysis by an amperometric biosensor based on a novel anti-interference bilayer. *Bioelectrochemistry, 129*, pp. 135–143.

Guo, X.-xia, Song, Z.-jun, Sun, J.-juan, and Song, J.-feng, 2011. Interaction of calf thymus dsDNA with anti-tumor drug tamoxifen studied by zero current potentiometry. *Biosensors and Bioelectronics, 26*(10), pp. 4001–4005.

Guo, Z., Hao, T., Wang, S., Gan, N., Li, X., and Wei, D., 2012. Electrochemiluminescence immunosensor for the determination of ag alpha fetoprotein based on energy scavenging of quantum dots. *Electrochemistry Communications, 14*(1), pp. 13–16.

Hasanzadeh, M., Rahimi, S., Solhi, E., Mokhtarzadeh, A., Shadjou, N., Soleymani, J., and Mahboob, S., 2018. Probing the antigen-antibody interaction towards ultrasensitive recognition of cancer biomarker in adenocarcinoma cell lysates using layer-by-layer assembled silver nano-cubics with porous structure on cysteamine caped GQDs. *Microchemical Journal, 143*, pp. 379–392.

Hasanzadeh, M., Shadjou, N., and de la Guardia, M., 2017. Early stage screening of breast cancer using electrochemical biomarker detection. *TrAC – Trends in Analytical Chemistry, 91*, pp. 67–76.

Hassanpour, S., and Hasanzadeh, M., 2021. Label-free electrochemical-immunoassay of cancer biomarkers: Recent progress and challenges in the efficient diagnosis of cancer employing electroanalysis and based on point of care (POC). *Microchemical Journal, 168*, p.106424.

Haupt, K., 2003. Imprinted polymers – tailor-made mimics of antibodies and receptors. *Chemical Communications*, (2), pp. 171–178.

He, L., Li, Z., Guo, C., Hu, B., Wang, M., Zhang, Z., and Du, M., 2019. Bifunctional bioplatform based on NiCo Prussian blue analogue: Label-free impedimetric aptasensor for the early detection of carcinoembryonic antigen and living cancer cells. *Sensors and Actuators, B: Chemical, 298*. doi: 10.1016/j.snb.2019.126852.

He, X., Yuan, R., Chai, Y., and Shi, Y., 2008. A sensitive amperometric immunosensor for carcinoembryonic antigen detection with porous nanogold film and nano-Au/chitosan composite as immobilization matrix. *Journal of Biochemical And Biophysical Methods, 70*(6), pp. 823–829.

Hermanson, G.T., 2008. *Bioconjugate Techniques; Pierce Biotechnology*. Thermo Fisher Scientific, Rockford, IL.

Ho, J.A.A., Lin, Y.C., Wang, L.S., Hwang, K.C., and Chou, P.T., 2009. Carbon nanoparticle-enhanced immunoelectrochemical detection for protein tumor marker with cadmium sulfide biotracers. *Analytical Chemistry, 81*(4), pp. 1340–1346.

Honda, K., Katzke, V., Hüsing, A., Okaya, S., Shoji, H., Onidani, K., Canzian, F., and Kaaks, R., 2018. Carbohydrate antigen 19-9 and apolipoprotein A2 isoform as early detection biomarkers for pancreatic cancer: A prospective evaluation by the EPIC study. *Annals of Oncology, 29*, p. viii41.

Huang, S.J., Kannaiyan, S., Venkatesh, K., Cheemalapati, S., Haidyrah, A.S., Ramaraj, S.K., Yang, C.C., and Karuppiah, C., 2021. Synthesis and fabrication of Ni-SiO2 nanosphere-decorated multilayer graphene nanosheets composite electrode for highly sensitive amperometric determination of guaifenesin drug. *Microchemical Journal, 167*. doi: 10.1016/j.microc.2021.106325.

Huang, Y., Tang, C., Liu, J., Cheng, J., Si, Z., Li, T., and Yang, M., 2017. Signal amplification strategy for electrochemical immunosensing based on a molybdophosphate induced enhanced redox current on the surface of hydroxyapatite nanoparticles. *Microchimica Acta, 184*(3), pp. 855–861.

Ibrahim, H., and Temerk, Y., 2021. A novel electrochemical sensor based on functionalized glassy carbon microparticles@CeO2 core–shell for ultrasensitive detection of breast anticancer drug exemestane in patient plasma and pharmaceutical dosage form. *Microchemical Journal, 167*, p. 106264.

Inbar, M., and Sachs, L., 1969. Interaction of the carbohydrate-binding protein concanavalin A with normal and transformed cells. *Proceedings of the National Academy of sciences, 63*(4), pp. 1418–1425.

Incebay, H., Aktepe, L., and Leblebici, Z., 2020. An electrochemical sensor based on green tea extract for detection of Cd(II) ions by differential pulse anodic stripping voltammetry. *Surfaces and Interfaces, 21*. doi: 10.1016/j.surfin.2020.100726.

Iqbal, N., and Iqbal, N., 2014. Human epidermal growth factor receptor 2 (HER2) in cancers: Overexpression and therapeutic implications. *Molecular Biology International, 2014.* doi: 10.1155/2014/852748.

Islam, M.N., Gorgannezhad, L., Masud, M.K., Tanaka, S., Hossain, M., Al, S., Yamauchi, Y., Nguyen, N.T., and Shiddiky, M.J., 2018. Graphene-oxide-loaded superparamagnetic iron oxide nanoparticles for ultrasensitive electrocatalytic detection of microRNA.*ChemElectroChem, 5*(17). pp. 2488–2495.

Islam, M.N., Masud, M.K., Nguyen, N.T., Gopalan, V., Alamri, H.R., Alothman, Z.A., Al Hossain, M.S., Yamauchi, Y., Lamd, A.K., and Shiddiky, M.J., 2018. Gold-loaded nanoporous ferric oxide nanocubes for electrocatalytic detection of microRNA at attomolar level. *Biosensors and Bioelectronics, 101,* pp. 275–281.

Iyama, S., Ono, M., Kawai-Nakahara, H., Husni, R.E., Dai, T., Shiozawa, T., Sakata, A., Kohrogi, H., and Noguchi, M., 2016. Drebrin: A new oncofetal biomarker associated with prognosis of lung adenocarcinoma. *Lung Cancer, 102,* pp. 74–81.

Jalalvand, A.R., Goicoechea, H.C., and Gu, H.W., 2019. An interesting strategy devoted to fabrication of a novel and high-performance amperometric sodium dithionite sensor. *Microchemical Journal, 144,* pp. 6–12.

Jia, Y., Qin, M., Zhang, H., Niu, W., Li, X., Wang, L., Li, X., Bai, Y., Cao, Y., and Feng, X., 2007. Label-free biosensor: A novel phage-modified Light Addressable Potentiometric Sensor system for cancer cell monitoring. *Biosensors and Bioelectronics, 22*(12), pp. 3261–3266.

Jia, Y.F., Gao, C.Y., He, J., Feng, D.F., Xing, K.L., Wu, M., Liu, Y., Cai, W.S., and Feng, X.Z., 2012. Unlabeled multi tumor marker detection system based on bioinitiated light addressable potentiometric sensor. *Analyst, 137*(16), pp. 3806–3813.

Jolly, P., Zhurauski, P., Hammond, J.L., Miodek, A., Liébana, S., Bertok, T., Tkáč, J., and Estrela, P., 2017. Self-assembled gold nanoparticles for impedimetric and amperometric detection of a prostate cancer biomarker. *Sensors and Actuators, B: Chemical, 251,* pp. 637–643.

Joo, J., Kwon, D., Yim, C., and Jeon, S., 2012. Highly sensitive diagnostic assay for the detection of protein biomarkers using microresonators and multifunctional nanoparticles. *ACS Nano, 6*(5), pp. 4375–4381.

Kangkamano, T., Numnuam, A., Limbut, W., Kanatharana, P., Vilaivan, T., and Thavarungkul, P., 2018. Pyrrolidinyl PNA polypyrrole/silver nanofoam electrode as a novel label-free electrochemical miRNA-21 biosensor. *Biosensors and Bioelectronics, 102,* pp. 217–225.

Kara, P., Erzurumlu, Y., Kirmizibayrak, P.B., and Ozsoz, M., 2016. Electrochemical aptasensor design for label free cytosensing of human non-small cell lung cancer. *Journal of Electroanalytical Chemistry, 775,* pp. 337–341.

Karakuş, E., Erden, P.E., and Pekyardımcı, S., 2006. Determination of creatine in commercial creatine powder with new potentiometric and amperometric biosensors. *Artificial Cells, Blood Substitutes, and Biotechnology, 34*(3), pp. 337–347.

Karimi-Maleh, H., Orooji, Y., Karimi, F., Alizadeh, M., Baghayeri, M., Rouhi, J., Tajik, S., Beitollahi, H., Agarwal, S., Gupta, V.K., Rajendran, S., Ayati, A., Fu, L., Sanati, A.L., Tanhaei, B., Sen, F., shabani-nooshabadi, M., Asrami, P.N., and Al-Othman, A., 2021. A critical review on the use of potentiometric based biosensors for biomarkers detection. *Biosensors and Bioelectronics.* doi: 10.1016/j.bios.2021.113252.

Khan, J.M., Khan, M.S., Ali, M.S., Al-Shabib, N.A., and Khan, R.H., 2016. Cetyltrimethylammonium bromide (CTAB) promote amyloid fibril formation in carbohydrate binding protein (concanavalin A) at physiological pH. *RSC Advances, 6*(44), pp. 38100–38111.

Killick, E., Morgan, R., Launchbury, F., Bancroft, E., Page, E., Castro, E., Kote-Jarai, Z., Aprikian, A., Blanco, I., Clowes, V., and Domchek, S., 2013. Role of Engrailed-2 (EN2) as a prostate cancer detection biomarker in genetically high risk men. *Scientific reports, 3*(1), pp. 1–5.

Kitade, T., Kitamura, K., Konishi, T., Takegami, S., Okuno, T., Ishikawa, M., Wakabayashi, M., Nishikawa, K., and Muramatsu, Y., 2004. Potentiometric immunosensor using artificial antibody based on molecularly imprinted polymers. *Analytical Chemistry, 76*(22), pp. 6802–6807.

Kojima, K., Hiratsuka, A., Suzuki, H., Yano, K., Ikebukuro, K., and Karube, I., 2003. Electrochemical protein chip with arrayed immunosensors with antibodies immobilized in a plasma-polymerized film. *Analytical Chemistry, 75*(5), pp. 1116–1122.

Komaki, Y., Komaki, F., Micic, D., Ido, A., and Sakuraba, A., 2017. Risk of colorectal cancer in chronic liver diseases: a systematic review and meta-analysis. *Gastrointestinal endoscopy, 86*(1), pp. 93–104.

Kongsuphol, P., Lee, G.C.F., Arya, S.K., Chiam, S.Y., and Park, M.K., 2017. Miniaturized electrophoresis electrochemical protein sensor (MEEPS) for multiplexed protein detections. *Sensors and Actuators, B: Chemical, 244,* pp. 823–830.

Korpan, Y.I., Raushel, F.M., Nazarenko, E.A., Soldatkin, A.P., Jaffrezic-Renault, N., and Martelet, C., 2006. Sensitivity and specificity improvement of an ion sensitive field effect transistors-based biosensor for potato glycoalkaloids detection. *Journal Of Agricultural And Food Chemistry*, *54*(3), pp. 707–712.

Krejcova, G., Kuca, K., and Sevelova, L., 2005. Cyclosarin – an organophosphate nerve agent. *Defence Science Journal*, *55*(2), p. 105.

Kuerer, H.M., 2006. Thomsen-Friedenreich and Tn antigens in nipple fluid: Carbohydrate biomarkers for breast cancer detection. *Breast Diseases*, *17*(2), p.156.

Kumar, P., Narwal, V., Jaiwal, R., and Pundir, C.S., 2018. Construction and application of amperometric sarcosine biosensor based on SOxNPs/AuE for determination of prostate cancer. *Biosensors and Bioelectronics*, *122*, pp. 140–146.

Kumar, S., Willander, M., Sharma, J.G., and Malhotra, B.D., 2015. A solution processed carbon nanotube modified conducting paper sensor for cancer detection. *Journal of Materials Chemistry B*, *3*(48), pp. 9305–9314.

Lai, R.Y., Plaxco, K.W., and Heeger, A.J., 2007. Aptamer-based electrochemical detection of picomolar platelet-derived growth factor directly in blood serum. *Analytical Chemistry*, *79*(1), pp. 229–233.

Lam, L., McAndrew, N., Yee, M., Fu, T., Tchou, J.C., and Zhang, H., 2012. Challenges in the clinical utility of the serum test for HER2 ECD. *Biochimica et Biophysica Acta (BBA)-Reviews on Cancer*, *1826*(1), pp. 199–208.

Lee, S., Jo, H., Her, J., Lee, H.Y., and Ban, C., 2015. Ultrasensitive electrochemical detection of engrailed-2 based on homeodomain-specific DNA probe recognition for the diagnosis of prostate cancer. *Biosensors and Bioelectronics*, *66*, pp. 32–38.

Lee, W.C., Kim, K.B., Gurudatt, N.G., Hussain, K.K., Choi, C.S., Park, D.S., and Shim, Y.B., 2019. Comparison of enzymatic and non-enzymatic glucose sensors based on hierarchical Au-Ni alloy with conductive polymer. *Biosensors and Bioelectronics*, *130*, pp. 48–54.

LeRoith, D., and Roberts Jr, C.T., 2003. The insulin-like growth factor system and cancer. *Cancer Letters*, *195*(2), pp. 127–137.

Levenson, V.V., and Melnikov, A.A., 2012. DNA methylation as clinically useful biomarkers – light at the end of the tunnel. *Pharmaceuticals*, *5*(1), pp. 94–113.

Li, H., He, J., Li, S., and Turner, A.P., 2013. Electrochemical immunosensor with N-doped graphene-modified electrode for label-free detection of the breast cancer biomarker CA 15-3. *Biosensors and Bioelectronics*, *43*, pp. 25–29.

Li, L., Liu, D., Wang, K., Mao, H., and You, T., 2017. Quantitative detection of nitrite with N-doped graphene quantum dots decorated N-doped carbon nanofibers composite-based electrochemical sensor. *Sensors and Actuators, B: Chemical*, *252*, pp. 17–23.

Li, Y., Chen, Y., Deng, D., Luo, L., He, H., and Wang, Z., 2017. Water-dispersible graphene/amphiphilic pyrene derivative nanocomposite: High AuNPs loading capacity for CEA electrochemical immunosensing. *Sensors and Actuators B: Chemical*, *248*, pp. 966–972.

Li, Y., Zhang, Z., Zhang, Y., Deng, D., Luo, L., Han, B., and Fan, C., 2016. Nitidine chloride-assisted bio-functionalization of reduced graphene oxide by bovine serum albumin for impedimetric immunosensing. *Biosensors and Bioelectronics*, *79*, pp. 536–542.

Liang, H., Xu, H., Zhao, Y., Zheng, J., Zhao, H., Li, G., and Li, C.P., 2019(a). Ultrasensitive electrochemical sensor for prostate specific antigen detection with a phosphorene platform and magnetic covalent organic framework signal amplifier. *Biosensors and Bioelectronics*, *144*. doi: 10.1016/j.bios.2019. 111691.

Liang, J., Wang, J., Zhang, L., Wang, S., Yao, C., and Zhang, Z., 2019(b). Conductometric immunoassay of alpha-fetoprotein in sera of liver cancer patients using bienzyme-functionalized nanometer-sized silica beads. *Analyst*, *144*(1), pp. 265–273.

Liang, X., Han, H., and Ma, Z., 2019(c). pH responsive amperometric immunoassay for carcinoma antigen 125 based on hollow polydopamine encapsulating methylene blue. *Sensors and Actuators, B: Chemical*, *290*, pp. 625–630.

Lim, J.M., Ryu, M.Y., Yun, J.W., Park, T.J., and Park, J.P., 2017. Electrochemical peptide sensor for diagnosing adenoma-carcinoma transition in colon cancer. *Biosensors and Bioelectronics*, *98*, pp. 330–337.

Lin, J., and Ju, H., 2005. Electrochemical and chemiluminescent immunosensors for tumor markers. *Biosensors and Bioelectronics*, *20*(8), pp. 1461–1470.

Liu, F., Xiang, G., Jiang, D., Zhang, L., Chen, X., Liu, L., Luo, F., Li, Y., Liu, C., and Pu, X., 2015. Ultrasensitive strategy based on PtPd nanodendrite/nano-flower-like@ GO signal amplification for the detection of long non-coding RNA. *Biosensors and Bioelectronics*, *74*, pp. 214–221.

Liu, F., Zhang, H., Wu, Z., Dong, H., Zhou, L., Yang, D., Ge, Y., Jia, C., Liu, H., Jin, Q., Zhao, J., Zhang, Q., and Mao, H., 2016. Highly sensitive and selective lateral flow immunoassay based on magnetic nanoparticles for quantitative detection of carcinoembryonic antigen. *Talanta, 161*, pp. 205–210.

Liu, J., Cai, J., Chen, H., Zhang, S., and Kong, J., 2016. A label-free impedimetric cytosensor based on galactosylated gold-nanoisland biointerfaces for the detection of liver cancer cells in whole blood. *Journal of Electroanalytical Chemistry, 781*, pp. 103–108.

Liu, J.X., Liang, X.L., Chen, F., and Ding, S.N., 2019. Ultrasensitive amperometric cytosensor for drug evaluation with monitoring early cell apoptosis based on Cu2O@PtPd nanocomposite as signal amplified label. *Sensors and Actuators, B: Chemical, 300*. doi: 10.1016/j.snb.2019.127046.

Liu, S., Zhang, X., Wu, Y., Tu, Y., and He, L., 2008. Prostate-specific antigen detection by using a reusable amperometric immunosensor based on reversible binding and leasing of HRP-anti-PSA from phenylboronic acid modified electrode. *Clinica Chimica Acta, 395*(1-2), pp. 51–56.

Liu, Y.M., Shahed-Al-Mahmud, M., Chen, X., Chen, T.H., Liao, K.S., Lo, J.M., Wu, Y.M., Ho, M.C., Wu, C.Y., Wong, C.H., and Jan, J.T., 2020. A carbohydrate-binding protein from the edible Lablab beans effectively blocks the infections of influenza viruses and SARS-CoV-2. *Cell Reports, 32*(6), p.108016.

Lotfi Zadeh Zhad, H.R., Rodríguez Torres, Y.M., and Lai, R.Y., 2017. A reagentless and reusable electrochemical aptamer-based sensor for rapid detection of Cd(II). *Journal of Electroanalytical Chemistry, 803*, pp. 89–94.

Lu, J., Wang, J., Hu, X., Gyimah, E., Yakubu, S., Wang, K., Wu, X., and Zhang, Z., 2019. Electrochemical biosensor based on tetrahedral DNA nanostructures and G-quadruplex–hemin conformation for the ultrasensitive detection of microRNA-21 in serum. *Analytical Chemistry, 91*(11), pp. 7353–7359.

Ludwig, J.A., and Weinstein, J.N., 2005. Biomarkers in cancer staging, prognosis and treatment selection. *Nature Reviews Cancer, 5*(11), pp. 845–856.

Luo, G., Jin, K., Deng, S., Cheng, H., Fan, Z., Gong, Y., Qian, Y., Huang, Q., Ni, Q., Liu, C., and Yu, X., 2021. Roles of CA19-9 in pancreatic cancer: Biomarker, predictor and promoter. *Biochimica et Biophysica Acta – Reviews on Cancer*. doi: 10.1016/j.bbcan.2020.188409.

Magner, E., 1998. Trends in electrochemical biosensors. *Analyst, 123*(10), pp. 1967–1970.

Mahony, J.O., Nolan, K., Smyth, M.R., and Mizaikoff, B., 2005. Molecularly imprinted polymers – potential and challenges in analytical chemistry. *Analytica Chimica Acta, 534*(1), pp. 31–39.

Maity, D., and Kumar, R.T.R., 2019. Highly sensitive amperometric detection of glutamate by glutamic oxidase immobilized Pt nanoparticle decorated multiwalled carbon nanotubes(MWCNTs)/polypyrrole composite. *Biosensors and Bioelectronics, 130*, pp. 307–314.

Mani, V., Chikkaveeraiah, B.V., and Rusling, J.F., 2011. Magnetic particles in ultrasensitive biomarker protein measurements for cancer detection and monitoring. *Expert Opinion on Medical Diagnostics, 5*(5), pp. 381–391.

Marques, R.C., Costa-Rama, E., Viswanathan, S., Nouws, H.P., Costa-García, A., Delerue-Matos, C., and González-García, M.B., 2018. Voltammetric immunosensor for the simultaneous analysis of the breast cancer biomarkers CA 15-3 and HER2-ECD. *Sensors and Actuators B: Chemical, 255*, pp. 918–925.

Marques, R.C., Viswanathan, S., Nouws, H.P., Delerue-Matos, C., and González-García, M.B., 2014. Electrochemical immunosensor for the analysis of the breast cancer biomarker HER2 ECD. *Talanta, 129*, pp. 594–599.

Marquette, C.A., Lawrence, M.F., and Blum, L.J., 2006. DNA covalent immobilization onto screen-printed electrode networks for direct label-free hybridization detection of p53 sequences. *Analytical Chemistry, 78*(3), pp. 959–964.

Marzouk, S.A., Ashraf, S.S., and Al Tayyari, K.A., 2007. Prototype amperometric biosensor for sialic acid determination. *Analytical Chemistry, 79*(4), pp. 1668–1674.

Mathur, A., Blais, S., Goparaju, C.M., Neubert, T., Pass, H., and Levon, K., 2013. Development of a biosensor for detection of pleural mesothelioma cancer biomarker using surface imprinting. *PLoS One, 8*(3). doi: 10.1371/journal.pone.0057681.

Maurya, P., Meleady, P., Dowling, P., and Clynes, M., 2007. Proteomic approaches for serum biomarker discovery in cancer. *Anticancer Research, 27*(3A), pp. 1247–1255.

McDowell, C.T., Klamer, Z., Hall, J., West, C.A., Wisniewski, L., Powers, T.W., Angel, P.M., Mehta, A.S., Lewin, D.N., Haab, B.B., and Drake, R.R., 2021. Imaging mass spectrometry and lectin analysis of n-linked glycans in carbohydrate antigen-defined pancreatic cancer tissues. *Molecular and Cellular Proteomics, 20*. doi: 10.1074/mcp.RA120.002256.

McGrath, S., Christidis, D., Perera, M., Hong, S.K., Manning, T., Vela, I., and Lawrentschuk, N., 2016. Prostate cancer biomarkers: Are we hitting the mark? *Prostate International*. doi: 10.1016/j.prnil.2016.07.002.

Medici, S., Peana, M., Coradduzza, D., and Zoroddu, M.A., 2021. Gold nanoparticles and cancer: Detection, diagnosis and therapy. *Seminars in Cancer Biology*. doi: 10.1016/j.semcancer.2021.06.017.

Mehrvar, M., and Abdi, M., 2004. Recent developments, characteristics, and potential applications of electrochemical biosensors. *Analytical Sciences*, 20(8), pp. 1113–1126.

Mizejewski, G.J., 1994. Alpha-fetoprotein binding proteins: Implications for transmembrane passage and subcellular localization. *Life Sciences*, 56(1), pp. 1–9.

Moazeni, M., Karimzadeh, F., and Kermanpur, A., 2018. Peptide modified paper based impedimetric immunoassay with nanocomposite electrodes as a point-of-care testing of Alpha-fetoprotein in human serum. *Biosensors and Bioelectronics*, 117, pp. 748–757.

Moreira, F.T.C., Ferreira, M.J.M.S., Puga, J.R.T., and Sales, M.G.F., 2016. Screen-printed electrode produced by printed-circuit board technology. Application to cancer biomarker detection by means of plastic antibody as sensing material. *Sensors and Actuators, B: Chemical*, 223, pp. 927–935.

Moreno-Guzmán, M., González-Cortés, A., Yáñez-Sedeño, P., and Pingarrón, J.M., 2012. Multiplexed ultrasensitive determination of adrenocorticotropin and cortisol hormones at a dual electrochemical immunosensor. *Electroanalysis*, 24(5), pp. 1100–1108.

Morgan, R., Boxall, A., Bhatt, A., Bailey, M., Hindley, R., Langley, S., Whitaker, H.C., Neal, D.E., Ismail, M., Whitaker, H., and Annels, N., 2011. Engrailed-2 (EN2): A tumor specific urinary biomarker for the early diagnosis of prostate cancer. *Clinical Cancer Research*, 17(5), pp. 1090–1098.

Mosbach, K., 1994. Molecular imprinting. *Trends in Biochemical Sciences*, 19(1), pp. 9–14.

Musa, A.M., Kiely, J., Luxton, R., and Honeychurch, K.C., 2021. Recent progress in screen-printed electrochemical sensors and biosensors for the detection of estrogens. *TrAC – Trends in Analytical Chemistry*.

Narwal, V., Kumar, P., Joon, P., and Pundir, C.S., 2018. Fabrication of an amperometric sarcosine biosensor based on sarcosine oxidase/chitosan/CuNPs/c-MWCNT/Au electrode for detection of prostate cancer. *Enzyme and Microbial Technology*, 113, pp. 44–51.

Negahdary, M., 2020. Aptamers in nanostructure-based electrochemical biosensors for cardiac biomarkers and cancer biomarkers: A review. *Biosensors and Bioelectronics*, 152. doi: 10.1016/j.bios.2020.112018.

Negi, S., Mittal, P., Kumar, B., and Juneja, P.K., 2019. Organic LED based light sensor for detection of ovarian cancer. *Microelectronic Engineering*, 218. doi: 10.1016/j.mee.2019.111154.

Nemčeková, K., and Labuda, J., 2021. Advanced materials-integrated electrochemical sensors as promising medical diagnostics tools: A review. *Materials Science and Engineering C*.

Newman, J.D., and Setford, S.J., 2006. Enzymatic biosensors. *Molecular biotechnology*, 32(3), pp. 249–268.

Nguyen, K.H., Nguyen, A.H., and Dabir, D.V., 2017. Clinical implications of augmenter of liver regeneration in cancer: A systematic review. *Anticancer Research*, 37(7), pp. 3379–3383.

Ouerghi, O., Touhami, A., Jaffrezic-Renault, N., Martelet, C., Ouada, H.B., and Cosnier, S., 2002. Impedimetric immunosensor using avidin–biotin for antibody immobilization. *Bioelectrochemistry*, 56(1-2), pp. 131–133.

Pacheco, J.G., Rebelo, P., Freitas, M., Nouws, H.P.A., and Delerue-Matos, C., 2018. Breast cancer biomarker (HER2-ECD) detection using a molecularly imprinted electrochemical sensor. *Sensors and Actuators, B: Chemical*, 273, pp. 1008–1014.

Parekh, A., Das, D., Das, S., Dhara, S., Biswas, K., Mandal, M., and Das, S., 2018. Bioimpedimetric analysis in conjunction with growth dynamics to differentiate aggressiveness of cancer cells. *Scientific Reports*, 8(1), pp. 1–10.

Park, S.H., Shin, J.H., Jung, K.U., and Lee, S.R., 2021. Prognostic value of carcinoembryonic antigen and carbohydrate antigen 19–9 in periampullary cancer patients receiving pancreaticoduodenectomy. *Asian Journal of Surgery*, 44(6), pp. 829–835.

Pascual, L., Campos, I., Vivancos, J.L., Quintás, G., Loras, A., Martínez-Bisbal, M.C., Martínez-Máñez, R., Boronat, F., and Ruiz-Cerdà, J.L., 2016. Detection of prostate cancer using a voltammetric electronic tongue. *Analyst*, 141(15), pp. 4562–4567.

Pauling, L., 1940. A theory of the structure and process of formation of antibodies. *Journal of the American Chemical Society*, 62(10), pp. 2643–2657.

Piletsky, S.A., Alcock, S., and Turner, A.P., 2001. Molecular imprinting: At the edge of the third millennium. *TRENDS in Biotechnology*, 19(1), pp. 9–12.

Pohanka, M., and Skládal, P., 2007. Serological diagnosis of tularemia in mice using the amperometric immunosensor. *Electroanalysis: An International Journal Devoted to Fundamental and Practical Aspects of Electroanalysis*, 19(24), pp. 2507–2512.

Pohanka, M., Jun, D., and Kuca, K., 2007. Amperometric biosensor for evaluation of competitive cholinesterase inhibition by the reactivator HI-6. *Analytical Letters*, 40(12), pp. 2351–2359.

Pradhan, R., Kalkal, A., Jindal, S., Packirisamy, G., and Manhas, S., 2021. Four electrode-based impedimetric biosensors for evaluating cytotoxicity of tamoxifen on cervical cancer cells. *RSC Advances*, *11*(2), pp. 798–806.

Promphet, N., Ummartyotin, S., Ngeontae, W., Puthongkham, P., and Rodthongkum, N., 2021. Non-invasive wearable chemical sensors in real-life applications. *Analytica Chimica Acta*, *338643*.

Pursey, J.P., Chen, Y., Stulz, E., Park, M.K., and Kongsuphol, P., 2017. Microfluidic electrochemical multiplex detection of bladder cancer DNA markers. *Sensors and Actuators, B: Chemical*, *251*, pp. 34–39.

Qian, L., Durairaj, S., Prins, S., and Chen, A., 2021. Nanomaterial-based electrochemical sensors and biosensors for the detection of pharmaceutical compounds. *Biosensors and Bioelectronics*, *175*. doi: 10.1016/j.bios.2020.112836.

Qian, L., Li, Q., Baryeh, K., Qiu, W., Li, K., Zhang, J., Yu, Q., Xu, D., Liu, W., Brand, R.E., Zhang, X., Chen, W., and Liu, G., 2019. Biosensors for early diagnosis of pancreatic cancer: A review. *Translational Research*. doi: 10.1016/j.trsl.2019.08.002

Rajaji, U., Muthumariyappan, A., Chen, S.M., Chen, T.W., and Ramalingam, R.J., 2019. A novel electrochemical sensor for the detection of oxidative stress and cancer biomarker (4-nitroquinoline N-oxide) based on iron nitride nanoparticles with multilayer reduced graphene nanosheets modified electrode. *Sensors and Actuators, B: Chemical*, *291*, pp. 120–129.

Rajapaksha, R.D.A.A., Hashim, U., Gopinath, S.C.B., Parmin, N.A., and Fernando, C.A.N., 2021. Nanoparticles in electrochemical bioanalytical analysis. *In: Nanoparticles in Analytical and Medical Devices* (pp. 83–112). Elsevier.

Raji, M.A., Amoabediny, G., Tajik, P., Hosseini, M., and Ghafar-Zadeh, E., 2015. An apta-biosensor for colon cancer diagnostics. *Sensors*, *15*(9), pp. 22291–22303.

Rama, E.C., and Costa-García, A., 2016. Screen-printed electrochemical immunosensors for the detection of cancer and cardiovascular biomarkers. *Electroanalysis*, *28*(8), pp. 1700–1715.

Ranjan, P., Parihar, A., Jain, S., Kumar, N., Dhand, C., Murali, S., Mishra, D., Sanghi, S.K., Chaurasia, J.P., Srivastava, A.K., and Khan, R., 2020. Biosensor-based diagnostic approaches for various cellular biomarkers of breast cancer: A comprehensive review. *Analytical Biochemistry*. doi: 10.1016/j.ab.2020.113996.

Rao, T.P., Kala, R., and Daniel, S., 2006. Metal ion-imprinted polymers – novel materials for selective recognition of inorganics. *Analytica Chimica Acta*, *578*(2), pp. 105–116.

Rasheed, P.A., and Sandhyarani, N., 2014. Graphene-DNA electrochemical sensor for the sensitive detection of BRCA1 gene. *Sensors and Actuators B: Chemical*, *204*, pp. 777–782.

Ravelli, A., Reuben, J.M., Lanza, F., Anfossi, S., Cappelletti, M.R., Zanotti, L., Gobbi, A., Senti, C., Brambilla, P., Milani, M., and Spada, D., 2015. Solid Tumor Working Party of European Blood and Marrow Transplantation Society (EBMT). *Breast cancer circulating biomarkers: advantages, drawbacks, and new insights. Tumour Biology*, *36*(9), pp. 6653–6665.

Raziq, A., Kidakova, A., Boroznjak, R., Reut, J., Öpik, A., and Syritski, V., 2021. Development of a portable MIP-based electrochemical sensor for detection of SARS-CoV-2 antigen. *Biosensors and Bioelectronics*, *178*. doi: 10.1016/j.bios.2021.113029.

Riegler, J., Ebert, A., Qin, X., Shen, Q., Wang, M., Ameen, M., Kodo, K., Ong, S.G., Lee, W.H., Lee, G., Neofytou, E., Gold, J.D., Connolly, A.J., and Wu, J.C., 2016. Comparison of magnetic resonance imaging and serum biomarkers for detection of human pluripotent stem cell-derived teratomas. *Stem Cell Reports*, *6*(2), pp. 176–187.

Rossetti, C., Pomati, F., and Calamari, D., 2001. Microorganisms' activity and energy fluxes in lake Varese (Italy): A field method. *Water Research*, *35*(5), pp. 1318–1324.

Ruiyi, L., Fangchao, C., Haiyan, Z., Xiulan, S., and Zaijun, L., 2018. Electrochemical sensor for detection of cancer cell based on folic acid and octadecylamine-functionalized graphene aerogel microspheres. *Biosensors and Bioelectronics*, *119*, pp. 156–162.

Saadati, A., Hassanpour, S., Hasanzadeh, M., Shadjou, N., and Hassanzadeh, A., 2019. Immunosensing of breast cancer tumor protein CA 15-3 (carbohydrate antigen 15.3) using a novel nano-bioink: A new platform for screening of proteins in human biofluids by pen-on-paper technology. *International Journal of Biological Macromolecules*, *132*, pp. 748–758.

Sadasivam, M., Sakthivel, A., Nagesh, N., Hansda, S., Veerapandian, M., Alwarappan, S., and Manickam, P., 2020. Magnetic bead-amplified voltammetric detection for carbohydrate antigen 125 with enzyme labels using aptamer-antigen-antibody sandwiched assay. *Sensors and Actuators, B: Chemical*, *312*. doi: 10.1016/j.snb.2020.127985.

Sadik, O.A., Aluoch, A.O., and Zhou, A., 2009. Status of biomolecular recognition using electrochemical techniques. *Biosensors and Bioelectronics*, *24*(9), pp. 2749–2765.

Salahandish, R., Ghaffarinejad, A., Omidinia, E., Zargartalebi, H., Majidzadeh-A, K., Naghib, S.M., and Sanati-Nezhad, A., 2018. Label-free ultrasensitive detection of breast cancer miRNA-21 biomarker employing electrochemical nano-genosensor based on sandwiched AgNPs in PANI and N-doped graphene. *Biosensors and Bioelectronics*, *120*, pp. 129–136.

Sánchez, S., Roldán, M., Pérez, S., and Fàbregas, E., 2008. Toward a fast, easy, and versatile immobilization of biomolecules into carbon nanotube/polysulfone-based biosensors for the detection of hCG hormone. *Analytical Chemistry*, *80*(17), pp. 6508–6514.

Sardesai, N.P., Barron, J.C., and Rusling, J.F., 2011. Carbon nanotube microwell array for sensitive electrochemiluminescent detection of cancer biomarker proteins. *Analytical Chemistry*, *83*(17), pp. 6698–6703.

Senel, M., Dervisevic, M., and Kokkokoğlu, F., 2019. Electrochemical DNA biosensors for label-free breast cancer gene marker detection. *Analytical and bioanalytical chemistry*, *411*(13), pp. 2925–2935.

Shahzad, F., Zaidi, S.A., and Koo, C.M., 2017. Highly sensitive electrochemical sensor based on environmentally friendly biomass-derived sulfur-doped graphene for cancer biomarker detection. *Sensors and Actuators, B: Chemical*, *241*, pp. 716–724.

Shandilya, R., Ranjan, S., Khare, S., Bhargava, A., Goryacheva, I.Y., and Mishra, P.K., 2021. Point-of-care diagnostics approaches for detection of lung cancer-associated circulating miRNAs. *Drug Discovery Today*. doi: 10.1016/j.drudis.2021.02.023.

Sharma, S.V., Bell, D.W., Settleman, J., and Haber, D.A., 2007. Epidermal growth factor receptor mutations in lung cancer. *Nature Reviews Cancer*, *7*(3), pp. 169–181.

Shekari, Z., Zare, H.R., and Falahati, A., 2021. Dual assaying of breast cancer biomarkers by using a sandwich–type electrochemical aptasensor based on a gold nanoparticles–3D graphene hydrogel nanocomposite and redox probes labeled aptamers. *Sensors and Actuators, B: Chemical*, *332*. doi: 10.1016/j.snb.2021.129515.

Silva, P.M.S., Lima, A.L.R., Silva, B.V.M., Coelho, L.C.B.B., Dutra, R.F., and Correia, M.T.S., 2016. Cratylia mollis lectin nanoelectrode for differential diagnostic of prostate cancer and benign prostatic hyperplasia based on label-free detection. *Biosensors and Bioelectronics*, *85*, pp. 171–177.

Singh, S., Gill, A.A.S., Nlooto, M., and Karpoormath, R., 2019. Prostate cancer biomarkers detection using nanoparticles based electrochemical biosensors. *Biosensors and Bioelectronics*, *137*. doi: 10.1016/j.bios.2019.03.065.

Sinibaldi, A., Sampaoli, C., Danz, N., Munzert, P., Sibilio, L., Sonntag, F., Occhicone, A., Falvo, E., Tremante, E., Giacomini, P., and Michelotti, F., 2017. Detection of soluble ERBB2 in breast cancer cell lysates using a combined label-free/fluorescence platform based on Bloch surface waves. *Biosensors and Bioelectronics*, *92*, pp. 125–130.

Solhi, E., and Hasanzadeh, M., 2020. Critical role of biosensing on the efficient monitoring of cancer proteins/biomarkers using label-free aptamer based bioassay. *Biomedicine and Pharmacotherapy*. doi: 10.1016/j.biopha.2020.110849.

Solovieva, S., Karnaukh, M., Panchuk, V., Andreev, E., Kartsova, L., Bessonova, E., Legin, A., Wang, P., Wan, H., Jahatspanian, I., and Kirsanov, D., 2019. Potentiometric multisensor system as a possible simple tool for non-invasive prostate cancer diagnostics through urine analysis. *Sensors and Actuators B: Chemical*, *289*, pp. 42–47.

Song, X., Lv, M.M., Lv, Q.Y., Cui, H.F., Fu, J., and Huo, Y.Y., 2021. A novel assay strategy based on isothermal amplification and cascade signal amplified electrochemical DNA sensor for sensitive detection of Helicobacter pylori. *Microchemical Journal*, *166*. doi: 10.1016/j.microc.2021.106243.

Sotiropoulou, S., and Chaniotakis, N.A., 2005. Lowering the detection limit of the acetylcholinesterase biosensor using a nanoporous carbon matrix. *Analytica Chimica Acta*, *530*(2), pp. 199–204.

Spanggaard, B., Gram, L., Okamoto, N., and Huss, H.H., 1994. Growth of the fish-pathogenic fungus, Ichthyophonus hoferi, measured by conductimetry and microscopy. *Journal of Fish Diseases*, *17*(2), pp. 145–153.

Su, B., Tang, J., Huang, J., Yang, H., Qiu, B., Chen, G., and Tang, D., 2010. Graphene and nanogold-functionalized immunosensing interface with enhanced sensitivity for one-step electro-chemical immunoassay of alpha-fetoprotein in human serum. *Electroanalysis*, *22*(22), pp. 2720–2728.

Subramani, I.G., Ayub, R.M., Gopinath, S.C.B., Perumal, V., Fathil, M.F.M., and Md Arshad, M.K., 2021. Lectin bioreceptor approach in capacitive biosensor for prostate-specific membrane antigen detection in diagnosing prostate cancer. *Journal of the Taiwan Institute of Chemical Engineers*, *120*, 9–16.

Tabish, T.A., Hayat, H., Abbas, A., and Narayan, R.J., 2021. Graphene quantum dot-based electrochemical biosensing for early cancer detection. *Current Opinion in Electrochemistry*, *100786*. doi: 10.1016/j.coelec.2021.100786.

Tabish, T.A., Hayat, H., Abbas, A., and Narayan, R.J., 2021. Graphene quantum dot-based electrochemical biosensing for early cancer detection. *Current Opinion in Electrochemistry, 100786*.

Takeda, K., Kusuoka, R., Inukai, M., Igarashi, K., Ohno, H., and Nakamura, N., 2021. An amperometric biosensor of L-fucose in urine for the first screening test of cancer. *Biosensors and Bioelectronics, 174*. doi: 10.1016/j.bios.2020.112831.

Tang, J., Huang, J., Su, B., Chen, H., and Tang, D., 2011. Sandwich-type conductometric immunoassay of alpha-fetoprotein in human serum using carbon nanoparticles as labels. *Biochemical Engineering Journal, 53*(2), pp. 223–228.

Tang, J., Tang, D., Niessner, R., Chen, G., and Knopp, D., 2011. Magneto-controlled graphene immunosensing platform for simultaneous multiplexed electrochemical immunoassay using distinguishable signal tags. *Analytical Chemistry, 83*(13), pp. 5407–5414.

Tang, Z., Fu, Y., and Ma, Z., 2017. Bovine serum albumin as an effective sensitivity enhancer for peptide-based amperometric biosensor for ultrasensitive detection of prostate specific antigen. *Biosensors and Bioelectronics, 94*, pp. 394–399.

Thangsunan, P., Lal, N., Tiede, C., Moul, S., Robinson, J.I., Knowles, M.A., Stockley, P.G., Beales, P.A., Tomlinson, D.C., McPherson, M.J., and Millner, P.A., 2021. Affimer-based impedimetric biosensors for fibroblast growth factor receptor 3 (FGFR3): A novel tool for detection and surveillance of recurrent bladder cancer. *Sensors and Actuators, B: Chemical, 326*. doi: 10.1016/j.snb.2020.128829.

Thévenot, D.R., Toth, K., Durst, R.A., and Wilson, G.S., 2001. Electrochemical biosensors: Recommended definitions and classification. *Analytical Letters, 34*(5), pp. 635–659.

Timur, S., and Telefoncu, A., 2004. Acetylcholinesterase (AChE) electrodes based on gelatin and chitosan matrices for the pesticide detection. *Artificial Cells, Blood Substitutes and Biotechnology, 32*(3), pp. 427–442.

Top, M., Er, O., Congur, G., Erdem, A., and Lambrecht, F.Y., 2016. Intracellular uptake study of radiolabeled anticancer drug and impedimetric detection of its interaction with DNA. *Talanta, 160*, pp. 157–163.

Tsai, H., Tsai, S.H., Deng, H.W., and Bor Fuh, C., 2013. Assessment of Cell Viability Using the Chronoamperometric Method Based on Screen-Printed Electrodes. *Electroanalysis, 25*(4), pp. 1005–1009.

Tsé, C., Gauchez, A.S., Jacot, W., and Lamy, P.J., 2012. HER2 shedding and serum HER2 extracellular domain: biology and clinical utility in breast cancer. *Cancer Treatment Reviews, 38*(2), pp. 133–142.

Uludag, Y., and Tothill, I.E., 2012. Cancer biomarker detection in serum samples using surface plasmon resonance and quartz crystal microbalance sensors with nanoparticle signal amplification. *Analytical Chemistry, 84*(14), pp. 5898–5904.

Verma, D., Yadav, A.K., Mukherjee, M. Das, and Solanki, P.R., 2021. Fabrication of a sensitive electrochemical sensor platform using reduced graphene oxide-molybdenum trioxide nanocomposite for BPA detection: An endocrine disruptor. *Journal of Environmental Chemical Engineering, 9* (4). doi: 10.1016/j.jece.2021.105504.

Walker, N.L., and Dick, J.E., 2021. Oxidase-loaded hydrogels for versatile potentiometric metabolite sensing. *Biosensors and Bioelectronics, 178*. doi: doi: 10.1016/j.bios.2021.112997.

Walker, N.L., Roshkolaeva, A.B., Chapoval, A.I., and Dick, J.E., 2021. Recent advances in potentiometric biosensing. *Current Opinion in Electrochemistry, 124*. doi: 10.1016/j.trac.2019.115803.

Wan, Y., Deng, W., Su, Y., Zhu, X., Peng, C., Hu, H., Peng, H., Song, S., and Fan, C., 2011. Carbon nanotube-based ultrasensitive multiplexing electrochemical immunosensor for cancer biomarkers. *Biosensors and Bioelectronics, 30*(1), pp. 93–99.

Wang, F., and Hu, S., 2009. Electrochemical sensors based on metal and semiconductor nanoparticles. *Microchimica Acta, 165*(1-2), pp. 1–22.

Wang, H., and Ma, Z., 2018. A cascade reaction signal-amplified amperometric immunosensor platform for ultrasensitive detection of tumour marker. *Sensors and Actuators, B: Chemical, 254*, pp. 642–647.

Wang, H., and Ma, Z., 2019. "Off-on" signal amplification strategy amperometric immunosensor for ultrasensitive detection of tumour marker. *Biosensors and Bioelectronics, 132*, pp. 265–270.

Wang, J., 2006. Electrochemical biosensors: Towards point-of-care cancer diagnostics. *Biosensors and Bioelectronics, 21*(10), pp. 1887–1892.

Wang, K., Zhou, L., Wang, Z., Cheng, Z., Dong, H., Wu, Z., Bai, Y., Jin, Q., Mao, H., and Zhao, J., 2018. Uniform distribution of microspheres based on pressure difference for carcinoma-embryonic antigen detection. *Sensors and Actuators, B: Chemical, 258*, pp. 558–565.

Wang, P., Wan, Y., Deng, S., Yang, S., Su, Y., Fan, C., Aldalbahi, A., and Zuo, X., 2016. Aptamer-initiated on-particle template-independent enzymatic polymerization (aptamer-OTEP) for electrochemical analysis of tumor biomarkers. *Biosensors and Bioelectronics, 86*, pp. 536–541.

Wang, Y., Zhang, Z., Jain, V., Yi, J., Mueller, S., Sokolov, J., Liu, Z., Levon, K., Rigas, B., and Rafailovich, M.H., 2010. Potentiometric sensors based on surface molecular imprinting: Detection of cancer biomarkers and viruses. *Sensors and Actuators B: Chemical, 146*(1), pp. 381–387.

Wang, Y., Zhang, Z., Jain, V., Yi, J., Mueller, S., Sokolov, J., Liu, Z., Levon, K., Rigas, B., and Rafailovich, M.H., 2010. Potentiometric sensors based on surface molecular imprinting: Detection of cancer biomarkers and viruses. *Sensors and Actuators B: Chemical, 146*(1), pp. 381–387.

Wei, F., Wang, J., Liao, W., Zimmermann, B.G., Wong, D.T., and Ho, C.M., 2008. Electrochemical detection of low-copy number salivary RNA based on specific signal amplification with a hairpin probe. *Nucleic Acids Research, 36*(11), pp. e65–e65.

Wei, Q., Mao, K., Wu, D., Dai, Y., Yang, J., Du, B., Yang, M., and Li, H., 2010. A novel label-free electrochemical immunosensor based on graphene and thionine nanocomposite. *Sensors and Actuators B: Chemical, 149*(1), pp. 314–318.

Wen, T., Wang, M., Luo, M., Yu, N., Xiong, H., and Peng, H., 2019. A nanowell-based molecularly imprinted electrochemical sensor for highly sensitive and selective detection of 17β-estradiol in food samples. *Food Chemistry, 297.* doi: doi: 10.1016/j.foodchem.2019.124968.

Wilson, M.S., 2005. Electrochemical immunosensors for the simultaneous detection of two tumor markers. *Analytical Chemistry, 77*(5), pp. 1496–1502.

Wong, R.J., Ahmed, A., and Gish, R.G., 2015. Elevated alpha-fetoprotein: differential diagnosis-hepatocellular carcinoma and other disorders. *Clinics in Liver Disease, 19*(2), pp. 309–323.

Wu, J., Yan, F., Zhang, X., Yan, Y., Tang, J., and Ju, H., 2008. Disposable reagentless electrochemical immunosensor array based on a biopolymer/sol-gel membrane for simultaneous measurement of several tumor markers. *Clinical Chemistry, 54*(9), pp. 1481–1488.

Xia, Y.M., Xia, M., Zhao, Y., Li, M.Y., Ou, X., and Gao, W.W., 2021. Photocatalytic electrochemical sensor based on three-dimensional graphene nanocomposites for the ultrasensitive detection of CYFRA21-1 gene. *Microchemical Journal, 166.* doi: 10.1016/j.microc.2021.106245.

Xiao, C., and Luong, J.H., 2003. On-line monitoring of cell growth and cytotoxicity using electric cell-substrate impedance sensing (ECIS). *Biotechnology Progress, 19*(3), pp. 1000–1005.

Xiao, X., Zhang, Z., Nan, F., Zhao, Y., Wang, P., He, F., and Wang, Y., 2021. Mesoporous CuCo2O4 rods modified glassy carbon electrode as a novel non-enzymatic amperometric electrochemical sensors with high-sensitive ascorbic acid recognition. *Journal of Alloys and Compounds, 852.* doi: 10.1016/j.jallcom.2020.157045.

Xing, Y., Zhou, S., Wu, G., Wang, C., Yuan, X., Feng, Q., Zhu, X., and Qu, J., 2021. A sensitive electrochemical sensor for bisphenol F detection and its application in evaluating cytotoxicity. *Microchemical Journal, 168,* pp. 106414.

Xu, H., Zheng, J., Liang, H., and Li, C.P., 2020. Electrochemical sensor for cancer cell detection using calix[8]arene/polydopamine/phosphorene nanocomposite based on host–guest recognition. *Sensors and Actuators, B: Chemical, 317.* doi: 10.1016/j.snb.2020.128193.

Xu, S., Zhang, R., Zhao, W., Zhu, Y., Wei, W., Liu, X., and Luo, J., 2017. Self-assembled polymeric nanoparticles film stabilizing gold nanoparticles as a versatile platform for ultrasensitive detection of carcino-embryonic antigen. *Biosensors and Bioelectronics, 92,* pp. 570–576.

Xu, T., Chi, B., Gao, J., Chu, M., Fan, W., Yi, M., Xu, H., and Mao, C., 2017. Novel electrochemical immune sensor based on Hep-PGA-PPy nanoparticles for detection of α-Fetoprotein in whole blood. *Analytica Chimica Acta, 977,* pp. 36–43.

Xu, X., Ji, J., Chen, P., Wu, J., Jin, Y., Zhang, L., and Du, S., 2020. Salt-induced gold nanoparticles aggregation lights up fluorescence of DNA-silver nanoclusters to monitor dual cancer markers carcinoembryonic antigen and carbohydrate antigen 125. *Analytica Chimica Acta, 1125,* pp. 41–49.

Xue, J., Yao, C., Li, N., Su, Y., Xu, L., and Hou, S., 2021. Construction of polydopamine-coated three-dimensional graphene-based conductive network platform for amperometric detection of dopamine. *Journal of Electroanalytical Chemistry, 886.* doi: 10.1016/j.jelechem.2021.115133.

Xue, Z., Wang, H., Rao, H., He, N., Wang, X., Liu, X., and Lu, X., 2017. Amperometric indicator displacement assay for biomarker monitoring: Indirectly sensing strategy for electrochemically inactive sarcosine. *Talanta, 167,* pp. 666–671.

Yadav, S., Kashaninejad, N., Masud, M.K., Yamauchi, Y., Nguyen, N.T., and Shiddiky, M.J.A., 2019. Autoantibodies as diagnostic and prognostic cancer biomarker: Detection techniques and approaches. *Biosensors and Bioelectronics.* doi: 10.1016/j.bios.2019.111315.

Yan, X., Song, Y., Liu, J., Zhou, N., Zhang, C., He, L., Zhang, Z., and Liu, Z., 2019. Two-dimensional porphyrin-based covalent organic framework: A novel platform for sensitive epidermal growth factor receptor and living cancer cell detection. *Biosensors and Bioelectronics, 126,* pp. 734–742.

Yang, Y., Yang, X., Yang, Y., and Yuan, Q., 2018. Aptamer-functionalized carbon nanomaterials electrochemical sensors for detecting cancer relevant biomolecules. *Carbon*, 129. doi: 10.1016/j.carbon.2017.12.013.

Yang, S., Zhang, F., Wang, Z., and Liang, Q., 2018. A graphene oxide-based label-free electrochemical aptasensor for the detection of alpha-fetoprotein. *Biosensors and Bioelectronics*, *112*, pp. 186–192.

Yazdi, M.K., Ghazizadeh, E., and Neshastehriz, A., 2020. Different liposome patterns to detection of acute leukemia based on electrochemical cell sensor. *Analytica Chimica Acta*, *1109*, pp. 122–129.

Yeh, C.H., Su, K.F., and Lin, Y.C., 2016. Development of an impedimetric immunobiosensor for measurement of carcinoembryonic antigen. *Sensors and Actuators, A: Physical*, *241*, pp. 203–211.

Yemini, M., Levi, Y., Yagil, E., and Rishpon, J., 2007. Specific electrochemical phage sensing for Bacillus cereus and Mycobacterium smegmatis. *Bioelectrochemistry*, *70*(1), pp. 180–184.

Yoshinobu, T., Iwasaki, H., Ui, Y., Furuichi, K., Ermolenko, Y., Mourzina, Y., Wagner, T., Näther, N., and Schöning, M.J., 2005. The light-addressable potentiometric sensor for multi-ion sensing and imaging. *Methods*, *37*(1), pp. 94–102.

Yuan, B., Xu, C., Zhang, R., Lv, D., Li, S., Zhang, D., Liu, L., and Fernandez, C., 2017. Glassy carbon electrode modified with 7,7,8,8-tetracyanoquinodimethane and graphene oxide triggered a synergistic effect: Low-potential amperometric detection of reduced glutathione. *Biosensors and Bioelectronics*, *96*, pp. 1–7.

Zandi, A., Gilani, A., Abbasvandi, F., Katebi, P., Tafti, S.R., Assadi, S., Moghtaderi, H., Parizi, M.S., Saghafi, M., Khayamian, M.A., and Hoseinpour, P., 2019. Carbon nanotube based dielectric spectroscopy of tumor secretion; electrochemical lipidomics for cancer diagnosis. *Biosensors and Bioelectronics*, *142*, p.111566.

Zeng, H., Agyapong, D.A.Y., Li, C., Zhao, R., Yang, H., Wu, C., Jiang, Y., and Liu, Y., 2015. A carcinoembryonic antigen optoelectronic immunosensor based on thiol-derivative-nanogold labeled anti-CEA antibody nanomaterial and gold modified ITO. *Sensors and Actuators B: Chemical*, *221*, pp. 22–27.

Zhang, C., Wang, C., Hao, T., Lin, H., Wang, Q., Wu, Y., Hu, Y., Wang, S., Huang, Y., and Guo, Z., 2021. Electrochemical sensor for the detection of ppq-level Cd2+ based on a multifunctional composite material by fast scan voltammetry. *Sensors and Actuators, B: Chemical*, *341*. doi: 10.1016/j.snb.2021.130037.

Zhang, H., Cai, X., Zhao, H., Sun, W., Wang, Z., and Lan, M., 2019. Enzyme-free electrochemical sensor based on ZIF-67 for the detection of superoxide anion radical released from SK-BR-3 cells. *Journal of Electroanalytical Chemistry*, *855*. doi: 10.1016/j.jelechem.2019.113653.

Zhang, S., Rong, F., Guo, C., Duan, F., He, L., Wang, M., Zhang, Z., Kang, M., and Du, M., 2021. Metal–organic frameworks (MOFs) based electrochemical biosensors for early cancer diagnosis in vitro. *Coordination Chemistry Reviews*. 10.1016/j.ccr.2021.213948.

Zhang, X., Liu, M., Zhang, X., Wang, Y., and Dai, L., 2020. Autoantibodies to tumor-associated antigens in lung cancer diagnosis. *In: Advances in Clinical Chemistry*. Academic Press Inc.

Zhang, X., Wu, D., Liu, Z., Cai, S., Zhao, Y., Chen, M., Xia, Y., Li, C., Zhang, J., and Chen, J., 2014. An ultrasensitive label-free electrochemical biosensor for microRNA-21 detection based on a 2′-O-methyl modified DNAzyme and duplex-specific nuclease assisted target recycling. *Chemical Communications*, *50*(82), pp. 12375–12377.

Zhao, X., Dai, X., Zhao, S., Cui, X., Gong, T., Song, Z., Meng, H., Zhang, X., and Yu, B., 2021. Aptamer-based fluorescent sensors for the detection of cancer biomarkers. *Spectrochimica Acta – Part A: Molecular and Biomolecular Spectroscopy*, *247*. doi: 10.1016/j.saa.2020.119038.

Zhao, Y., Cai, X., Zhu, C., Yang, H., and Du, D., 2020. A novel fluorescent and electrochemical dual-responsive immunosensor for sensitive and reliable detection of biomarkers based on cation-exchange reaction. *Analytica Chimica Acta*, *1096*, pp. 61–68.

Zhou, E., Li, Y., Wu, F., Guo, M., Xu, J., Wang, S., Tan, Q., Ma, P., Song, S., and Jin, Y., 2021. Circulating extracellular vesicles are effective biomarkers for predicting response to cancer therapy. *EBioMedicine*, 67, p.103365. doi:10.1016/j.ebiom.2021.103365

Zhou, J., Du, L., Zou, L., Zou, Y., Hu, N., and Wang, P., 2014. An ultrasensitive electrochemical immunosensor for carcinoembryonic antigen detection based on staphylococcal protein A—Au nanoparticle modified gold electrode. *Sensors and Actuators B: Chemical*, *197*, pp. 220–227.

Zhou, Y., Yu, B., and Levon, K., 2005. Potentiometric sensor for dipicolinic acid. *Biosensors and Bioelectronics*, *20*(9), pp. 1851–1855.

Zouari, M., Campuzano, S., Pingarrón, J.M., and Raouafi, N., 2020a. Determination of miRNAs in serum of cancer patients with a label-and enzyme-free voltammetric biosensor in a single 30-min step. *Microchimica Acta*, *187*(8), pp. 1–11.

Zouari, M., Campuzano, S., Pingarrón, J.M., and Raouafi, N., 2020b. Femtomolar direct voltammetric determination of circulating miRNAs in sera of cancer patients using an enzymeless biosensor. *Analytica Chimica Acta*, *1104*, pp. 188–198.

6 Mechanisms of Different Anticancer Drugs

6.1 INTRODUCTION

Anticancer drugs are classified not only based on their mechanism of action but also based on their anticancerous selectivity towards different types of cancer. These anticancer drugs are classified according to chemical structure, their mechanism of action, and cytotoxic effects associated with the cell cycle are well studied (Calman et al., 1980; Sun et al., 2017; Advani et al., 2020; Behl et al., 2020; Mohammed and Hanoon, 2020, 2021a, 2021b; Arora et al., 2021; Liang et al., 2021; Mameri et al., 2021; Mohammadi et al., 2021; Palmieri and Macpherson, 2021). Anticancer therapies aim to check the growth of cancer and its spread. Anticancer therapies are divided into different ways. Traditionally, it is based on their procedure type like surgery, chemotherapy, radiotherapy, immunotherapy, targeted, or cell therapy (Alam et al., 2018; Sarah and David, 2018; Bo Chiang et al., 2021; Cristóvão et al., 2021; dehghan banadaki et al., 2021; Erol et al., 2021; J. Wang et al., 2021; Kalita et al., 2021; Kelbert et al., 2021; Kim and Choi, 2021; Pranzini et al., 2021; Rahimi and Solimannejad, 2021; Safaei and Shishehbore, 2021; Shen and Noguchi, 2021; Tian et al., 2021). However, these classifications or types can be unclear due to combination therapy, which has more interest. With the arrival of immune checkpoint therapy, it has been predicted that cancer can become a chronic rather than deadly disease soon. Immune therapies involve cell therapies such as CAR-T cells and are currently being employed in more than 3,000 phase 2 and 3 of clinical studies (FDA approved) in different cancer types like the combination therapies (ClinicalTrials.gov).

All the cancer treatments are mainly divided based on the cancer properties they target and mechanism during targeting, either inhibitory or interfering (Agborbesong et al., 2020; B. Li et al., 2020; Bhat et al., 2020; Chakravarty et al., 2020; Champ and Klement, 2020; Farayola et al., 2020; Fathi Maroufi et al., 2020; Juste et al., 2020; Kim and Clavijo, 2020; Maroufi et al., 2020; Nguyen et al., 2020; Shan et al., 2020; Sun et al., 2020; Thomas et al., 2020; Yao et al., 2020; YeKedüz et al., 2020; Sweilam et al., 2021). However, all the molecular pathways involved in a particular type of cancer, genes and the epigenetic factors responsible for malignancy, proliferation and metastasis are unclear. But, to get an idea for the genetic and epigenetic sites of distinct tumor types, enough studies are available. There are worldwide schemes available to record cancer genetics and epigenetics and even their proteomic landscape. Such data is available in databases such as the Cancer Genomics, provided by the National Cancer Institute (NCI) in 1997. This is an online reference tool with data on benign tissues and pre- and post-cancerous genomes. Additionally, it has the tools for analyzing data and browsing the genes involved in different tumor progressions. In an essential study of Hanahan and Weinberg (Hanahan and Weinberg, 2011; Hainaut and Plymoth, 2013), it became evident that a minimum of a eukaryotic cell malignancy is caused by a minimum of one mutation in each ten characteristics that are hallmarks of cancer.

We are now aware that cancer stem cells (CSCs) can develop cancer by accumulating such variations over a long period. Although some changes are unique for particular cancer types, many different cancers share some common changes also. Therefore, these types of cancer can be treated through a combination therapy capable of targeting familiar and definite cancer hallmarks (Dembic, 2020).

DOI: 10.1201/9781003201946-6

6.2   CANCER RESEARCH AND ANTICANCER DRUG HISTORY

The first documented origin of cancer chemotherapy started from 1861 when Robert Bentley demonstrated the antitumor effect of extract of *Podophyllum peltatum* roots at Kings College London. Certainly, this was even mentioned in the first U.S. Pharmacopoeia named "Materia Medica," issued in 1820. In 1946, the tumor suppression mechanism was explained, when it was shown that podophyllotoxin can inhibit mitotic spindle formation, leading to holding the dividing cells at metaphase. This discovery consequently encouraged the synthesis of structural analogs – etoposide and teniposide –developed in the late 1960s and early 1970s at Sandoz. Remarkably, these two analogs have also been shown to obstruct the action of an important DNA regulatory enzyme, i.e., topoisomerase II. These analogs have also shown activity against the small-cell lung cancer, lymphomas and testicular cancer, and etoposide is also found to inhibit various brain tumors (Jones, 2014; DeVita and Chu, 2008; Papac, 2001).

Over the past three decades, surgery is the oldest way of cancer treatment with a varying success rate (Wyld et al., 2015; Falzone et al., 2018). In the second half of the nineteenth century, breast cancer cases in women decreased due to ovariectomy (Battey, 1873; Hegar, 1878; Love and Philips, 2002). After finding x-rays, surgeons or doctors started using them to treat different skin lesions and basal cell carcinoma (Jolly, 1983; Pusey, 1983; Sharpe, 1900; Fox, 1924). Firstly, the use of x-rays for cancer therapy was reported in 1896 in breast cancer cases by Emil H. Grubbe (Grubbe, 1902; Hummler et al., 1997; Schulze-Rath et al., 2008; Slater, 2012) and again in 1899, Anton U. Sjøgren has used this method to treat mouth epi-thelioma (Nakayama and Bonasso, 2016; Ren and Hu, 2019). It was first discovered in 1899 that radioactivity could be a solution for cancer treatment (Curie and Curie, 1899; Kułakowski, 2011; Dembic, 2020). A list of tumors treated by radiotherapy is available in "Kassabian S Medical Manual." It was listed in 1907 in Philadelphia (Kassabian, 1907; Kułakowski, 2011). Later in 1922, French radiologist Henry Coutard presented and published data at the International Congress of Oncology in Paris, according to which radiotherapy could be used to treat buccal cavity and larynx and pharynx tumors without any deleterious effects (Coutard, 1937; Martins, 2018). This study was also supported by Claudius Regaud, who discovered radiation fractionation to treat various cancers reducing the side effects (Deloch et al., 2016; Frey et al., 2017; Moulder and Seymour, 2018; Rückert et al., 2018). These studies have been proven as the beginning of modern radiotherapy. In the early 1940s, chemotherapy came into the picture to treat cancer (Gilman, 1946). They were synthesized from the toxins used for chemical warfare, like nitrogen mustards and antifolate drugs (Golomb, 1963; Cheung-Ong et al., 2013; Gustafson and Page, 2013). In 1942, a lymphosarcoma (x-ray resistant) patient was treated by Gustav Lindskog with nitrogen mustard (Gilman, 1963; Chen et al., 2018). This was the first evidence of the cancer regression property of chemotherapy. Hence, nitrogen mustard mustine was approved as the first chemotherapy by U.S. Food and Drug Administration (FDA) in 1949 (Hirsch, 2006; Christakis, 2011). Soon afterward, other ni-trogen mustards such as cyclophosphamide, melphalan, chlorambucil, and uramustine were developed (Mattes et al., 1986; Hartley et al., 1992). All the drugs have gained therapeutic importance with time except the discontinued mustine due to extensive toxicity. There is a timeline for anticancer medicine approvals until the present (Chabner and Roberts, 2005; Jayashree et al., 2015; Kumar et al., 2017; Sun et al., 2017; Wang et al., 2019; Falzone et al., 2018; Olgen, 2018). Collectively, cancers are responsible for most of the deaths worldwide and have the capability to become the primary killer in the early parts of this period. Chemotherapy is playing an important role in malignancies management, in either direct form or indirect forms such as adjuvant to surgery or radiotherapy. There is a range of agents that have been approved clinically, as combinational chemotherapy "cocktails." Understanding of many agents' mechanisms has become clear recently. The cancer chemotherapies history eras back into the nineteenth century.

6.3 MUSTARD GAS

The history of organized clinical chemotherapy era started from World War I, though, it includes the warfare agents together known as mustard gases. In a study done by Edward B. Krumbhaar and group, the effects of mustard gas were seen on fighters who were killed due to exposure of this agent (Jarrell et al., 2020). The results, obtained from 75 post-mortems, it was revealed that an extreme decrease of leucocyte cells had occurred along with other phenomena. Simultaneous research was conducted at the U.S. Chemical Warfare Service on mustard gases. In this study, rabbits were injected with a lethal limit of doses and a prominent drop in leucocyte numbers was observed immediately along with bone marrow damage (Jones, 2014). Cancer cells characteristically multiply faster than the healthy cells, with a few exceptions; therefore, cytotoxins that control cell division have more effect on cancers than normal cells. The realization that leucopenia was due to the mustard gases imposing damage to fast-dividing cells and its cancer controlling ability was then legally followed after World War II (an era of first clinical chemotherapy trials) (Einhorn, 1985; Ghanei and Vosoghi, 2002). Originally, b-chloroethyl sulfides and its analogs, b-chloroethyl amines, were considered which later switched to dozens of mustard gas analogs. This eventually directed the final assimilation of "nitrogen mustards" in standard chemotherapy regimens. Mechlorethamine, the simplest and foremost candidate of this class, is still used for the treatment of non-Hodgkin lymphoma, and its structural relatives are used for other cancer treatment such as cyclophosphamide for leukemia, breast, lung cancer, and ovarian cancer, chlorambucil for chronic lymphocytic leukemia treatment, and melphalan for multiple myeloma and ovarian cancer management (Baguley and Kerr, 2001).

6.4 SYNTHETIC AGENTS

The introduction of synthetic agents is majorly responsible for bringing advances in this field (Parameswaran et al., 2019; Zargar et al., 2019; Alsehli, 2020; Cao and Liu, 2020; Mondal et al., 2020; Říhová et al., 2000; Sunwoo et al., 2020; Xu et al., 2020; Amin et al., 2021). Chemists who contributed in this discovery led the development of insecticide DDT after researching mitotane, a new class of antimetabolite, i.e., methotrexate, Charles Heidelberger's stimulated design of antimetabolite 5-fluorouracil; and Southern Research Institute's work that directed to the compound, N,N'-bis(2-chloroethyl)-N-nitrosourea, conventionally known as BCNU (Jones, 2014). In recent times, the current research is focused towards the development of synthetic and semi-synthetic drugs for treatment of different types of cancers (Gupta et al., 2021; H.W. Cheng et al., 2021; Jain et al., 2021; Jiang et al., 2021; Mosaddad et al., 2021; Rana and Bhatnagar, 2021; Sadat et al., 2021; Saeed et al., 2021; Shubhra et al., 2021; Thng et al., 2021; Truong Hoang et al., 2021; Xiao et al., 2021; Zhi et al., 2021).

6.5 NATURAL PRODUCTS

A significant conventional source of anticancer agents is fermentation products. A very efficient and clinically proven agent known as mitomycin C (mutamycin) was developed in Japan at the Kitasato Institute in 1956 from *Streptomyces caespitosu* culture. This agent is effective in combinational chemotherapy. Other important fermentation products comprise the structurally associated anthracycline antibiotics i.e., daunorubicin and doxorubicin, developed from *Streptomyces peucetius* in 1960s (Jones, 2014; Dias et al., 2012). However, some other manufactured analogs of these natural agents have been assessed and only a few of them had shown any extra benefit over the original structures, which still manages the acute lymphocytic and myelocytic leukemia. Compounds from plant and marine sources also exhibit anticancer activity. Among them, the best examples are vinblastine and vincristine (Gordaliza, 2007). This study was invented from reports that a tea prepared from the *Vinca rosea* leaves was effective for diabetes mellitus treatment. Further, research has shown that when this material was injected into rats, their cellular blood components were extensively decreased. Gordon H. Svoboda was successful in isolating vin-blastine in 1965, and later other substances related to it including vincristine. Vincristine was eventually found to be exceptionally effective for the treatment of acute leukemia in children, while vinblastine was shown to be advantageous against the varied type of tumors.

To date, thousands of plant species have been identified with remarkable anticancer activity (Mukherjee et al., 2001; Kajiwara et al., 2007; Kuete and Efferth, 2011; Wang et al., 2012; Zhong et al., 2012; Kim et al., 2015; Hung et al., 2017; Lin et al., 2017; Mohammadi, Mansoori and Baradaran, 2017; Bazrafshani et al., 2019; Chando et al., 2019; Han et al., 2019; So et al., 2019; Tsai et al., 2019; Zhai et al., 2019; Kong et al., 2020; Qi et al., 2020; Sidhu and Zafar, 2020; Yeh et al., 2020; Xu et al., 2021; Yin et al., 2021; Liu et al., 2021); for example, vinca alkaloids isolation such as isolation of vinblastine (Balunas, and Kinghorn, 2005) from Madagascar periwinkle (*Catharanthus roseus*). Plant constituents of the family Apocynaceae have also been used along with the vincristine and vinblastine for the range of such as leukemias, lymphomas, breast cancers, and Kaposi's sarcoma (Cragg and Newman, 2005; Chando et al., 2019) etc. Paclitaxel (Taxol) (Butler, 2004) obtained from the bark of *Taxus brevifolia* (Taxaceae), is another main drug discovery in natural products. The use of different parts of *Taxus brevifolia* for the discovery of Paclitaxel by Native American Tribes spark the idea of native information-based therapeutic plants (Cragg and Newman, 2005; Chando et al., 2019). Another potent compound extracted from the Chinese tree *Cephalotaxus harringtonia* was Homoharringtonine (Farnsworth et al., 1985) and has been used effectively for acute myelogenous leukemia treatment (Cragg and Newman, 2005). Elliptinium, an ellipticine derivative that was dis-covered from *Bleekeria vitensis* (a Fijian medicinal plant), was transported to France for breast cancer treatment (Cragg and Newman, 2005; Chando et al., 2019).

6.6   TARGETING THE GROWTH OF TUMORS

6.6.1   INHIBITION OF CANCER PROLIFERATION

Proliferation is the essential trait of cancer (Biddle et al., 2011). In normal conditions, the growth of tissues and cell division are firmly regulated which involves growth factors production and regulates cell receptors and signaling pathways leading to cell division. Homeostasis of each tissue is maintained by the equilibrium between cell death, restitution, reconstruction, and proliferation (Hine and Martin, 2015; Hall and Hall 2020). The growth factor signals may transmit locally to control the cells' numbers and positions in the tissues (Wolff et al., 2018). It is assumed to occur in a three-dimensional and time-based manner among a cell and its neighboring tissues. Well-known methods involve the binding of growth factors to the cell surface receptors, especially tyrosine kinase domains. Other intracellular proteins having tyrosine or threonine phosphorylation sites also get attracted by additional intracellular receptors. All of these together transduce the signal started by binding the nucleus factor and the growth factor; thus, it helps in the cellular response setting for growth and division (H.W. Cheng et al., 2019; Huang et al., 2019; Ishigaki et al., 2019; Kurdyukov et al., 2019; Li et al., 2019; Takahashi et al., 2019; Thu et al., 2019; Buranrat et al., 2020; Rattanaburee et al., 2020; Vasquez et al., 2020; Wang et al., 2020; Z. Li et al., 2020; Zhang et al., 2020; Zhao et al., 2020; Zúñiga et al., 2020; Jin et al., 2020; Kim et al., 2020; Gallo et al., 2021; Hamza et al., 2021; Yoshizawa et al., 2021).

Cancer cells can sustain proliferative signaling by producing their growth factors followed by autocrine stimulation through the ligand-specific receptors or via paracrine growth enhancement of the surrounding tissues. Additionally, growth factor receptors can be changed or decontrolled such that they could be hypersensitive without a ligand (Clara et al., 2020; Hanahan and Weinberg, 2000). Examples include signaling molecules such as proto-oncogenes or "oncogenes" MYC, RAS, etc., whose continuous activation by mutation or any epigenetic variations describes them as oncogenes or activated oncogenes. Ras proteins are small GTPases that help in transducing the transient signals related to the cell membrane. There are three types of RAS genes in humans: NRAS, HRAS, and KRAS. Ras superfamily consists of the largest group with 154 enzyme candidates among 220 with affinity to bind with guanosine diphosphate (GDP) and guanosine triphosphate (GTP). It hydrolyses the bound GTP to GDP and releases orthophosphate. Mutation in codons 12, 13, and 61 of the RAS gene leads to conversion of these genes to oncogenes. From the transient state, proteins are bound with GTP constantly and remain active. Permanent activation of Ras proteins due to mutation is the most common among humans and is responsible for around 30% of all cancers in humans and approximately 75% in specific cancers such as pancreatic cancer (Karnoub and Weinberg, 2008; Pylayeva-Gupta et al., 2011). KRAS is most dominantly deregulated in lung cancer, pancreatic cancer, colorectal, and adenocarcinomas (Bailey et al., 2016; Campbell et al., 2016; Giannakis et al., 2016; Jordan et al., 2017).

6.6.2 TARGETING THE SUPPRESSORS RESPONSIBLE FOR TUMOR GROWTH

p53 is also known as the guardian of the genome and responsible for cell cycle regulation. Mutation in hereditary copies of *p53* and other tumor suppressor genes such as retinoblastoma (RB1) is required to become oncogenic and for cancer development (Murphree and Benedict, 1984; Surget et al., 2014; Toufektchan and Toledo, 2018).

The planning of cancer therapy involves permitting the regular function of these alleles in cancer cells with non-functional Rb or *p53* proteins. This might be attained by gene therapy. It could be insufficient as long as a cancer treatment because its progeny would ultimately conquer the "rescued" Rb cancer stem cell. Hence, the inserted RB1 allele will be inactivated due to the genetic instability of cancer cells. Inactivation of both copies of the RB1 gene may not be an excellent option for cancer treatment. Fascinatingly, the pharmacological renaissance of *p53* (another tumor suppressor gene)- Gendicine was accepted as gene therapy for cancer in China in 2003 for neck and head cancers. Retinoblastoma is a multifunctional protein. It can interact with over 100 cellular proteins (Burkhart and Sage, 2008; Vélez-Cruz and Johnson, 2017; Dick et al., 2018). Besides, if RB protein is not inactivated at the time of carcinogenesis, molecules acting as tumor suppressors or the mimics of Rb would be possible treatments for cancer (Burkhart and Sage, 2008). The most crucial step is the binding and suppression of the E2F transcription factor, which prevents cell access into the S-phase of the cell cycle. This step is characterized by DNA synthesis for replication during cell division (Goodrich et al., 1991; Meng et al., 2016). RB1 ensures the inactivation of the E2F target by keeping the cell in the G1 phase (Burkhart and Sage, 2008).

6.6.3 INHIBITING ANTIAPOPTOTIC BEHAVIOR OF TUMOR

Cancer can evade apoptosis, which is an essential factor for tumor progression and resistance to chemotherapy. In patients of chronic lymphatic leukemia (CLL) and acute myeloid leukemia (AML), leukemic cells have shown a higher level of B-cell lymphoma-2 (Bcl-2) protein. Thus, they can avoid apoptosis (Campos et al., 1993; Konopleva et al., 2006; Lagadinou et al., 2013). Research conducted on specific tumor cell lines have shown that by increasing the expression of intracellular Bax protein (Bcl-2-associated-X), these properties of cells can be reversed. Bax induces the proapoptotic factors release from the mitochondria like cytochrome C and can alert the breast tumor cells to anticancer agents or therapies. Pterostilbene (3,5-dimethoxy-4-hydroxystilbene) is a naturally occurring drug having antioxidant properties. It can inhibit the proliferation of cancer cells and tempt apoptosis in breast cancer cells (Moon et al., 2013).

Venetoclax is a novel anticancer compound that blocks the antiapoptotic Bcl-2 protein from programmed cell death (Roberts and Huang, 2017). It showed promising results in AML patients in combination with azacitidine and scientists are hoping that it will also be effective for other cancer types in the future. More prolonged survival and reduction in incidence were observed compared to those treated with azacitidine alone (DiNardo et al., 2020). Apart from the stimulation of G1/S transition gene expression, E2F also induces the proapoptotic gene expression. However, many tumor cells can develop signaling pathways to avoid cell death, countering the E2F. These cancer-shielding unique pathways are targets for novel treatments. *p53* is the most crucial tumor suppressor protein involved in apoptosis. More than 50% of cancers have mutated *p53* genes (Schmitt et al., 2002).

6.6.4 TARGETING ABILITY OF TUMORS TO DIVIDE INDEFINITELY BY EXTENDING TELOMERES

The ends of chromosomes are called telomeres, consisting of a repetitive sequence of TTAGGG. The telomerase enzyme adds these specific DNA elements. This enzyme is a ribonucleoprotein complex and responsible for maintaining the length of telomeres in human stem cells and 85% of cancer cells (Shay and Bacchetti, 1997). Another telomere expansion procedure is used in 15% of cancers (Heaphy et al., 2011). During cell division, telomere shortening occurs in normal somatic cells for growth seizure, called cellular senescence. Hence, its long-term stability offers cancer cells infinite proliferation ability and cancer cells can grow indeterminately due to higher telomerase expression (Hanahan and Weinberg, 2011). In this targeting process, a telomere crisis occurs in RB- and *p53*-deficient cells due to ongoing telomere shortening. This phenomenon can cause different genomic aberrations. Telomerase activation permits cancer cells to escape from disaster and damage genome integrity costs (Maciejowski and deLange, 2017).

Telomerase inhibition is a feasible option for antitumor drugs as therapeutics. Currently, medicines used to inhibit telomerase activity are antagonists of telomerase, i.e., GRN163L (imetelstat, Geron). This telomerase template inhibitor is under screening for (refractory myelofibrosis since 2019 (FDA). It can control tumorigenicity and invasiveness of tumor cells and act as anticancer drugs for cancer therapy (Harley, 2008).

6.6.5 TARGETING ANGIOGENESIS PROPERTY OF CANCER CELLS

Drugs used to check cancer angiogenesis are inhibitors of cancer processes, including invasiveness, angiogenesis, and metastasis. Enzymes responsible for tissue homeostasis like matrix metalloproteinases (MMPs) are involved in these processes. This enzyme belongs to zinc-dependent proteinases and is carried out by the degradation of the extracellular matrix. MMPs play a significant part in physiologic progressions like wound healing; although they can also ease the process of tumor growth, proliferation, angiogenesis, and metastases (Folkman, 1995; Hidalgo and Eckhardt, 2001). Its overexpression in some cancers has been associated with a worse prognosis (Murray et al., 1996; Yamamoto et al., 1999). In different clinical researches, it has been observed that by inhibiting the MMPs expression alongside other regulatory molecules, metastatic and tumor growth in breast cancer models were reduced (Sledge et al., 1995; Fallah et al., 2019; Fei et al., 2020; Lee et al., 2020; Saravanan et al., 2020; Wu et al., 2020; Zhu et al., 2020; Al-Ostoot et al., 2021; Armani et al., 2021; Berger et al., 2021; Chen et al., 2021; Guo et al., 2021; K. J. Cheng et al., 2021; Ma et al., 2021; Razavi et al., 2021; S. Wang et al., 2021; Yetkin-Arik et al., 2021; Zhang et al., 2021; Zhou et al., 2021). This finding leads to developing various MMP inhibitors, i.e., batimastat and marimastat, having the low molecular weight and peptidomimetic inhibitors (Zucker et al., 2002).

6.6.6 TARGETING INFLAMMATION-PROMOTING TUMORS

Inflammatory agents can cause DNA and tissue damage if produced persistently. Also, it can lead to developing cancer. Mutated tumor cells can also instigate the inflammatory response leading to the development of cancer. Recently, cyclic GMP-AMP synthase that can stimulate interferon genes (STING) with a downstream effector has shown antimicrobial and antitumor immunity. The cytokines generated in this pathway can relate innate immunity with developed adaptive immunity against cancer. Activation of chronic STING can result in generating a tumor promotor phenotype and malignancy. Therefore, by inhibiting the proinflammatory effects of cancers, the growth and development of cancers can be seized (Mercadante et al., 2002; D'Arrigo et al., 2014; Marques et al., 2014; Magee et al., 2019; Augustin et al., 2020; Di Lorenzo et al., 2020; Zheng et al., 2020; Bai et al., 2021). The most commonly used anti-inflammatory drugs among anti-inflammatory drugs are aspirin, diclofenac, colecoxib, sulindac, ibuprofen, piroxicam, dexamethasone, hydrocortisone, and prednisone. The antitumor effects of aspirin and non-steroidal anti-inflammatory drugs are well known (Thun et al., 2002; Dettorre et al., 2021; Dougan et al., 2021; Fantini and Guadagni, 2021; K. J. Cheng et al., 2021; Motolani et al., 2021; Shariare et al., 2021; Tewari et al., 2021; Xing et al., 2021; M. Zhang et al., 2021). They have anti-invasiveness properties, anti-metastasis, and can also increase the apoptosis rate and sensitivity to chemotherapy (Thun et al., 2002; Zappavigna et al., 2020).

6.7 DIFFERENT CLASSES OF ANTICANCER DRUGS

Anticancer drugs can be classified based on different criteria including their structure, origin, and mode of action (Figure 6.1).

6.7.1 CHEMICAL STRUCTURE-BASED CLASSIFICATION

The classification of anticancer drugs based on chemical structure is quite satisfactory. Based on structural features, anticancer drugs are mainly classified into two groups: (a) alkylating agents and (b) nucleic acid antimetabolites (Huang et al., 2017; Sun et al., 2017).

Alkylating agents react covalently with various cell components such as enzymes, nucleic acids, and proteins (Asselin and Rizzari, 2015) under physiological settings. In the alkylation process of an organic molecule, the alkyl group replaces a hydrogen atom. Therefore, standard alkylating agents possess an alkyl group with the ability to interact with negatively charged centered molecules. These agents interact mainly with thiol groups and ionized acidic groups of amino acids/proteins and nucleic acids (Konstantinov and Berger, 2008), respectively.

The second grouping based on the chemical structure of anticancer drugs is nucleic acid antimetabolites. These chemical agents are similar to nucleosides, which are responsible for building

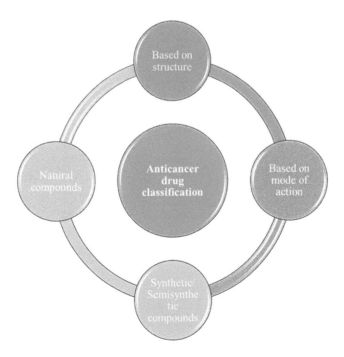

FIGURE 6.1 Classification of anticancer drugs.

DNA and RNA. Any compound with the ability to compete with the natural metabolite utilization because of possessing structural similarity is termed an antimetabolite (Calman et al., 1980; Asselin and Rizzari, 2015). Currently, precursors of nucleic acid and folic acid analogs are used for cancer chemotherapy.

6.7.2   Classification Based on Mechanism of Action

On the basis of the mechanism of action (Figures 6.2 and 6.4), anticancer drugs identify in three different assay groups (Trevor et al., 2010; Calman et al., 1980; https://training.seer.cancer.gov/).

1. **Class I - Cell cycle (non-specific):** In this class, drugs kill the cancerous cells, whether they are present in the cell cycle (Trevor et al., 2010) or not, e.g., nitrogen mustard, gamma irradiation.
2. **Class II - Cell cycle (phase-) specific:** Drugs from this class are toxic only to the cells present in the particular cell cycle phase. Bone marrow and the tumor cells that are not present in this phase escape toxicity at the time of treatment (Trevor et al., 2010), e.g., vinblastine, vincristine, methotrexate.
3. **Class III - Cell cycle (non-phase-) specific** (https://training.seer.cancer.gov/): Drugs of this class kill the cells in proportion to increasing dose. However, bone marrow cells show less sensitivity than cancerous cells, e.g., cyclophosphamide, actinomycin D.

6.7.3   Common Classification Based on Different Groups

6.7.3.1   Antimetabolites

As discussed earlier, antimetabolites can compete with natural metabolite utilization. Although these antimetabolites are available in large numbers, only a few are recognized as anticancer agents (Table 6.1, Figure 6.3).

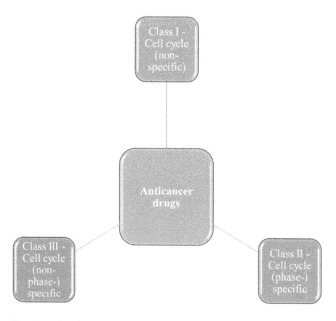

FIGURE 6.2 Classification of anticancer drugs based on the mechanism of action.

TABLE 6.1

Different Types of Anticancer Drugs and their Action Mechanism

S. No.	Type of Anticancer Drug	Mechanism of Action	References
1.	Antimetabolites e.g., Methotrexate, 5-Fluorouracil, Azacytidine, Azauridine, Azathioprine, 6-Mercaptopurine, 6-Thioguanine, etc.	Compete with the natural metabolites	Hess and Khasawneh, 2015; Zhang et al., 2008; Waller and Sampson, 2017; Ilowite and Laxer, 2011; Choughule et al., 2014; Wang et al., 2016; Munshi et al., 2014
2.	Alkylating agents e.g., Cyclophosphamide, Chlorambucil, Melphalan, Nitrogen mustard	Bind to different cell components via forming covalent bonds	Asselin and Rizzari, 2015; Weber, 2015; Cheng et al., 2016; Weber, 2015; Pocasap et al., 2020; Chen et al., 2018
3.	Natural products e.g., Vincristine, Vinblastine, Vindesine, Actinomycin D, L-Asparaginase	Inhibition of cell division, protein, and nucleic acid synthesis	Lichota and Gwozdzinski, 2018; Keglevich et al., 2012; Park et al., 2019; Li et al., 2018; Marinello et al., 2018; Asselin and Rizzari, 2015
4.	Hormones	Alteration of hormonal conjugation and metabolism responsible for tumor cell growth	Hadoke et al., 2009; Ke et al., 2017

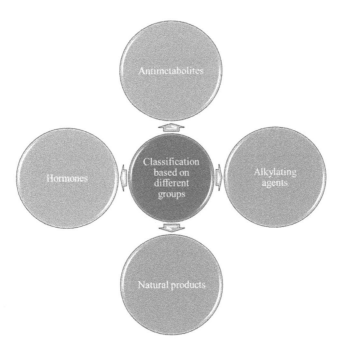

FIGURE 6.3   Classification of anticancer drugs based on different groups.

6.7.3.1.1 Methotrexate

Methotrexate action involves inhibition of folate metabolism intracellularly via the interaction with dihydrofolate reductase enzyme in the cancer cells (Hess and Khasawneh, 2015). In normal conditions, this enzyme leads to the production of tetrahydrofolates that are responsible for the synthesis and metabolism of nucleic acids and amino acids (Hagner and Joerger, 2010),

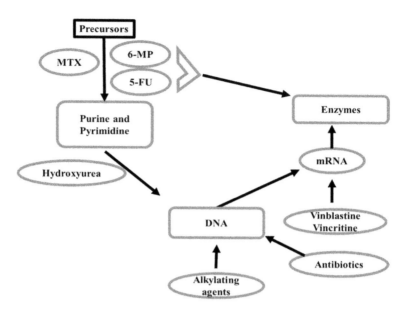

FIGURE 6.4   Action mechanism of some common anticancer drugs.

respectively. Methotrexate affinity is affected by coenzyme concentrations and intracellular pH. In some studies, weak binding of methotrexate with other enzymes of pyrimidine is also reported.

6.7.3.1.2 5-Fluorouracil

Fluorouracil is broadly metabolized and it possesses two separate and specific action mechanisms. In one mechanism, DNA synthesis is prevented as fluorouracil is metabolized to fluorodeoxyuridine monophosphate, which stops thymidylate synthetase (Zhang et al., 2008). In another mechanism, it can also interfere with the synthesis of RNA and protein. The specific mechanism of fluorouracil lethalness differs probably between tissues according to the effects of fluorouracil on the metabolism of RNA and biosynthesis of thymidine (Longley et al., 2003). Metabolism of fluorouracil is also influenced by the simultaneous administration of allopurinol and thymidine (Calman, 2015). It is depicted that the rate response of this drug is associated with a high dose during treatment.

6.7.3.1.3 Azacytidine and Azauridine

These drugs are used as pyrimidine antimetabolites. Azacytidine plays a significant role in acute myeloblastic leukaemia treatment while Azauridine inhibits the activity of the enzyme, orotidylate decarboxylase (Calman, 2015), and used in the treatment of gestational carcinoma.

6.7.3.1.4 6-Mercaptopurine (6-Mp)

6-Mercaptopurine is an analog of hypoxanthine, a purine base (Waller and Sampson, 2017). It has been used for acute leukemia treatment. 6-Mp, after metabolizing into thioinosinic acid, blocks the DNA biosynthesis via inhibiting the denovo purine metabolism (Ilowite and Laxer, 2011).

6.7.3.1.5 Azathioprine

Azathioprine is an extensively used antimetabolite of a purine base, 6-Mercaptopurine (Polifka and Friedman, 2002), after metabolizing into the liver, converts in this form (Choughule et al., 2014). Therefore, it is also considered a dormant form of 6-Mercaptopurine. Azathioprine is broadly used as an immunosuppressive drug (Diehl et al., 2017). It is advantageous over 6-Mercaptopurine due to its ease of administration and pharmacological properties.

6.7.3.1.6 6-Thioguanine (6-TG)

6-Thioguanine is a guanine analog, blocks various purine biosynthesis steps (Wang et al., 2016), and is widely incorporated in DNA itself. It is already reported that there is a close association between 6-TG incorporation into DNA and cell sensitivity. 6-TG is most commonly involved in acute leukemia treatment (Munshi et al., 2014).

6.7.3.2 Alkylating Agents

Alkylating agents can bind to different cell components covalently (Asselin and Rizzari, 2015). There are numerous synthesized alkylating agents that have been investigated as antitumor drugs. Despite having different chemical structures, alkylating agents have a common action mechanism. They are positively charged in physiological conditions and react with the negatively charged molecules (Weber, 2015). Therefore, for alkylating agents, DNA is considered a principal target site. The synthesis of DNA is blocked by these alkylating agents in a dose-dependent manner. Mostly alkylating agents are of dual functionality with two alkylation arms and cross-linking reaction that is responsible for cytotoxicity (Weber, 2015; Cheng et al., 2016).

6.7.3.2.1 Cyclophosphamide

Cyclophosphamide was developed as a dormant form of nitrogen mustard. It can be hydrolyzed via enzyme phosphoramidase into tumor cells (Wermuth, 2008). However, its mechanism of action in the case of cancer is not related to hydrolysis by phosphoramidase. Its mode of action involves the metabolism of cyclophosphamide into an active form of metabolites, i.e., 4-hydroxy derivative (Frangou et al., 2016) (in equilibrium with aldophosphamide) and their release into the circulation.

6.7.3.2.2 Chlorambucil

For ovarian cancer and lymphocytic leukemia treatment, chlorambucil has been used for the past 20 years (IARC working group, 2012). It is available commercially for oral use by absorption of the gastrointestinal tract. This drug is readily transported to cancerous cells through facilitated diffusion (Pocasap et al., 2020). It has the same action mechanism spectrum as other alkylating drugs for anticancer activity; rarely, it causes alopecia or toxicity of the bladder.

6.7.3.2.3 Melphalan

Melphalan, also called as L-phenylalanine mustard, was first synthesized in 1953. Phenylalanine was a precursor for melanin synthesis; therefore, melphalan was initially used for the treatment of malignant melanoma. Its effect has been observed in other types of cancer such as ovarian cancer and multiple myeloma (Thirumaran et al., 2007). Melphalan can be administered orally as well as parenterally.

6.7.3.2.4 Nitrogen mustard

The first alkylating drug in the 1940s used for disease treatment was nitrogen mustard. It is broadly used for the treatment of Hodgkin's disease and malignancy. It is a highly labile alkylating drug and available only in an injectable form (Chen et al., 2018). Nitrogen mustard is actively transported and absorbed in tumor cells.

6.7.3.3 Natural Products

Many natural products are also available that can be used as anticancer drugs, including mitotic inhibitors, enzymes, antibiotics, etc.

6.7.3.3.1 Mitotic Inhibitors

Under this antitumor class, only vinca-alkaloids can be used, generally. Derivatives of vindesine and epipodophyllotoxin are still in the clinical trial stage.

6.7.3.3.2 Vincristine and vinblastine

Both of these antitumor agents share similar structures but they differ in their biological properties. They have broad anticancerous properties. During the mitotic metaphase, cell growth is arrested by vinca alkaloids via spindle body breakage. Along with increased disorganization of microtubules, ribosomes also occur (Lichota and Gwozdzinski, 2018). In this way, vinca alkaloids inhibit the DNA biosynthesis in tumor cells, which can be prevented by using aspartate and glutamate *in vivo* (Calman et al., 1980). The antitumor spectrum of both compounds is the same but the neurotoxicity of vincristine is more than vinblastine.

Vindesine is a derivative of vinblastine, also known as desacetyl-vinblastine. Recently, it has shown its antitumor effects in different clinical trials (Keglevich et al., 2012). Its pharmacology is similar to vinblastine. Vindesine therapy involves intravenous injections. This drug also exerts neurotoxic and myelosuppressive effects.

6.7.3.3.3 Antibiotics

Actinomycins were the first anticancer antibiotics that have been clinically evaluated. Fermentation products obtained from bacteria and fungi have added various compounds or chemical agents for experimental clinical studies (Park et al., 2019).

- **Actinomycin D:** Actinomycin D was first isolated and used for clinical use in 1954. It has a very broad spectrum of anticancerous activity. It is broadly used as a drug for the treatment of gestational trophoblastic tumors, tumors of germ cells, and Wilms's tumors. The mode of action involves the binding of actinomycin D with dsDNA and inhibition of RNA synthesis (Li et al., 2018). Its specificity is determined by its penetration and ability to retain inside the tissues. It is not metabolized widely.

  Daunorubicin and adriamycin are the anticancerous antibiotics that belong to the anthracycline group. These antibiotics are structurally similar and isolated from the fungus culture, i.e., *Streptomyces peucetius* (Edwardson et al., 2015). Both of these drugs have shown their clinical potential against lymphomas, leukemia, and a number of other tumors. These antitumor drugs show collective dose-dependent cardiomyopathy. Both drugs show a similar mode of action, i.e., blockage of nucleic acid synthesis. Their high-affinity binding with DNA by intercalation has been shown in different studies. These antitumor drugs show maximum toxicity during the synthetic phase of DNA (Marinello et al., 2018).

6.7.3.3.4 Enzymes

- **L-Asparaginase:** This drug was first identified as an antitumor agent in guinea pig serum after showing antitumor activity in mice (Avramis and Tiwari, 2006). The clinical practice of asparaginase is restricted to acute lymphoblastic leukemia and malignant lymphomas. Different studies on tumor-bearing animals have shown that asparaginase shows its anti-tumor effect directly via the reduction of asparagine (Asselin and Rizzari, 2015). By using asparaginase, glutamine and asparagine levels can be decreased, which leads to immediate inhibition of protein and nucleic acid synthesis.

6.7.3.4   Steroid Hormones

Exogenous administration of steroid hormones changes the hormonal balance in cancer patients and alter the cancer growth of tissues, susceptible to hormonal changes. The exact mechanism of the antitumor property of hormones is not clear yet, but it is believed to be related to RNA and protein synthesis.

- **O,p'DDD (Mitotane):** This steroid hormone possesses a unique antitumor mechanism for inhibition of glucocorticoids and 17-ketosteroids production and alteration of their conjugation and metabolism (Hadoke et al., 2009). By doing so, it alleviates the clinical indices of the production of steroid hormones useful for carcinoma of the adrenal cortex (Lehmann et al., 2013). Several anticancer drugs are under investigation or not established yet for cancer therapy.

- **Steroidal Pyrimidines and Dihydrotriazine:** In recent studies, steroidal pyrimidines have shown their promising potential for the design of antitumor drugs. In a study (Ke et al., 2017), dehydroepiandrosterone-derived novel steroidal [17,16-d] pyrimidines were evaluated for inhibitory actions against gastric and liver cancer (Ke et al., 2017). They have found that few heterocyclic steroidal pyrimidines had antitumor activities against Huh-7 and HepG2 cell lines using 5-fluorouracil as reference.

  The structure of a dihydrotriazine ring consists of two acetimidamides and one steroid molecule. Steroidal dihydrotriazines have been proven more potent against MCF-7 cancer cells in comparison to steroid-bonded pyrimidines. Natural hormones such as estrone and 3β-acetoxyandrostene were changed into corresponding improved steroidal pyrimidines and dihydrotriazines by utilizing the Vilsmeier–Haack reaction process and condensation through amidines. The new modified compounds have shown significant anticancer activity against prostate and breast cancer cells.

REFERENCES

Advani, D., Gupta, R., Tripathi, R., Sharma, S., Ambasta, R.K., and Kumar, P., 2020. Protective role of anticancer drugs in neurodegenerative disorders: A drug repurposing approach. *Neurochemistry International*. doi: 10.1016/j.neuint.2020.104841.

Agborbesong, O., Helmer, S.D., Reyes, J., Strader, L.A., and Tenofsky, P.L., 2020. Breast cancer treatment in the elderly: Do treatment plans that do not conform to NCCN recommendations lead to worse outcomes?. *American Journal of Surgery*, 220, pp. 381–384.

Alam, A., Farooq, U., Singh, R., Dubey, V.P., Kumar, S., Kumari, R., Kumar, K., Naik, B.D., and Dhar, K.L., 2018. Chemotherapy treatment and strategy schemes: A review. *Open Access Journal of Toxicology*, 2, p. 555600.

Al-Ostoot, F.H., Salah, S., Khamees, H.A., and Khanum, S.A., 2021. Tumor angiogenesis: Current challenges and therapeutic opportunities. *Cancer Treatment and Research Communications*, 28, p. 100422.

Alsehli, M., 2020. Polymeric nanocarriers as stimuli-responsive systems for targeted tumor (cancer) therapy: Recent advances in drug delivery. *Saudi Pharmaceutical Journal*. doi: 10.1016/j.jsps. 2020.01.004.

Amin, H., Khan, A., Makeen, H.A., Rashid, H., Amin, I., Masoodi, M.H., Khan, R., Arafah, A., and Rehman, M.U., 2021. Nanosized delivery systems for plant-derived therapeutic compounds and their synthetic derivative for cancer therapy. *Phytomedicine*. pp. 655–675. doi: 10.1016/B978-0-12-824109-7.00020-0.

Armani, G., Pozzi, E., Pagani, A., Porta, C., Rizzo, M., Cicognini, D., Rovati, B., Moccia, F., Pedrazzoli, P., and Ferraris, E., 2021. The heterogeneity of cancer endothelium: The relevance of angiogenesis and endothelial progenitor cells in cancer microenvironment. *Microvascular Research*, 138, pp. 104189.

Arora, S., Surakiatchanukul, T., Arora, T., Errera, M.H., Agrawal, H., Lupidi, M., and Chhablani, J., 2021. Retinal toxicities of systemic anticancer drugs. *Survey of Ophthalmology*. doi: 10.1016/j.survophthal.2021.05.007.

Asselin, B., and Rizzari, C., 2015. Asparaginase pharmacokinetics and implications of therapeutic drug monitoring. *Leukemia & Lymphoma*, 56(8), pp. 2273–2280.

Augustin, Y., Staines, H.M., and Krishna, S., 2020. Artemisinins as a novel anticancer therapy: Targeting a global cancer pandemic through drug repurposing. *Pharmacology and Therapeutics*. doi: 10.1016/j.pharmthera.2020.107706.

Avendaño, C., and Menendez, J.C., 2015. DNA Alkylating Agents. *In*: *Medicinal Chemistry of Anticancer Drugs*, 2nd edition. *Elsevier*.

Avramis, V.I., and Tiwari, P.N., 2006. Asparaginase (native ASNase or pegylated ASNase) in the treatment of acute lymphoblastic leukemia. *International Journal of Nanomedicine*, *1*(3), p. 241.

Baguley, B.C., and Kerr, D.J. eds., 2001. *Anticancer Drug Development. Elsevier.*

Bai, R., Yao, C., Zhong, Z., Ge, J., Bai, Z., Ye, X., Xie, T., and Xie, Y., 2021. Discovery of natural anti-inflammatory alkaloids: Potential leads for the drug discovery for the treatment of inflammation. *European Journal of Medicinal Chemistry.* doi: 10.1016/j.ejmech.2021.113165.

Bailey, P., Chang, D.K., Nones, K., Johns, A.L., Patch, A.M., Gingras, M.C., Miller, D.K., Christ, A.N., Bruxner, T.J., Quinn, M.C., and Nourse, C., 2016. Genomic analyses identify molecular subtypes of pancreatic cancer. *Nature*, *531*(7592), pp. 47–52.

Balunas, M.J., and Kinghorn, A.D., 2005. Drug discovery from medicinal plants. *Life Sciences*, *78*(5), pp. 431–441.

Battey, R., 1873. *Normal Ovariotomy*. Herald Publishing Company.

Bazrafshani, M.S., Khandani, B.K., Pardakhty, A., Tajadini, H., Pour Afshar, R.M., Moazed, V., Nemati, A., Nasiri, N., and Sharifi, H., 2019. The prevalence and predictors of using herbal medicines among Iranian cancer patients. *Complementary Therapies in Clinical Practice*, *35*, pp. 368–373.

Behl, A., Parmar, V.S., Malhotra, S., and Chhillar, A.K., 2020. Biodegradable diblock copolymeric PEG-PCL nanoparticles: Synthesis, characterization and applications as anticancer drug delivery agents. *Polymer*, *207*. doi: 10.1016/j.polymer.2020.122901.

Berger, A.A., Dao, F., and Levine, D.A., 2021. Angiogenesis in endometrial carcinoma: Therapies and biomarkers, current options, and future perspectives. *Gynecologic Oncology.* doi: 10.1016/j.ygyno.2020.12.016.

Bhat, D., Heiman, A.J., Talwar, A.A., Dunne, M., Amanjee, K., and Ricci, J.A., 2020. Access to breast cancer treatment and reconstruction in rural populations: Do women have a choice? *Journal of Surgical Research*, *254*, pp. 223–231.

Biddle, A., Liang, X., Gammon, L., Fazil, B., Harper, L.J., Emich, H., Costea, D.E., and Mackenzie, I.C., 2011. Cancer stem cells in squamous cell carcinoma switch between two distinct phenotypes that are preferentially migratory or proliferative. *Cancer research*, *71*(15), pp. 5317–5326.

Bo Chiang, J.C., Goldstein, D., Park, S.B., Krishnan, A. V., and Markoulli, M., 2021. Corneal nerve changes following treatment with neurotoxic anticancer drugs. *The Ocular Surface.* doi: 10.1016/j.jtos.2021.06.007.

Buranrat, B., Noiwetch, S., Suksar, T., and Ta-ut, A., 2020. Inhibition of cell proliferation and migration by Oroxylum indicum extracts on breast cancer cells via Rac1 modulation. *Journal of Pharmaceutical Analysis*, *10* (2), pp. 187–193.

Burkhart, D.L., and Sage, J., 2008. Cellular mechanisms of tumour suppression by the retinoblastoma gene. *Nature Reviews Cancer*, *8*(9), pp. 671–682.

Butler, M.S., 2004. The role of natural product chemistry in drug discovery. *Journal of Natural Products*, *67*(12), pp. 2141–2153.

Calman, K.C., 2015. Basic Principles of Cancer Chemotherapy. *In*: *Macmillan International Higher Education*. Macmillan.

Calman, K.C., Smyth, J.F., and Tattersall, M.H., 1980. Mechanism of Action of Anti-Cancer Drugs. *In*: *Basic Principles of Cancer Chemotherapy* (pp. 49–78). Palgrave, London.

Campbell, J.D., Alexandrov, A., Kim, J., Wala, J., Berger, A.H., Pedamallu, C.S., Shukla, S.A., Guo, G., Brooks, A.N., Murray, B.A., and Imielinski, M., 2016. Distinct patterns of somatic genome alterations in lung adenocarcinomas and squamous cell carcinomas. *Nature Genetics*, *48*(6), p. 607.

Campos, L., Rouault, J.P., Sabido, O., Oriol, P., Roubi, N., Vasselon, C., Archimbaud, E., Magaud, J.P., and Guyotat, D., 1993. High expression of bcl-2 protein in acute myeloid leukemia cells is associated with poor response to chemotherapy. *Blood*, 81. pp. 3091–3096.

Cao, Z., and Liu, J., 2020. Bacteria and bacterial derivatives as drug carriers for cancer therapy. *Journal of Controlled Release*, *326*, pp. 396–407.

Chabner, B.A., and Roberts, T.G., 2005. Chemotherapy and the war on cancer. *Nature Reviews Cancer*, *5*(1), pp. 65–72.

Chakravarty, D., Huang, L., Kahn, M., and Tewari, A.K., 2020. Immunotherapy for metastatic prostate cancer: Current and emerging treatment options. *Urologic Clinics of North America.* doi: 10.1016/j.ucl.2020.07.010.

Champ, C.E., and Klement, R.J., 2020. Assessing successful completion of calorie restriction studies for the prevention and treatment of cancer. *Nutrition*, *78*. doi: doi: 10.1016/j.nut.2020.110829.

Chando, R.K., Hussain, N., Rana, M.I., Sayed, S., Alam, S., Fakir, T.A., Sharma, S., Tanu, A.R., Mobin, F., Apu, E.H., and Hasan, M.K., 2019. CDK4 as a phytochemical based anticancer drug target. *bioRxiv.* doi: 10.1101/859595.

Chen, Y., Jia, Y., Song, W., and Zhang, L., 2018. Therapeutic potential of nitrogen mustard based hybrid molecules. *Frontiers in pharmacology*, 9, p. 1453.

Chen, Y., Zhang, W., Liang, C., Zheng, D., Wang, Y., Li, X., Shi, Y., Wang, F., Dong, W., and Yang, Z., 2021. Supramolecular nanofibers with superior anti-angiogenesis and antitumor properties by enzyme-instructed self-assembly (EISA). *Chemical Engineering Journal*, 425, p. 130531.

Cheng, H., Li, X., Wang, C., Chen, Y., Li, S., Tan, J., Tan, B., and He, Y., 2019. Inhibition of tankyrase by a novel small molecule significantly attenuates prostate cancer cell proliferation. *Cancer Letters*, 443, pp. 80–90.

Cheng, H.W., Chiang, C.S., Ho, H.Y., Chou, S.H., Lai, Y.H., Shyu, W.C., and Chen, S.Y., 2021. Dextran-modified Quercetin-Cu(II)/hyaluronic acid nanomedicine with natural poly(ADP-ribose) polymerase inhibitor and dual targeting for programmed synthetic lethal therapy in triple-negative breast cancer. *Journal of Controlled Release*, 329, pp. 136–147.

Cheng, J., Ye, F., Dan, G., Zhao, Y., Wang, B., Zhao, J., Sai, Y., and Zou, Z., 2016. Bifunctional alkylating agent-mediated MGMT-DNA cross-linking and its proteolytic cleavage in 16HBE cells. *Toxicology and Applied Pharmacology*, 305, pp. 267–273.

Cheng, K.J., Mejia Mohammed, E.H., Khong, T.L., Mohd Zain, S., Thavagnanam, S., and Ibrahim, Z.A., 2021. IL-1α and colorectal cancer pathogenesis: Enthralling candidate for anticancer therapy. *Critical Reviews in Oncology/Hematology*, 163, pp. 103398.

Cheung-Ong, K., Giaever, G., and Nislow, C., 2013. DNA-damaging agents in cancer chemotherapy: Serendipity and chemical biology. *Chemistry & Biology*, 20(5), pp. 648–659.

Choughule, K.V., Barnaba, C., Joswig-Jones, C.A., and Jones, J.P., 2014. In vitro oxidative metabolism of 6-mercaptopurine in human liver: Insights into the role of the molybdoflavoenzymes aldehyde oxidase, xanthine oxidase, and xanthine dehydrogenase. *Drug Metabolism and Disposition*, 42(8), pp. 1334–1340.

Christakis, P., 2011. Bicentennial: The birth of chemotherapy at Yale: Bicentennial Lecture Series: Surgery Grand Round. *The Yale Journal of Biology and Medicine*, 84(2), p. 169.

Clara, J.A., Monge, C., Yang, Y., and Takebe, N., 2020. Targeting signalling pathways and the immune microenvironment of cancer stem cells – a clinical update. *Nature Reviews Clinical Oncology*, 17(4), pp. 204–232.

ClinicalTrials.gov. Clinical Studies for Neoplasms (phase 2 and 3). Available online: https://www.clinicaltrials.gov/ct2/results?cond=Neoplasms&recrs=d&age_v=&gndr=&type=Intr&rslt=&phase=1&phase=2&Search=Apply (accessed on May 1, 2021).

Coutard, H., 1937. The results and methods of treatment of cancer by radiation. *Annals of Surgery*, 106(4), p. 584.

Cragg, G.M., and Newman, D.J., 2005. Plants as a source of anticancer agents. *Journal of Ethnopharmacology*, 100(1-2), pp. 72–79.

Cristóvão, M.B., Bento-Silva, A., Bronze, M.R., Crespo, J.G., and Pereira, V.J., 2021. Detection of anticancer drugs in wastewater effluents: Grab versus passive sampling. *Science of the Total Environment*, 786. doi: 10.1016/j.scitotenv.2021.147477.

Curie, P., and Curie, M., 1899. *Sur la radioactivité provoquée par les rayons de Becquerel*. Gauthier-Villars.

D'Arrigo, G., Navarro, G., Di Meo, C., Matricardi, P., and Torchilin, V., 2014. Gellan gum nanohydrogel containing anti-inflammatory and anticancer drugs: A multi-drug delivery system for a combination therapy in cancer treatment. *European Journal of Pharmaceutics and Biopharmaceutics*, 87(1), pp. 208–216.

dehghan banadaki, M., Aghaie, M., and Aghaie, H., 2021. Folic acid functionalized boron nitride oxide as targeted drug delivery system for fludarabine and cytarabine anticancer drugs: A DFT study. *Journal of Molecular Liquids*, 116753. doi: 10.1016/j.molliq.2021.116753.

Deloch, L., Derer, A., Hartmann, J., Frey, B., Fietkau, R., and Gaipl, U.S., 2016. Modern radiotherapy concepts and the impact of radiation on immune activation. *Frontiers in Oncology*, 6, p. 141.

Dembic, Z., 2020. Antitumor drugs and their targets. *Molecules*, 25(23), p. 5776.

Dettorre, G.M., Patel, M., Gennari, A., Pentheroudakis, G., Romano, E., Cortellini, A., and Pinato, D.J., 2021. The systemic pro-inflammatory response: targeting the dangerous liaison between COVID-19 and cancer. *ESMO Open*, 6(3), p. 100123.

DeVita, V.T., and Chu, E., 2008. A history of cancer chemotherapy. *Cancer Research*, 68(21), pp. 8643–8653.

Di Lorenzo, G., Di Trolio, R., Kozlakidis, Z., Busto, G., Ingenito, C., Buonerba, L., Ferrara, C., Libroia, A., Ragone, G., Ioio, C. dello, Savastano, B., Polverino, M., De Falco, F., Iaccarino, S., and Leo, E., 2020. COVID 19 therapies and anticancer drugs: A systematic review of recent literature. *Critical Reviews in Oncology/Hematology*. doi: doi: 10.1016/j.critrevonc.2020.102991.

Dias, D.A., Urban, S., and Roessner, U., 2012. A historical overview of natural products in drug discovery. *Metabolites*, 2(2), pp. 303–336.

Dick, F.A., Goodrich, D.W., Sage, J., and Dyson, N.J., 2018. Non-canonical functions of the RB protein in cancer. *Nature Reviews Cancer*, 18(7), pp. 442–451.

Diehl, R., Ferrara, F., Müller, C., Dreyer, A.Y., McLeod, D.D., Fricke, S., and Boltze, J., 2017. Immunosuppression for in vivo research: state-of-the-art protocols and experimental approaches. *Cellular & Molecular Immunology*, 14(2), pp. 146–179.

DiNardo, C.D., Jonas, B.A., Pullarkat, V., Thirman, M.J., Garcia, J.S., Wei, A.H., Konopleva, M., Döhner, H., Letai, A., Fenaux, P., and Koller, E., 2020. Azacitidine and venetoclax in previously untreated acute myeloid leukemia. *New England Journal of Medicine*, 383(7), pp. 617–629.

Dougan, M., Luoma, A.M., Dougan, S.K., and Wucherpfennig, K.W., 2021. Understanding and treating the inflammatory adverse events of cancer immunotherapy. *Cell*. doi: doi: 10.1016/j.cell.2021.02.011.

Edwardson, D.W., Narendrula, R., Chewchuk, S., Mispel-Beyer, K., PJ Mapletoft, J., and M Parissenti, A., 2015. Role of drug metabolism in the cytotoxicity and clinical efficacy of anthracyclines. *Current Drug Metabolism*, 16(6), pp. 412–426.

Einhorn, J., 1985. Nitrogen mustard: the origin of chemotherapy for cancer. *International Journal of Radiation Oncology, Biology, Physics*, 11(7), pp. 1375–1378.

Erol, A., Akpınar, F., and Muti, M., 2021. Electrochemical determination of anticancer drug Bendamustine and its interaction with double strand DNA in the absence and presence of quercetin. *Colloids and Surfaces B: Biointerfaces*, 205, p. 111884.

Fallah, A., Heidari, H.R., Bradaran, B., Sisakht, M.M., Zeinali, S., and Molavi, O., 2019. A gene-based anti-angiogenesis therapy as a novel strategy for cancer treatment. *Life Sciences*, 239. doi: 10.1016/j.lfs.2019.117018.

Falzone, L., Salomone, S., and Libra, M., 2018. Evolution of cancer pharmacological treatments at the turn of the third millennium. *Frontiers in Pharmacology*, 9, p. 1300.

Fantini, M.C., and Guadagni, I., 2021. From inflammation to colitis-associated colorectal cancer in inflammatory bowel disease: Pathogenesis and impact of current therapies. *Digestive and Liver Disease*. doi: 10.1016/j.dld.2021.01.012.

Farayola, M.F., Shafie, S., Siam, F.M., and Khan, I., 2020. Mathematical modeling of radiotherapy cancer treatment using Caputo fractional derivative. *Computer Methods and Programs in Biomedicine*, 188. doi: 10.1016/j.cmpb.2019.105306.

Farnsworth, N.R., Akerele, O., Bingel, A.S., Soejarto, D.D., and Guo, Z., 1985. Medicinal plants in therapy. *Bulletin of The World Health Organization*, 63(6), p. 965.

Fathi Maroufi, N., Rashidi, M.R., Vahedian, V., Akbarzadeh, M., Fattahi, A., and Nouri, M., 2020. Therapeutic potentials of Apatinib in cancer treatment: Possible mechanisms and clinical relevance. *Life Sciences*. doi: 10.1016/j.lfs.2019.117106.

Fei, L., Huimei, H., and Dongmin, C., 2020. Pivalopril improves anticancer efficiency of cDDP in breast cancer through inhibiting proliferation, angiogenesis and metastasis. *Biochemical and Biophysical Research Communications*, 533(4), pp. 853–860.

Folkman, J., 1995. Angiogenesis in cancer, vascular, rheumatoid and other disease. *Nature Medicine*, 1(1), pp. 27–30.

Fox, H., 1924. The Röntgen Ray in the treatment of skin diseases. *Archives of Dermatology and Syphilology*, 9(1), pp. 13–37.

Frangou, E.A., Bertsias, G., and Boumpas, D.T., 2016. Cytotoxic-immunosuppressive drug treatment. In*Systemic Lupus Erythematosus* (pp. 533–541). Academic Press.

Frey, B., Rückert, M., Weber, J., Mayr, X., Derer, A., Lotter, M., Bert, C., Rödel, F., Fietkau, R., and Gaipl, U.S., 2017. Hypofractionated irradiation has immune stimulatory potential and induces a timely restricted infiltration of immune cells in colon cancer tumors. *Frontiers in Immunology*, 8, p. 231.

Gallo, E., Kelil, A., Haughey, M., Cazares-Olivera, M., Yates, B.P., Zhang, M., Wang, N.-Y., Blazer, L., Carderelli, L., Adams, J.J., Kossiakoff, A.A., Wells, J.A., Xie, W., and Sidhu, S.S., 2021. Inhibition of cancer cell adhesion, migration and proliferation by a bispecific antibody that targets two distinct epitopes on αv integrins. *Journal of Molecular Biology*, 433(15), p. 167090.

Ghanei, M., and Vosoghi, A.A., 2002. An epidemiologic study to screen for chronic myelocytic leukemia in war victims exposed to mustard gas. *Environmental Health Perspectives*, 110(5), pp. 519–521.

Giannakis, M., Mu, X.J., Shukla, S.A., Qian, Z.R., Cohen, O., Nishihara, R., Bahl, S., Cao, Y., Amin-Mansour, A., Yamauchi, M., and Sukawa, Y., 2016. Genomic correlates of immune-cell infiltrates in colorectal carcinoma. *Cell Reports*, 15(4), pp. 857–865.

Gilman, A., 1946, June. Therapeutic applications of chemical warfare agents. *In: Federation Proceedings (Vol. 5*, pp. 285–292).

Gilman, A., 1963. The initial clinical trial of nitrogen mustard. *The American Journal of Surgery, 105*(5), pp. 574–578.

Golomb, F.M., 1963. Agents used in cancer chemotherapy. *The American Journal of Surgery, 105*(5), pp. 579–590.

Goodrich, D.W., Wang, N.P., Qian, Y.W., Eva, Y.H.L., and Lee, W.H., 1991. The retinoblastoma gene product regulates progression through the G1 phase of the cell cycle. *Cell, 67*(2), pp. 293–302.

Gordaliza, M., 2007. Natural products as leads to anticancer drugs. *Clinical and Translational Oncology, 9*(12), pp. 767–776.

Grubbe, E.H., 1902. X-rays in the treatment of cancer and other malignant diseases. *Medical Record (1866–1922), 62*(18), p. 692.

Guo, J., Zeng, H., Liu, Y., Shi, X., Liu, Y., Liu, C., and Chen, Y., 2021. Multicomponent thermosensitive lipid complexes enhance desmoplastic tumor therapy through boosting anti-angiogenesis and synergistic strategy. *International Journal of Pharmaceutics, 601*. doi: 10.1016/j.ijpharm.2021.120533.

Gupta, P.K., Gahtori, R., Govarthanan, K., Sharma, V., Pappuru, S., Pandit, S., Mathuriya, A.S., Dholpuria, S., and Bishi, D.K., 2021. Recent trends in biodegradable polyester nanomaterials for cancer therapy. *Materials Science and Engineering: C, 127*, p. 112198.

Gustafson, D.L., and Page, R.L., 2013. Cancer Chemotherapy. *In: Withrow and MacEwen's Small Animal Clinical Oncology* (pp. 157–179). Elsevier.

Hadoke, P.W., Iqbal, J., and Walker, B.R., 2009. Therapeutic manipulation of glucocorticoid metabolism in cardiovascular disease. *British Journal of Pharmacology, 156*(5), pp. 689–712.

Hagner, N., and Joerger, M., 2010. Cancer chemotherapy: targeting folic acid synthesis. *Cancer Management and Research, 2*, p. 293.

Hainaut, P., and Plymoth, A., 2013. Targeting the hallmarks of cancer: towards a rational approach to next-generation cancer therapy. *Current opinion in oncology, 25*(1), pp. 50–51.

Hall, J.E., and Hall, M.E., 2020. *Guyton and Hall Textbook of Medical Physiology e-Book.* Elsevier Health Sciences.

Hamza, A.A., Heeba, G.H., Hamza, S., Abdalla, A., and Amin, A., 2021. Standardized extract of ginger ameliorates liver cancer by reducing proliferation and inducing apoptosis through inhibition oxidative stress/ inflammation pathway. *Biomedicine and Pharmacotherapy, 134*. doi: 10.1016/j.biopha.2020.111102.

Han, G., Lee, J., Seong, S., and Kim, S., 2019. Herbal Medicines or Natural Products for Immune Checkpoint Therapy in Cancer: a Systematic Review of Experimental Studies. *Advances in Integrative Medicine, 6*, p. S134.

Hanahan, D., and Weinberg, R.A., 2000. The hallmarks of cancer. *Cell, 100*(1), pp. 57–70.

Hanahan, D., and Weinberg, R.A., 2011. Hallmarks of cancer: The next generation. *Cell, 144*(5), pp. 646–674.

Harley, C.B., 2008. Telomerase and cancer therapeutics. *Nature Reviews Cancer, 8*(3), pp. 167–179.

Hartley, J.A., Bingham, J.P., and Souhami, R.L., 1992. DNA sequence selectivity of guanine-N7 alkylation by nitrogen mustards is preserved in intact cells. *Nucleic Acids Research, 20*(12), pp. 3175–3178.

Heaphy, C.M., Subhawong, A.P., Hong, S.M., Goggins, M.G., Montgomery, E.A., Gabrielson, E., Netto, G.J., Epstein, J.I., Lotan, T.L., Westra, W.H., and Shih, I.M., 2011. Prevalence of the alternative lengthening of telomeres telomere maintenance mechanism in human cancer subtypes. *The American Journal of Pathology, 179*(4), pp. 1608–1615.

Hegar, A., 1878. *Die Castration der Frauen vom physiologischen und chirugischen Standpunkte aus.* Breitkopf und Härtel.

Hess, J.A., and Khasawneh, M.K., 2015. Cancer metabolism and oxidative stress: Insights into carcinogenesis and chemotherapy via the non-dihydrofolate reductase effects of methotrexate. *BBA Clinical, 3*, pp. 152–161.

Hidalgo, M., and Eckhardt, S.G., 2001. Development of matrix metalloproteinase inhibitors in cancer therapy. *Journal of the National Cancer Institute, 93*(3), pp. 178–193.

Hine, R., and Martin, E. eds., 2015. *A Dictionary of Biology.* Oxford University Press, USA.

Hirsch, J., 2006. An anniversary for cancer chemotherapy. *JAMA, 296*(12), pp. 1518–1520.

Huang, C.Y., Ju, D.T., Chang, C.F., Reddy, P.M., and Velmurugan, B.K., 2017. A review on the effects of current chemotherapy drugs and natural agents in treating non–small cell lung cancer. *Biomedicine, 7*(4). doi: 10.1051/bmdcn/2017070423.

Huang, S.H., Tseng, J.C., Lin, C.Y., Kuo, Y.Y., Wang, B.J., Kao, Y.H., Muller, C.J.F., Joubert, E., and Chuu, C.P., 2019. Rooibos suppresses proliferation of castration-resistant prostate cancer cells via inhibition of Akt signaling. *Phytomedicine, 64.* doi: 10.1016/j.phymed.2019.153068.

Hummler, E., Barker, P., Talbot, C., Wang, Q., Verdumo, C., Grubb, B., Gatzy, J., Burnier, M., Horisberger, J.D., Beermann, F., and Boucher, R., 1997. A mouse model for the renal salt-wasting syndrome pseudohypoaldosteronism. *Proceedings of the National Academy of Sciences, 94*(21), pp. 11710–11715.

Hung, K.F., Hsu, C.P., Chiang, J.H., Lin, H.J., Kuo, Y.T., Sun, M.F., and Yen, H.R., 2017. Complementary Chinese herbal medicine therapy improves survival of patients with gastric cancer in Taiwan: A nationwide retrospective matched-cohort study. *Journal of Ethnopharmacology, 199*, pp. 168–174.

IARC Working Group on the Evaluation of Carcinogenic Risk to Humans. Pharmaceuticals. Lyon (FR): International Agency for Research on Cancer; 2012. (IARC Monographs on the Evaluation of Carcinogenic Risks to Humans, No. 100A.) CHLORAMBUCIL. Available from: https://www.ncbi.nlm.nih.gov/books/NBK304324/.

Ilowite, N.T., and Laxer, R.M., 2011. Pharmacology and Drug Therapy. *In: Textbook of Pediatric Rheumatology* (pp. 71–126). Elsevier Inc.

Ishigaki, H., Minami, T., Morimura, O., Kitai, H., Horio, D., Koda, Y., Fujimoto, E., Negi, Y., Nakajima, Y., Niki, M., Kanemura, S., Shibata, E., Mikami, K., Takahashi, R., Yokoi, T., Kuribayashi, K., and Kijima, T., 2019. EphA2 inhibition suppresses proliferation of small-cell lung cancer cells through inducing cell cycle arrest. *Biochemical and Biophysical Research Communications, 519*(4), pp. 846–853.

Jain, P., Kathuria, H., and Momin, M., 2021. Clinical therapies and nano drug delivery systems for urinary bladder cancer. *Pharmacology and Therapeutics.* doi: 10.1016/j.pharmthera.2021.107871.

Jarrell, D.K., Drake, S., and Brown, M.A., 2020. Advancing therapies for cancer – from mustard gas to CAR T. *Science, 2*(2), p. 42.

Jayashree, B.S., Nigam, S., Pai, A., Patel, H.K., Reddy, N.D., Kumar, N., and Rao, C.M., 2015. Targets in anticancer research – a review.

Jiang, X., He, C., and Lin, W., 2021. Supramolecular metal-based nanoparticles for drug delivery and cancer therapy. *Current Opinion in Chemical Biology.* doi: 10.1016/j.cbpa.2021.01.005.

Jin, W., He, X., Zhu, J., Fang, Q., Wei, B., Sun, J., Zhang, W., Zhang, Z., Zhang, F., Linhardt, R.J., Wang, H., and Zhong, W., 2020. Inhibition of glucuronomannan hexamer on the proliferation of lung cancer through binding with immunoglobulin G. *Carbohydrate Polymers, 248.*

Jolly, H.W., 1983. Commentary: Roentgen rays in the treatment of skin disease and for the removal of hair. *Archives of Dermatology, 119*(2), pp. 176–177.

Jones, Graham B., 2014. History of Anticancer Drugs. *In:* eLS, John Wiley & Sons, Ltd, Chichester.

Jordan, E.J., Kim, H.R., Arcila, M.E., Barron, D., Chakravarty, D., Gao, J., Chang, M.T., Ni, A., Kundra, R., Jonsson, P., and Jayakumaran, G., 2017. Prospective comprehensive molecular characterization of lung adenocarcinomas for efficient patient matching to approved and emerging therapies. *Cancer Discovery, 7*(6), pp. 596–609.

Juste, B., Miró, R., Morató, S., Verdú, G., and Peris, S., 2020. Prostate cancer Monte Carlo dose model with 177Lutetium and 125Iodine treatments. *Radiation Physics and Chemistry, 174.*

Kajiwara, M., Oki, M., Mutaguchi, K., and Moriyama, H., 2007. POS-03.58: Prevalence and distress due to hot flushes in patients receiving endocrine therapy for prostate cancer and therapeutic efficacy of Japanese herbal medicine. *Urology, 70*(3), p. 290.

Kalita, B., Saviola, A.J., and Mukherjee, A.K., 2021. From venom to drugs: A review and critical analysis of Indian snake venom toxins envisaged as anticancer drug prototypes. *Drug Discovery Today.* doi 10.1016/j.drudis.2020.12.021.

Karnoub, A.E., and Weinberg, R.A., 2008. Ras oncogenes: Split personalities. *Nature reviews Molecular cell biology, 9*(7), pp. 517–531.

Kassabian, M.K., 1907. *Roentgen rays and electro-therapeutics: With chapters on radium and phototherapy.* JB Lippincott Comapny.

Ke, S., Shi, L., Zhang, Z., and Yang, Z., 2017. Steroidal [17, 16-d] pyrimidines derived from dehydroepiandrosterone: a convenient synthesis, antiproliferation activity, structure-activity relationships, and role of heterocyclic moiety. *Scientific Reports, 7*(1), pp. 1–7.

Keglevich, P., Hazai, L., Kalaus, G., and Szántay, C., 2012. Modifications on the basic skeletons of vinblastine and vincristine. *Molecules, 17*(5), pp. 5893–5914.

Kelbert, M., Pereira, C.S., Daronch, N.A., Cesca, K., Michels, C., de Oliveira, D., and Soares, H.M., 2021. Laccase as an efficacious approach to remove anticancer drugs: A study of doxorubicin degradation, kinetic parameters, and toxicity assessment. *Journal of Hazardous Materials*, *409*. doi: 10.1016/j.jhazmat.2020.124520.

Kim, C.W., and Choi, K.C., 2021. Effects of anticancer drugs on the cardiac mitochondrial toxicity and their underlying mechanisms for novel cardiac protective strategies. *Life Sciences*. doi: 10.1016/j.lfs.2021.119607.

Kim, P., and Clavijo, R.I., 2020. Management of male sexual dysfunction after cancer treatment. *Urologic Oncology: Seminars and Original Investigations*. doi: 10.1016/j.urolonc.2020.08.006.

Kim, S., Kim, W., Kim, D.H., Jang, J.H., Kim, S.J., Park, S.A., Hahn, H., Han, B.W., Na, H.K., Chun, K.S., Choi, B.Y., and Surh, Y.J., 2020. Resveratrol suppresses gastric cancer cell proliferation and survival through inhibition of PIM-1 kinase activity. *Archives of Biochemistry and Biophysics*, *689*. doi: 10.1016/j.abb.2020.108413.

Kim, W., Lee, W.B., Lee, J.W., Min, B. Il, Baek, S.K., Lee, H.S., and Cho, S.H., 2015. Traditional herbal medicine as adjunctive therapy for breast cancer: A systematic review. *Complementary Therapies in Medicine*, *23*(4), pp. 626–632.

Kong, M. yan, Li, L. yan, Lou, Y. mei, Chi, H. yu, and Wu, J. jun, 2020. Chinese herbal medicines for prevention and treatment of colorectal cancer: From molecular mechanisms to potential clinical applications. *Journal of Integrative Medicine*. doi: 10.1016/j.joim.2020.07.005.

Konopleva, M., Contractor, R., Tsao, T., Samudio, I., Ruvolo, P.P., Kitada, S., Deng, X., Zhai, D., Shi, Y.X., Sneed, T., and Verhaegen, M., 2006. Mechanisms of apoptosis sensitivity and resistance to the BH3 mimetic ABT-737 in acute myeloid leukemia. *Cancer Cell*, *10*(5), pp. 375–388.

Konstantinov, S.M., Berger, M.R., 2008. Alkylating Agents. In: Offermanns S., Rosenthal W. (eds) Encyclopedia of Molecular Pharmacology. Springer.

Kuete, V., and Efferth, T., 2011. Pharmacogenomics of Cameroonian traditional herbal medicine for cancer therapy. *Journal of Ethnopharmacology*, *137*(1), pp. 752–766.

Kułakowski, A., 2011. The contribution of Marie Skłodowska-Curie to the development of modern oncology.

Kumar, B., Singh, S., Skvortsova, I., and Kumar, V., 2017. Promising targets in anticancer drug development: Recent updates. *Current Medicinal Chemistry*, *24*(42), pp. 4729–4752.

Kurdyukov, D.A., Eurov, D.A., Shmakov, S. V., Kirilenko, D.A., Kukushkina, J.A., Smirnov, A.N., Yagovkina, M.A., Klimenko, V. V., Koniakhin, S. V., and Golubev, V.G., 2019. Fabrication of doxorubicin-loaded monodisperse spherical micro-mesoporous silicon particles for enhanced inhibition of cancer cell proliferation. *Microporous and Mesoporous Materials*, *281*, pp. 1–8.

Lagadinou, E.D., Sach, A., Callahan, K., Rossi, R.M., Neering, S.J., Minhajuddin, M., Ashton, J.M., Pei, S., Grose, V., O'Dwyer, K.M., and Liesveld, J.L., 2013. BCL-2 inhibition targets oxidative phosphorylation and selectively eradicates quiescent human leukemia stem cells. *Cell Stem Cell*, *12*(3), pp. 329–341.

Lee, J.Y., Ham, J., Lim, W., and Song, G., 2020. Apomorphine facilitates loss of respiratory chain activity in human epithelial ovarian cancer and inhibits angiogenesis in vivo: Apomorphine suppresses ovarian cancer via multiple intracellular mechanisms. *Free Radical Biology and Medicine*, *154*, pp. 95–104.

Lehmann, T.P., Wrzesiński, T., and Jagodziński, P.P., 2013. The effect of mitotane on viability, steroidogenesis and gene expression in NCI-H295R adrenocortical cells. *Molecular Medicine Reports*, *7*(3), pp. 893–900.

Li, B., Yang, L., Peng, X., Fan, Q., Wei, S., Yang, S., Li, X., Jin, H., Wu, B., Huang, M., Tang, S., Liu, J., and Li, H., 2020. Emerging mechanisms and applications of ferroptosis in the treatment of resistant cancers. *Biomedicine and Pharmacotherapy*. doi: 10.1016/j.biopha.2020.110710.

Li, L., Wan, X., Feng, F., Ren, T., Yang, J., Zhao, J., Jiang, F., and Xiang, Y., 2018. Pulse actinomycin D as first-line treatment of low-risk post-molar non-choriocarcinoma gestational trophoblastic neoplasia. *BMC Cancer*, *18*(1), p. 585.

Li, S., Yao, M., Niu, C., Liu, D., Tang, Z., Gu, C., Zhao, H., Ke, J., Wu, S., Wang, X., and Wu, F., 2019. Inhibition of MCF-7 breast cancer cell proliferation by a synthetic peptide derived from the C-terminal sequence of Orai channel. *Biochemical and Biophysical Research Communications*, *516*(4), p. 1066–1072.

Li, Z., Li, Y., Jia, Y., Ding, B., and Yu, J., 2020. Rab1A knockdown represses proliferation and promotes apoptosis in gastric cancer cells by inhibition of mTOR/p70S6K pathway. *Archives of Biochemistry and Biophysics*, *685*. doi: 10.1016/j.abb.2020.108352.

Liang, X., Wu, P., Yang, Q., Xie, Y., He, C., Yin, L., Yin, Z., Yue, G., Zou, Y., Li, L., Song, X., Lv, C., Zhang, W., and Jing, B., 2021. An update of new small-molecule anticancer drugs approved from 2015 to 2020. *European Journal of Medicinal Chemistry*. doi: 10.1016/j.ejmech.2021.113473.

Lichota, A., and Gwozdzinski, K., 2018. Anticancer activity of natural compounds from plant and marine environment. *International Journal of Molecular Sciences*, *19*(11), p. 3533.

Lin, T.H., Yen, H.R., Chiang, J.H., Sun, M.F., Chang, H.H., and Huang, S.T., 2017. The use of Chinese herbal medicine as an adjuvant therapy to reduce incidence of chronic hepatitis in colon cancer patients: A Taiwanese population-based cohort study. *Journal of Ethnopharmacology*, *202*, pp. 225–233.

Liu, H., Wang, Z.Y., Zhou, Y.C., Song, W., Ali, U., and Sze, D.M.Y., 2021. Immunomodulation of Chinese Herbal Medicines on NK cell populations for cancer therapy: A systematic review. *Journal of Ethnopharmacology*. doi: 10.1016/j.jep.2020.113561.

Longley, D.B., Harkin, D.P., and Johnston, P.G., 2003. 5-fluorouracil: mechanisms of action and clinical strategies. *Nature Reviews Cancer*, *3*(5), pp. 330–338.

Love, R.R., and Philips, J., 2002. Oophorectomy for breast cancer: History revisited. *Journal of the National Cancer Institute*, *94*(19), pp. 1433–1434.

Ma, W., Ou, T., Cui, X., Wu, K., Li, H., Li, Y., Peng, G., Xia, W., and Wu, S., 2021. HSP47 contributes to angiogenesis by induction of CCL2 in bladder cancer. *Cellular Signalling*, *85*, p. 110044.

Maciejowski, J., and de Lange, T., 2017. Telomeres in cancer: tumour suppression and genome instability. *Nature reviews Molecular Cell Biology*, *18*(3), p. 175.

Magee, D.J., Jhanji, S., Poulogiannis, G., Farquhar-Smith, P., and Brown, M.R.D., 2019. Nonsteroidal anti-inflammatory drugs and pain in cancer patients: a systematic review and reappraisal of the evidence. *British Journal of Anaesthesia*. doi: 10.1016/j.bja.2019.02.028.

Mameri, A., Bournine, L., Mouni, L., Bensalem, S., and Iguer-Ouada, M., 2021. Oxidative stress as an underlying mechanism of anticancer drugs cytotoxicity on human red blood cells' membrane. *Toxicology in Vitro*, *72*. doi: 10.1016/j.tiv.2021.105106.

Marinello, J., Delcuratolo, M., and Capranico, G., 2018. Anthracyclines as topoisomerase II poisons: from early studies to new perspectives. *International Journal of Molecular Sciences*, *19*(11), p. 3480.

Maroufi, N.F., Vahedian, V., Hemati, S., Rashidi, M.R., Akbarzadeh, M., Zahedi, M., Pouremamali, F., Isazadeh, A., Taefehshokr, S., Hajazimian, S., Seraji, N., and Nouri, M., 2020. Targeting cancer stem cells by melatonin: Effective therapy for cancer treatment. *Pathology Research and Practice*. doi: 10.1016/j.prp.2020.152919.

Marques, J.G., Gaspar, V.M., Costa, E., Paquete, C.M., and Correia, I.J., 2014. Synthesis and characterization of micelles as carriers of non-steroidal anti-inflammatory drugs (NSAID) for application in breast cancer therapy. *Colloids and Surfaces B: Biointerfaces*, *113*, pp. 375–383.

Martins, P.N., 2018. A brief history about radiotherapy. *International Journal of Latest Research in Engineering and Technology (IJLRET)*, *4*. doi: 10.5281/zenodo.3824294.

Mattes, W.B., Hartley, J.A., and Kohn, K.W., 1986. DNA sequence selectivity of guanine–N7 alkylation by nitrogen mustards. *Nucleic Acids Research*, *14*(7), pp. 2971–2987.

Meng, F., Qian, J., Yue, H., Li, X., and Xue, K., 2016. SUMOylation of Rb enhances its binding with CDK2 and phosphorylation at early G1 phase. *Cell Cycle*, *15*(13), pp. 1724–1732.

Mercadante, S., Fulfaro, F., and Casuccio, A., 2002. A randomised controlled study on the use of anti-inflammatory drugs in patients with cancer pain on morphine therapy: Effects on dose-escalation and a pharmacoeconomic analysis. *European Journal of Cancer*, *38*(10), pp. 1358–1363.

Mohammadi, A., Mansoori, B., and Baradaran, B., 2017. Regulation of miRNAs by herbal medicine: An emerging field in cancer therapies. *Biomedicine and Pharmacotherapy*. doi: 10.1016/j.biopha.2016.12.023.

Mohammadi, E., Tabatabaei, M., Habibi-Anbouhi, M., and Tafazzoli-Shadpour, M., 2021. Chemical inhibitor anticancer drugs regulate mechanical properties and cytoskeletal structure of non-invasive and invasive breast cancer cell lines: Study of effects of Letrozole, Exemestane, and Everolimus. *Biochemical and Biophysical Research Communications*, *565*, pp. 14–20.

Mohammed, M.H., and Hanoon, F.H., 2020. Theoretical prediction of delivery and adsorption of various anticancer drugs into pristine and metal-doped graphene nanosheet. *Chinese Journal of Physics*, *68*, pp. 578–595.

Mohammed, M.H., and Hanoon, F.H., 2021a. Application of zinc oxide nanosheet in various anticancer drugs delivery: Quantum chemical study. *Inorganic Chemistry Communications*, *127*. doi: 10.1016/j.inoche.2021.108522.

Mohammed, M.H., and Hanoon, F.H., 2021b. First-principles study on the physical properties for various anticancer drugs using density functional theory (DFT). *Solid State Communications*, *325*. doi: 10.1016/j.ssc.2020.114160.

Mondal, P., Natesh, J., Penta, D., and Meeran, S.M., 2020. Progress and promises of epigenetic drugs and epigenetic diets in cancer prevention and therapy: A clinical update. *Seminars in Cancer Biology*.

Moon, D., McCormack, D., McDonald, D., and McFadden, D., 2013. Pterostilbene induces mitochondrially derived apoptosis in breast cancer cells in vitro. *Journal of Surgical Research, 180*(2), pp. 208–215.

Mosaddad, S.A., Beigi, K., Doroodizadeh, T., Haghnegahdar, M., Golfeshan, F., Ranjbar, R., and Tebyanian, H., 2021. Therapeutic applications of herbal/synthetic/bio-drug in oral cancer: An update. *European Journal of Pharmacology.* doi: 10.1016/j.ejphar.2020.173657.

Motolani, A., Martin, M., Sun, M., and Lu, T., 2021. NF-κB and Cancer Therapy Drugs. *In: Reference Module in Biomedical Sciences.* Elsevier.

Moulder, J.E., and Seymour, C., 2018. Radiation fractionation: The search for isoeffect relationships and mechanisms. *International Journal of Radiation Biology, 94*(8), pp. 743–751.

Mukherjee, A.K., Basu, S., Sarkar, N., and Ghosh, A.C., 2001. Advances in cancer therapy with plant based natural products. *Current Medicinal Chemistry, 8*(12), pp. 1467–1486.

Munshi, P.N., Lubin, M., and Bertino, J.R., 2014. 6-thioguanine: a drug with unrealized potential for cancer therapy. *The oncologist, 19*(7), p. 760.

Murphree, A.L., and Benedict, W.F., 1984. Retinoblastoma: Clues to human oncogenesis. *Science, 223*(4640), pp. 1028–1033.

Murray, G.I., Duncan, M.E., O'Neil, P., Melvin, W.T., and Fothergill, J.E., 1996. Matrix metalloproteinase–1 is associated with poor prognosis in colorectal cancer. *Nature Medicine, 2*(4), pp. 461–462.

Nakayama, D.K., and Bonasso, P.C., 2016. The history of multimodal treatment of Wilms' tumor. *The American Surgeon, 82*(6), pp. 487–492.

Nguyen, T.T., Nguyen, T.T.D., Ta, Q.T.H., and Vo, V.G., 2020. Advances in non and minimal-invasive transcutaneous delivery of immunotherapy for cancer treatment. *Biomedicine and Pharmacotherapy, 131.* doi: 10.1016/j.biopha.2020.110753.

Okamoto, K., and Seimiya, H., 2019. Revisiting telomere shortening in cancer. *Cells, 8*(2), p. 107.

Olgen, S., 2018. Overview on anticancer drug design and development. *Current Medicinal Chemistry, 25*(15), pp. 1704–1719.

Palmieri, C., and Macpherson, I.R., 2021. A review of the evidence base for utilizing Child-Pugh criteria for guiding dosing of anticancer drugs in patients with cancer and liver impairment. *ESMO Open, 6*(3), p. 100162.

Papac, R.J., 2001. Origins of cancer therapy. *The Yale Journal of Biology and Medicine, 74*(6), p. 391.

Parameswaran, S., Kundapur, D., Vizeacoumar, F.S., Freywald, A., Uppalapati, M., and Vizeacoumar, F.J., 2019. A road map to personalizing targeted cancer therapies using synthetic lethality. *Trends in Cancer.* doi: 10.1016/j.trecan.2018.11.001.

Park, S.R., Yoon, Y.J., Pham, J.V., Yilma, M.A., Feliz, A., Majid, M.T., Maffetone, N., Walker, J.R., Kim, E., Reynolds, J.M., and Song, M.C., 2019. A review of the microbial production of bioactive natural products and biologics. *Frontiers in Microbiology, 10*, p. 1404.

Pocasap, P., Weerapreeyakul, N., Timonen, J., Järvinen, J., Leppänen, J., Kärkkäinen, J., and Rautio, J., 2020. Tyrosine–Chlorambucil Conjugates Facilitate Cellular Uptake through L-Type Amino Acid Transporter 1 (LAT1) in Human Breast Cancer Cell Line MCF-7. *International Journal of Molecular Sciences, 21*(6), p. 2132.

Polifka, J.E., and Friedman, J.M., 2002. Teratogen update: azathioprine and 6-mercaptopurine. *Teratology, 65*(5), pp. 240–261.

Pranzini, E., Pardella, E., Paoli, P., Fendt, S.M., and Taddei, M.L., 2021. Metabolic reprogramming in anticancer drug resistance: A focus on amino acids. *Trends in Cancer.* doi: 10.1016/j.trecan. 2021.02.004.

Pusey, A., 1983. Roentgen-rays in the treatment of skin diseases and for the removal of hair. 1. *Archives of Dermatology, 119*(2), pp. 162–175.

Pylayeva-Gupta, Y., Grabocka, E., and Bar-Sagi, D., 2011. RAS oncogenes: weaving a tumorigenic web. *Nature Reviews Cancer, 11*(11), pp. 761–774.

Qi, L., Zhang, Y., Song, F., and Ding, Y., 2020. Chinese herbal medicine promote tissue differentiation in colorectal cancer by activating HSD11B2. *Archives of Biochemistry and Biophysics, 695.* doi: 10.1016/j.abb.2020.108644.

Rahimi, R., and Solimannejad, M., 2021. First-principles survey on the pristine BC2N monolayer as a promising vehicle for delivery of β-lapachone anticancer drug. *Journal of Molecular Liquids, 321.*

Rana, A., and Bhatnagar, S., 2021. Advancements in folate receptor targeting for anticancer therapy: A small molecule-drug conjugate approach. *Bioorganic Chemistry, 112.* doi: 10.1016/j.bioorg.2021. 104946.

Rattanaburee, T., Tipmanee, V., Tedasen, A., Thongpanchang, T., and Graidist, P., 2020. Inhibition of CSF1R and AKT by (±)-kusunokinin hinders breast cancer cell proliferation. *Biomedicine and Pharmacotherapy*, *129*. doi: 10.1016/j.biopha.2020.110361.

Razavi, Z.S., Asgarpour, K., Mahjoubin-Tehran, M., Rasouli, S., Khan, H., Shahrzad, M.K., Hamblin, M.R., and Mirzaei, H., 2021. Angiogenesis-related non-coding RNAs and gastrointestinal cancer. *Molecular Therapy – Oncolytics*. doi: 10.1016/j.omto.2021.04.002.

Ren, P., and Hu, M., 2019. A three long non-coding RNA signature to improve survival prediction in patients with Wilms' tumor. *Oncology Letters*, *18*(6), pp. 6164–6170.

Říhová, B., Jelínková, M., Strohalm, J., Šubr, V., Plocová, D., Hovorka, O., Novák, M., Plundrová, D., Germano, Y., and Ulbrich, K., 2000. Polymeric drugs based on conjugates of synthetic and natural macromolecules. II. Anti-cancer activity of antibody or (Fab')2-targeted conjugates and combined therapy with immunomodulators. *Journal of Controlled Release*, 64. pp. 241–261.

Roberts, A.W., and Huang, D.C.S., 2017. Targeting BCL2 with BH3 mimetics: basic science and clinical application of venetoclax in chronic lymphocytic leukemia and related B cell malignancies. *Clinical Pharmacology & Therapeutics*, *101*(1), pp. 89–98.

Rückert, M., Deloch, L., Fietkau, R., Frey, B., Hecht, M., and Gaipl, U.S., 2018. Immune modulatory effects of radiotherapy as basis for well-reasoned radioimmunotherapies. *Strahlentherapie und Onkologie*, *194*(6), pp. 509–519.

Sadat, S.M.A., Paiva, I.M., Shire, Z., Sanaee, F., Morgan, T.D.R., Paladino, M., Karimi-Busheri, F., Mani, R.S., Martin, G.R., Jirik, F.R., Hall, D.G., Weinfeld, M., and Lavasanifar, A., 2021. A synthetically lethal nanomedicine delivering novel inhibitors of polynucleotide kinase 3'-phosphatase (PNKP) for targeted therapy of PTEN-deficient colorectal cancer. *Journal of Controlled Release*, *334*, pp. 335–352.

Saeed, A.F.U.H., Su, J., and Ouyang, S., 2021. Marine-derived drugs: Recent advances in cancer therapy and immune signaling. *Biomedicine and Pharmacotherapy*. doi: 10.1016/j.biopha.2020.111091.

Safaei, M., and Shishehbore, M.R., 2021. A review on analytical methods with special reference to electroanalytical methods for the determination of some anticancer drugs in pharmaceutical and biological samples. *Talanta*. doi: 10.1016/j.talanta.2021.122247.

Sarah, P., and David, M., 2018. Mechanisms of Anticancer Drugs. *In: Scott-Brown's Otorhinolaryngology Head and Neck Surgery* (pp. 39–50). CRC Press.

Saravanan, S., Vimalraj, S., Pavani, K., Nikarika, R., and Sumantran, V.N., 2020. Intussusceptive angiogenesis as a key therapeutic target for cancer therapy. *Life Sciences*. doi: 10.1016/j.lfs.2020.117670.

Schmitt, C.A., Fridman, J.S., Yang, M., Baranov, E., Hoffman, R.M., and Lowe, S.W., 2002. Dissecting p53 tumor suppressor functions in vivo. *Cancer Cell*, *1*(3), pp. 289–298.

Schulze-Rath, R., Hammer, G.P., and Blettner, M., 2008. Are pre-or postnatal diagnostic X-rays a risk factor for childhood cancer? A systematic review. *Radiation and Environmental Biophysics*, *47*(3), pp. 301–312.

Shan, X., Li, S., Sun, B., Chen, Q., Sun, J., He, Z., and Luo, C., 2020. Ferroptosis-driven nanotherapeutics for cancer treatment. *Journal of Controlled Release*. doi: 10.1016/j.jconrel.2020.01.008.

Shariare, M.H., Noor, H.B., Khan, J.H., Uddin, J., Ahamad, S.R., Altamimi, M.A., Alanazi, F.K., and Kazi, M., 2021. Liposomal drug delivery of Corchorus olitorius leaf extract containing phytol using design of experiment (DoE): In-vitro anticancer and in-vivo anti-inflammatory studies. *Colloids and Surfaces B: Biointerfaces*, *199*. doi: 10.1016/j.colsurfb.2020.111543.

Sharpe, M.M., 1900. The X-ray treatment of skin diseases. *Archives of the Roentgen Ray*, *4*(3), pp. 52–60.

Shay, J.W., and Bacchetti, S., 1997. A survey of telomerase activity in human cancer. *European journal of cancer*, *33*(5), pp. 787–791.

Shen, Y., and Noguchi, H., 2021. Impacts of anticancer drug parity laws on mortality rates. *Social Science and Medicine*, *272*. doi: 10.1016/j.socscimed.2021.113714.

Shubhra, Q.T.H., Guo, K., Liu, Y., Razzak, M., Manir, M.S., and Alam, A.K.M., 2021. Dual targeting smart drug delivery system for multimodal synergistic combination cancer therapy with reduced cardiotoxicity. *Acta Biomaterialia*. doi: 10.1016/j.actbio.2021.06.016.

Sidhu, J.S., and Zafar, T.A., 2020. Indian Herbal Medicine and Their Functional Components in Cancer Therapy and Prevention. *In: Functional Foods in Cancer Prevention and Therapy* (pp. 169–194). Elsevier.

Slater, J.M., 2012. From X-Rays to Ion Beams: A Short History of Radiation Therapy. *In: Ion Beam Therapy* (pp. 3–16). Springer, Berlin, Heidelberg.

Sledge Jr, G.W., Qulali, M., Goulet, R., Bone, E.A., and Fife, R., 1995. Effect of matrix metalloproteinase inhibitor batimastat on breast cancer regrowth and metastasis in athymic mice. *JNCI: Journal of the National Cancer Institute*, *87*(20), pp. 1546–1551.

So, T.H., Chow, Z., Chan, K.S., Lam, K.O., Lee, V.H.F., and Choi, H.H., 2019. A pilot cross-sectional study on incidence of liver toxicity in cancer patients on western anticancer drug therapy with or without concurrent Chinese herbal medicine. *Annals of Oncology*, *30*, p. ix146.

Sparano, J.A., Bernardo, P., Stephenson, P., Gradishar, W.J., Ingle, J.N., Zucker, S., and Davidson, N.E., 2004. Randomized phase III trial of marimastat versus placebo in patients with metastatic breast cancer who have responding or stable disease after first-line chemotherapy: Eastern Cooperative Oncology Group trial E2196. *Journal of Clinical Oncology*, *22*(23), pp. 4683–E4690.

Sun, J., Wei, Q., Zhou, Y., Wang, J., Liu, Q., and Xu, H., 2017. A systematic analysis of FDA-approved anticancer drugs. *BMC Systems Biology*, *11*(5), p. 87.

Sun, M., He, L., Fan, Z., Tang, R., and Du, J., 2020. Effective treatment of drug-resistant lung cancer via a nanogel capable of reactivating cisplatin and enhancing early apoptosis. *Biomaterials*, *257*. doi: 10.1016/j.biomaterials.2020.120252.

Sunwoo, K., Won, M., Ko, K.P., Choi, M., Arambula, J.F., Chi, S.G., Sessler, J.L., Verwilst, P., and Kim, J.S., 2020. Mitochondrial relocation of a common synthetic antibiotic: A non-genotoxic approach to cancer therapy. *Chem*, *6*(6), pp. 1408–1419.

Surget, S., Khoury, M.P., and Bourdon, J.C., 2014. Uncovering the role of p53 splice variants in human malignancy: a clinical perspective. *OncoTargets and Therapy*, *7*, p. 57.

Sweilam, N.H., AL-Mekhlafi, S.M., Albalawi, A.O., and Tenreiro Machado, J.A., 2021. Optimal control of variable-order fractional model for delay cancer treatments. *Applied Mathematical Modelling*, *89*, pp. 1557–1574.

Takahashi, T., Ichikawa, H., Morimoto, Y., Tsuneyama, K., and Hijikata, T., 2019. Inhibition of EP2/EP4 prostanoid receptor-mediated signaling suppresses IGF-1-induced proliferation of pancreatic cancer BxPC-3 cells via upregulating γ-glutamyl cyclotransferase expression. *Biochemical and Biophysical Research Communications*, *516*(2), pp. 388–396.

Tewari, D., Bawari, S., Sharma, S., DeLiberto, L.K., and Bishayee, A., 2021. Targeting the crosstalk between canonical Wnt/β-catenin and inflammatory signaling cascades: A novel strategy for cancer prevention and therapy. *Pharmacology & Therapeutics*, *107876*. doi: 10.1016/j.pharmthera.2021.107876.

Thirumaran, R., Prendergast, G.C., and Gilman, P.B., 2007. Cytotoxic Chemotherapy in Clinical Treatment of Cancer. *In: Cancer Immunotherapy* (pp. 101–116). Academic Press.

Thng, D.K.H., Toh, T.B., and Chow, E.K.H., 2021. Capitalizing on synthetic lethality of MYC to treat cancer in the digital age. *Trends in Pharmacological Sciences*. doi: 10.1016/j.tips.2020.11.014.

Thomas, G., Eisenhauer, E., Bristow, R.G., Grau, C., Hurkmans, C., Ost, P., Guckenberger, M., Deutsch, E., Lacombe, D., and Weber, D.C., 2020. The European Organisation for Research and Treatment of Cancer, State of Science in Radiation Oncology and Priorities for Clinical Trials Meeting Report. *In: European Journal of Cancer* (pp. 76–88). Elsevier Ltd.

Thu, P.M., Zheng, Z.G., Zhou, Y.P., Wang, Y.Y., Zhang, X., Jing, D., Cheng, H.M., Li, J., Li, P., and Xu, X., 2019. Phellodendrine chloride suppresses proliferation of KRAS mutated pancreatic cancer cells through inhibition of nutrients uptake via macropinocytosis. *European Journal of Pharmacology*, *850*, pp. 23–34.

Thun, M.J., Henley, S.J., and Patrono, C., 2002. Nonsteroidal anti-inflammatory drugs as anticancer agents: mechanistic, pharmacologic, and clinical issues. *Journal of the National Cancer Institute*, *94*(4), pp. 252–266.

Tian, B., Liu, Y., and Liu, J., 2021. Chitosan-based nanoscale and non-nanoscale delivery systems for anticancer drugs: A review. *European Polymer Journal*, *154*, p. 110533.

Toufektchan, E., and Toledo, F., 2018. The guardian of the genome revisited: p53 downregulates genes required for telomere maintenance, DNA repair, and centromere structure. *Cancers*, *10*(5), p. 135.

Trevor, A.J., Katzung, B.G., Masters, S.B., and Kruidering-Hall, M., 2010. Cancer chemotherapy. *Pharmacology Examination & Board Review*. New York: McGraw-Hill Medical.

Truong Hoang, Q., Lee, D.Y., Choi, D.G., Kim, Y.C., and Shim, M.S., 2021. Efficient and selective cancer therapy using pro-oxidant drug-loaded reactive oxygen species (ROS)-responsive polypeptide micelles. *Journal of Industrial and Engineering Chemistry*, *95*, pp. 101–108.

Tsai, F.J., Liu, X., Chen, C.J., Li, T.M., Chiou, J.S., Chuang, P.H., Ko, C.H., Lin, T.H., Liao, C.C., Huang, S.M., Liang, W.M., and Lin, Y.J., 2019. Chinese herbal medicine therapy and the risk of overall mortality for patients with liver cancer who underwent surgical resection in Taiwan. *Complementary Therapies in Medicine*, *47*.

Vasquez, J.L., Lai, Y., Annamalai, T., Jiang, Z., Zhang, M., Lei, R., Zhang, Z., Liu, Y., Tse-Dinh, Y.C., and Agoulnik, I.U., 2020. Inhibition of base excision repair by natamycin suppresses prostate cancer cell proliferation. *Biochimie*, *168*, pp. 241–250.

Vélez-Cruz, R., and Johnson, D.G., 2017. The retinoblastoma (RB) tumor suppressor: Pushing back against genome instability on multiple fronts. *International journal of molecular sciences*, *18*(8), p. 1776.

Waller, D.G., and Sampson, T., 2017. *Medical Pharmacology and Therapeutics E-Book*. Elsevier Health Sciences.

Wang, D., Liu, W., Wang, L., Wang, Y., Liao, C.K., Chen, J., Hu, P., Hong, W., Huang, M., Chen, Z., and Xu, P., 2020. Suppression of cancer proliferation and metastasis by a versatile nanomedicine integrating photodynamic therapy, photothermal therapy, and enzyme inhibition. *Acta Biomaterialia*, *113*, pp. 541–553.

Wang, J., Yang, J., Cao, M., Zhao, Z., Cao, B., and Yu, S., 2021. The potential roles of Nrf2/Keap1 signaling in anticancer drug interactions. *Current Research in Pharmacology and Drug Discovery*, *2*, p. 100028.

Wang, S., Wu, X., Tan, M., Gong, J., Tan, W., Bian, B., Chen, M., and Wang, Y., 2012. Fighting fire with fire: Poisonous Chinese herbal medicine for cancer therapy. *Journal of Ethnopharmacology*. doi: 10.1016/j.jep.2011.12.041.

Wang, S., Zhang, R.-H., Zhang, H., Wang, Y.-C., Yang, D., Zhao, Y.-L., Yan, G.-Y., Xu, G.-B., Guan, H.-Y., Zhou, Y.-H., Cui, D.-B., Liu, T., Li, Y.-J., Liao, S.-G., and Zhou, M., 2021. Design, synthesis, and biological evaluation of 2,4-diamino pyrimidine derivatives as potent FAK inhibitors with anticancer and anti-angiogenesis activities. *European Journal of Medicinal Chemistry*, *222*, p. 113573.

Wang, S., Zhou, D., Xu, Z., Song, J., Qian, X., Lv, X., and Luan, J., 2019. Anti-tumor drug targets analysis: Current insight and future prospect. *Current Drug Targets*, *20*(11), pp. 1180–1202.

Wang, Y., Wang, W., Xu, L., Zhou, X., Shokrollahi, E., Felczak, K., Van Der Laan, L.J., Pankiewicz, K.W., Sprengers, D., Raat, N.J., and Metselaar, H.J., 2016. Cross talk between nucleotide synthesis pathways with cellular immunity in constraining hepatitis E virus replication. *Antimicrobial Agents and Chemotherapy*, *60*(5), pp. 2834–2848.

Weber, G.F., 2015. DNA Damaging Drugs. *In*: *Molecular Therapies of Cancer* (pp. 9–112). Springer, Cham.

Wermuth, C.G., 2008. Designing Prodrugs and Bioprecursors. *In*: *The Practice of Medicinal Chemistry* (pp. 721–746). Academic Press.

Wolff, A.C., Hammond, M.E.H., Allison, K.H., Harvey, B.E., Mangu, P.B., Bartlett, J.M., Bilous, M., Ellis, I.O., Fitzgibbons, P., Hanna, W., and Jenkins, R.B., 2018. Human epidermal growth factor receptor 2 testing in breast cancer: American Society of Clinical Oncology/College of American Pathologists clinical practice guideline focused update. *Archives of Pathology & Laboratory Medicine*, *142*(11), pp. 1364–1382.

Wu, J.Q., Fan, R.Y., Zhang, S.R., Li, C.Y., Shen, L.Z., Wei, P., He, Z.H., and He, M.F., 2020. A systematical comparison of anti-angiogenesis and anticancer efficacy of ramucirumab, apatinib, regorafenib and cabozantinib in zebrafish model. *Life Sciences*, *247*. doi: 10.1016/j.lfs.2020.117402.

Wyld, L., Audisio, R.A., and Poston, G.J., 2015. The evolution of cancer surgery and future perspectives. *Nature Reviews Clinical Oncology*, *12*(2), p. 115.

Xiao, Y., Gu, Y., Qin, L., Chen, L., Chen, X., Cui, W., Li, F., Xiang, N., and He, X., 2021. Injectable thermosensitive hydrogel-based drug delivery system for local cancer therapy. *Colloids and Surfaces B: Biointerfaces*. doi: 10.1016/j.colsurfb.2021.111581.

Xing, L., Yang, C.X., Zhao, D., Shen, L.J., Zhou, T.J., Bi, Y.Y., Huang, Z.J., Wei, Q., Li, L., Li, F., and Jiang, H.L., 2021. A carrier-free anti-inflammatory platinum (II) self-delivered nanoprodrug for enhanced breast cancer therapy. *Journal of Controlled Release*, *331*, pp. 460–471.

Xu, C.H., Ye, P.J., Zhou, Y.C., He, D.X., Wei, H., and Yu, C.Y., 2020. Cell membrane-camouflaged nanoparticles as drug carriers for cancer therapy. *Acta Biomaterialia*. https://doi.org/10.1016/j.actbio.2020.01.036.

Xu, X., Chen, R., Chen, Q., An, K., Ding, L., Zhang, L., Wang, F., and Deng, Y., 2021. Efficacy of traditional herbal medicine versus transcatheter arterial chemoembolization in postsurgical patients with hepatocellular carcinoma: A retrospective study. *Complementary Therapies in Clinical Practice*, *43*. 10.1016/j.ctcp.2021.101359.

Yamamoto, H., Adachi, Y., Itoh, F., Iku, S., Matsuno, K., Kusano, M., Arimura, Y., Endo, T., Hinoda, Y., Hosokawa, M., and Imai, K., 1999. Association of matrilysin expression with recurrence and poor prognosis in human esophageal squamous cell carcinoma. *Cancer Research*, *59*(14), pp. 3313–3316.

Yao, H.P., Hudson, R., and Wang, M.H., 2020. Progress and challenge in development of biotherapeutics targeting MET receptor for treatment of advanced cancer. *Biochimica et Biophysica Acta – Reviews on Cancer.* doi: 10.1016/j.bbcan.2020.188425.

Yeh, M.H., Wu, H.C., Lin, N.W., Hsieh, J.J., Yeh, J.W., Chiu, H.P., Wu, M.C., Tsai, T.Y., Yeh, C.C., and Li, T.M., 2020. Long-term use of combined conventional medicine and Chinese herbal medicine decreases the mortality risk of patients with lung cancer. *Complementary Therapies in Medicine, 52.* doi: 10.1016/j.ctim.2020.102427.

Yekedüz, E., Utkan, G., and Ürün, Y., 2020. A systematic review and meta-analysis: the effect of active cancer treatment on severity of COVID-19. *European Journal of Cancer, 141*, pp. 92–104.

Yetkin-Arik, B., Kastelein, A.W., Klaassen, I., Jansen, C.H.J.R., Latul, Y.P., Vittori, M., Biri, A., Kahraman, K., Griffioen, A.W., Amant, F., Lok, C.A.R., Schlingemann, R.O., and van Noorden, C.J.F., 2021. Angiogenesis in gynecological cancers and the options for anti-angiogenesis therapy. *Biochimica et Biophysica Acta – Reviews on Cancer.* doi: 10.1016/j.bbcan.2020. 188446.

Yin, X., Cai, S., Tao, L., Chen, L., Zhang, Z., Xiao, S., Fan, A.Y., and Zou, X., 2021. Recovery of a patient with severe COVID-19 by acupuncture and Chinese herbal medicine adjuvant to standard care. *Journal of Integrative Medicine.* doi: 10.1016/j.joim.2021.06.001.

Yoshizawa, A., Takahara, K., Saruta, M., Zennami, K., Nukaya, T., Fukaya, K., Ichino, M., Fukami, N., Niimi, A., Sasaki, H., Kusaka, M., Suzuki, M., Sumitomo, M., and Shiroki, R., 2021. Combined α-methylacyl-CoA racemase inhibition and docetaxel treatment reduce cell proliferation and decrease expression of heat shock protein 27 in androgen receptor-variant-7–positive prostate cancer cells. *Prostate International, 9*(1), pp. 18–24.

Zappavigna, S., Cossu, A.M., Grimaldi, A., Bocchetti, M., Ferraro, G.A., Nicoletti, G.F., Filosa, R., and Caraglia, M., 2020. Anti-inflammatory drugs as anticancer agents. *International Journal of Molecular Sciences, 21*(7), p. 2605.

Zargar, A., Chang, S., Kothari, A., Snijders, A.M., Mao, J.-H., Wang, J., Hernández, A.C., Keasling, J.D., and Bivona, T.G., 2019. Overcoming the challenges of cancer drug resistance through bacterial-mediated therapy. *Chronic Diseases and Translational Medicine, 5*(4), pp. 258–266.

Zhai, B., Zhang, N., Han, X., Li, Q., Zhang, M., Chen, X., Li, G., Zhang, R., Chen, P., Wang, W., Li, C., Xiang, Y., Liu, S., Duan, T., Lou, J., Xie, T., and Sui, X., 2019. Molecular targets of β-elemene, a herbal extract used in traditional Chinese medicine, and its potential role in cancer therapy: A review. *Biomedicine and Pharmacotherapy.* doi: 10.1016/j.biopha.2019.108812.

Zhang, D., Wu, P., Zhang, Z., An, W., Zhang, C., Pan, S., Tan, Y., and Xu, H., 2020. Overexpression of negative regulator of ubiquitin-like proteins 1 (NUB1) inhibits proliferation and invasion of gastric cancer cells through upregulation of p27Kip1 and inhibition of epithelial-mesenchymal transition. *Pathology Research and Practice, 216*(8). doi: 10.1016/j.prp.2020.153002.

Zhang, M., Chen, X., and Radacsi, N., 2021. New tricks of old drugs: Repurposing non-chemo drugs and dietary phytochemicals as adjuvants in anti-tumor therapies. *Journal of Controlled Release.* doi: 10.1016/j.jconrel.2020.11.047.

Zhang, N., Yin, Y., Xu, S.J., and Chen, W.S., 2008. 5-Fluorouracil: mechanisms of resistance and reversal strategies. *Molecules, 13*(8), pp. 1551–1569.

Zhang, S., Cao, M., Hou, Z., Gu, X., Chen, Y., Chen, L., Luo, Y., Chen, L., Liu, D., Zhou, H., Zhu, K., Wang, Z., Zhang, X., Zhu, X., Cui, Y., Li, H., Guo, H., and Zhang, T., 2021. Angiotensin-converting enzyme inhibitors have adverse effects in anti-angiogenesis therapy for hepatocellular carcinoma. *Cancer Letters, 501*, pp. 147–161.

Zhao, B., Aggarwal, A., Marshall, J.A., and Nehs, M.A., 2020. Inhibition of Glucose Metabolism with Metformin Decreases Tumor Proliferation and Migration in Anaplastic Thyroid Cancer. *Journal of the American College of Surgeons, 231*(4), p. S77.

Zheng, J., Mo, J., Zhu, T., Zhuo, W., Yi, Y., Hu, S., Yin, J., Zhang, W., Zhou, H., and Liu, Z., 2020. Comprehensive elaboration of the cGAS-STING signaling axis in cancer development and immunotherapy. *Molecular Cancer, 19*(1), pp. 1–19.

Zhi, D., Yang, T., Zhang, T., Yang, M., Zhang, S., and Donnelly, R.F., 2021. Microneedles for gene and drug delivery in skin cancer therapy. *Journal of Controlled Release, 335*, pp. 158–177.

Zhong, L.L.D., Chen, H.Y., Cho, W.C.S., Meng, X. ming, and Tong, Y., 2012. The efficacy of Chinese herbal medicine as an adjunctive therapy for colorectal cancer: A systematic review and meta-analysis. *Complementary Therapies in Medicine.* doi: 10.1016/j.ctim.2012.02.004.

Zhou, L., Yin, R., Gao, N., Sun, H., Chen, D., Cai, Y., Ren, L., Yang, L., Zuo, Z., Zhang, H., and Zhao, J., 2021. Oligosaccharides from fucosylated glycosaminoglycan prevent breast cancer metastasis in mice by inhibiting heparanase activity and angiogenesis. *Pharmacological Research*, *166*. doi: 10.1016/j.phrs.2021.105527.

Zhu, Y., Yang, L., Xu, J., Yang, X., Luan, P., Cui, Q., Zhang, P., Wang, F., Li, R., Ding, X., Jiang, L., Lin, G., and Zhang, J., 2020. Discovery of the anti-angiogenesis effect of eltrombopag in breast cancer through targeting of HuR protein. *Acta Pharmaceutica Sinica B*, *10*(8), pp. 1414–1425.

Zucker, S., Cao, J., and Molloy, C.J., 2002. Role of matrix metalloproteinases and plasminogen activators in cancer invasion and metastasis: Therapeutic strategies. *Anticancer Drug Development*, *6*, pp. 91–122.

Zúñiga, R., Concha, G., Cayo, A., Cikutović-Molina, R., Arevalo, B., González, W., Catalán, M.A., and Zúñiga, L., 2020. Withaferin A suppresses breast cancer cell proliferation by inhibition of the two-pore domain potassium (K2P9) channel TASK-3. *Biomedicine and Pharmacotherapy*, *129*. doi: 10.1016/j.biopha.2020.110383.

7 *In Silico* Approach to Cancer Therapy

7.1 INTRODUCTION

The ultimate aim of modern oncology relies on personalized medicine and therapy individualization approaches (Ghetiu et al., 2011; Toss and Cristofanilli, 2015; Amin et al., 2017; Mehta et al., 2019; Shamsi et al., 2019; Bianco, Goodsell and Forli, 2020; Paredes-Ramos et al., 2020; Zoudani and Soltani, 2020; Balasubramanian, 2021; Wadanambi and Mannapperuma, 2021). Clinicians aim to diagnose the emergence of diseases as well as prediction of response to treatment. However, a reliable and accurate prognostic approximation is the basic issue for the therapy course. The ability to estimate disease effects more precisely would allow clinicians to make decisions for surgical procedure extension and the potential need for adjuvant therapy (Park et al., 2013; Chari et al., 2021; Dong et al., 2021; Mohanty et al., 2021; Moradi Kashkooli et al., 2021; Poustforoosh et al., 2021). It is very challenging to objectively measure the impact of different parameters important for disease development, to analyze them, and to make satisfactory choices of therapy. Computer-aided methods (Figure 7.1) that work on different artificial intelligence–based classification techniques can help specialists substantially (Cruz and Wishart, 2007; Kourou et al., 2015) though the use of these computer-based methods in medical practices is strongly based on proving predictions reliably and demonstration of developed knowledge.

7.2 MACHINE LEARNING (ML) APPROACHES IN BREAST CANCER PROGNOSIS PREDICTION

For classification based on increased recurrence risk and metastases, prognostic calculators are used for patients (Kalinli et al., 2013; Kourou et al., 2015; Burki, 2016; Brieu et al., 2018; Alizadeh Savareh et al., 2020; Andjelkovic Cirkovic, 2020; Praiss et al., 2020; Shao et al., 2020; Wang et al., 2020; Alabi, Mäkitie et al., 2021). The first predictive method that was introduced in medical examinations was called the Gail model. This model was used for the prediction of breast cancer development risk and confirmed in different studies (Gail et al., 1989; Costantino et al., 1999; Jacobi et al., 2009; Mealiffe et al., 2010). The Gail model is an interactive tool consisting of six factors including age, menarche, earlier breast biopsy, tissue atypia, age at the time of first childbirth, and family history. Various literature are showing the prediction models for the prognosis of breast and other types of cancers and the development of databases (Alabi, Youssef et al., 2021; Chai et al., 2021; Eun et al., 2021; Fahami et al., 2021; Gopal et al., 2021; Guo et al., 2021; Jung et al., 2021; Kaushik et al., 2021; Masuda et al., 2021; Panahi et al., 2021; Shaikh and Rao, 2021). In a study (Park et al. 2013), the surveillance epidemiology end results (SEER) database was used for comparison of artificial neural networks (ANNs), semi-supervised learning (SSL), and support vector machines (SVMs) for the prediction of 5-year survival of breast cancer patients. In a study (Kim et al., 2012), a model was developed for the prediction of recurrence of breast cancer by comparing the SVM and ANN classifiers in the database of the Korean Breast Cancer Centre hospital. Wisconsin Prognostic Breast Cancer (WPBC) was used for the testing of ANN models (Newman et al., 1998) (Figure 7.2). In a study of Burke et al. (1997), it was concluded that the

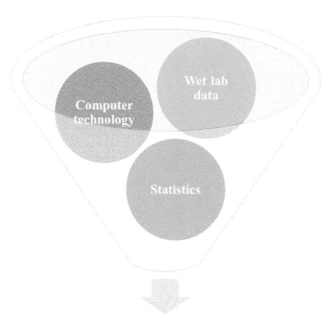

Computational models for
drug design

FIGURE 7.1 Computer-based approach for drug design.

5-year survival rate obtained through ANN predictions was more precise than the TNM staging system. In this study, patient care evaluation (PCE) and SEER breast carcinoma data sets were used. Moreover, in another study (Strumbelj et al. 2010), to predict breast cancer recurrence, a standard machine learning methods study was presented by using the database of the Institute of Oncology, Lju-bljana.

7.2.1 METHODOLOGICAL OUTLINE FOR MACHINE LEARNING METHODS

Machine learning signifies the set of procedures with the ability to learn from data. It is an artificial intelligence branch. Several mathematical models have been executed to approximate multifaceted relationships in data for predictive goals (Figure 7.3). Hence, ML is quite a useful method for many domains where relevant, real data collections could not be surveyed (Chunyu et al., 2020; D'Souza et al., 2020; Kowalewski and Ray, 2020; Réda, Kaufmann and Delahaye-Duriez, 2020; Zhao et al., 2020; Bannigan et al., 2021; Rajput et al., 2021; Sasahara et al., 2021a, 2021b). The methodological framework was followed in different studies reported at different times (Andjelkovic-Cirkovic et al., 2015; Elbadawi et al., 2021; Farizhandi et al., 2021; Gaur et al., 2021; Hathout, 2021; Issa et al., 2021; Kamerzell and Middaugh, 2021; Woillard et al., 2021; Yeung et al., 2021; Zhu et al., 2021).

7.2.1.1 Preprocessing

Preprocessing of the data is done for the reduction of number variables and to get clean data. Specifically, variables with missing values in high proportion must be omitted from further processing, although variables with a low proportion of missing variables could be analyzed. The standard approach to deal with this concern is the replacement of missing values by feature means (in case of numerical variables) and modes (nominal variables).

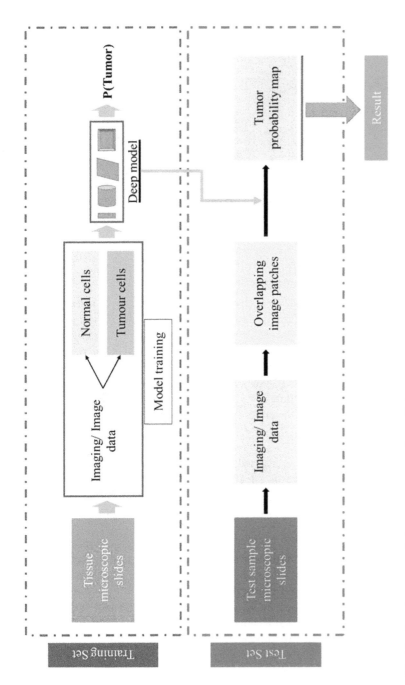

FIGURE 7.2 Machine-learning approach for cancer prognosis prediction.

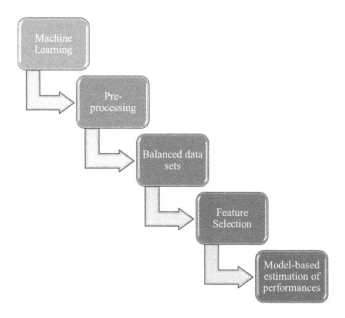

FIGURE 7.3 Outline of models based on machine learning.

7.2.1.2   The Requirement of Balanced Data Sets

Generally, the 5-year survival and recurrence rate of disease among breast cancer patients might lead to imbalanced data classification distribution (Brewster et al., 2008). Usually, trials with lesser frequency can affect the ML models negatively, and its inappropriate classification might lead to adverse decisions and lead to heavy costs. Therefore, for model assessment, a performance measurement metric that exploits class distribution must be adopted. The SMOTE (Synthetic Minority Over-sampling TEchnique) algorithm (Chawla et al., 2002) has been proven as a suitable choice in the case of a data set with a fewer number of minority cases. This system is based on the k-Nearest Neighbours (k-NN) method.

7.2.1.3   Features Selection

After organizing the data, feature values can be inserted directly into classification algorithms, or this step can be exploited for the maintenance of a feature subset contributing mostly to classification accuracy and ease of the task of classification (Figure 7.4).

In a study (Filipovic et al., 2020) for feature election, three filter algorithms and wrapper methods were used.

1. **mRMR (minimum redundancy-maximum relevance):** This algorithm of feature selection inclines for the selection of the features subset with the maximum relevance with the yield and the minimum correlation among the features themselves.
2. **Relief F:** This algorithm was developed by Kira and Rendell (1992) and upgraded by Kononenko (1993) and Robnik-Sikonja and Kononenko (1997). This algorithm evaluates the efficacy of the feature rendering to their dissimilar/similar values for neighbors' occurrences from the same or different classes. The algorithm correctly estimates the quality of features in queries, with strong dependencies between the features.
3. **Information gain (IG):** This algorithm evaluates a distinct feature at a time. This filter divides a set of cases into separate subsets rendering attribute values.
4. **Wrapper algorithm:** This algorithm was joined with the first search and IWSS (Incremental Wrapper Subset Selection) algorithm (Bermejo et al., 2009) following advance selection.

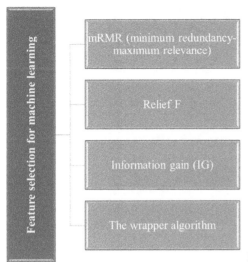

FIGURE 7.4   Selection of features for machine learning.

The assessment of feature division was completed by estimating the classifier's AUC (area under the curve) factors contained by the 10-fold cross-examination process.

7.2.1.4   Model-Based Estimation of Performances

The next step to pre-processing data is the learning task by various algorithms (Figure 7.5). In a study (Filipovic et al., 2020), the ML algorithm for the prognosis of breast cancer was described:

1. **Naive Bayes (NB):** This algorithm classification is based on Bayes's theorem. It can perform remarkably in real cases, specifically in the field of medicine (Kononenko, 1993).
2. **Logistic regression (LR):** This algorithm is used entirely for two-class cases. It exploits maximum probability estimation for evaluation of the possibility of class association.
3. **Random forest (RF):** It is a collective classifier consisting of several verdict trees and yields the class by utilizing the "voting" principle. To reduce the variance with the production of decision trees, the bootstrap-combining method along with the tree-learning algorithm was used.
4. **Support vector machine (SVM):** It belongs to a set of best algorithms effective for binary classification. SVM inclines to exploit all accessible features, despite being insignificant. This system places classifying hyperplanes in the unique or altered feature area.
5. **Artificial neural network (ANN):** It is a multilayer insight with a single unknown layer. It is best suitable for binary classification and prediction of survival (Zhang, 2000). Due to being motivated by the human brain, this algorithm contains several numbers of vastly interconnected nodes structured into layers for architecture preparation of data processing. This architecture is skilled to identify complex and nonlinear configurations between the input data and output variables.

7.3   ENZYME-MEDIATED CANCER IMAGING AND THERAPY (EMCIT) CONCEPT

This concept (Ho et al., 2002; Chen et al., 2006; Pospisil et al., 2007; Wang et al., 2007; Kassis et al., 2008) includes enzyme overexpression on the cancer cells' surface specifically. The enzyme can hydrolyze the radioisotopically labeled, water-soluble prodrug to a water-insoluble drug. This enzyme-based site-specific reaction of hydrolysis offers a noninvasive method for cancer imaging and therapy.

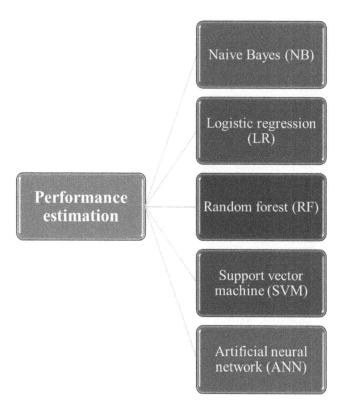

FIGURE 7.5   Alogorithms used in performance estimation of machine learning models.

In a study, the use of the EMCIT approach was done in the case of pancreatic cancer. To identify an EMCIT-suitable hydrolase, they have recognized five phosphatases and one extracellular sulfatase (SULF1), considered an excellent target for EMCIT. SULF1 is known for its specific endoglucosamine-6-sulfatase and (Morimoto-Tomita et al., 2002) arylsulfatase activity, due to causing desulfurization from the aromatic phenyl ring. Therefore, arylsulfatase activity might be linked to arylphosphatase action that is known for alkaline phosphatase (ALP) and prostatic acid phosphatase (PAP) activity for hydrolysis of IQ_{2-P} (ammonium 2-(2′-phosphoryloxyphenyl)-6-iodo-4-(3H)-quinazolinone) (Ho et al., 2002; Chen et al., 2006; Pospisil et al., 2007; Wang et al., 2007; Kassis et al., 2008). IQ_{2-P} was modified to IQ_{2-S} (2-(2′-sulfooxyphenyl)-6-iodo-4-(3H)-quinazolinone), a sulphate-based analog. This IQ_{2-S} was docked into the neighboring homologous 3-D configuration of SULF1, arylsulfatase A (ARSA). A docking study of IQ_{2-S} and IQ_{2-P} with ARSA, ALP, and PAP revealed that these three enzymes of an alkaline-phosphatase family can bind para-nitrophenyl sulfates or phosphates, exhibiting sulfatidic and phosphatidic actions (O'Brien and Herschlag, 1999; Catrina et al., 2007). Overall, the docking study predicts that the replacement of phosphate with sulfate on an iodoquinazolinone skeleton would shift this EMCIT concept to sulfatases from phosphatases, leading to a new prodrug target scheme generation. Particularly in the case of pancreatic cancer, IQ_{2-S} (sulfate derived prodrug), has been proven as a potential prodrug for ARSA and later its neighboring analog SULF1.

It is expected that these quinazolinone-based pharmaceuticals will ultimately result in the novel noninvasive method development for imaging (^{123}I-SPECT; [124]I-PET) and further in treatment (^{131}I) of pancreatic cancer. The EMCIT concept has shown sulfatase and phosphatase-facilitated hydrolysis of quinazolinone and its derivatives, resulting in the production of an active insoluble drug, 2-(2′-hydroxyphenyl)-6-iodo-4-(3H)-quinazolinone (IQ_{2-OH}) (Pospisil and Kassis, 2014), that can move significant mediation to a much earlier stage in the pancreatic cancer progression path before the disease becomes deadly.

7.4   TOOLS AND DATABASES FOR THE STUDY OF CANCER PATHWAYS

There are different databases and tools available for the study of pathways involved in cancer signaling.

7.4.1   TARGET PATHWAYS FOR TREATMENT OF CANCER

Tumorigenesis is a multistep activity in humans and involves alterations at the genetic level that leads to the advanced transformation of normal human cells to malignant entities. Moreover, mutations can occur early in specific oncogenes and tumor suppressor genes and late in others. As a result, tumor cells gain different hallmarks, such as apoptosis resistance, nonstop angiogenesis, proliferation, and infinite replication potential during the time of tumor progression (Hanahan and Weinberg, 2000). Signaling pathways and the processes required for the growth and survival of cells are generally activated in cancer because of genetic changes or mutations involving amplifications (Figure 10.3). Rationally designed drugs and molecules are increasing day by day in clinical trials for cancer treatment. These molecules specifically target the signaling or survival pathways like *JAK-STAT*, *PI3K-AKT-mTOR*, and the mitogen-activated protein (*MAP*) kinase pathway to inhibit cancer growth (Van Ummersen et al., 2004; Hennessy et al., 2005; McCubrey et al., 2008). The *PI3K/AKT* signaling pathway is the main aim for cancer treatment (Vivanco and Sawyers, 2002). For example, in different cancer cases, perifosine, an *AKT* inhibitor, was used in preclinical and clinical studies like prostate cancer (Van Ummersen et al., 2004). This drug works by interrupting the communication between *PIP3* and *AKT*'s pleckstrin homology (PH) domain, leading to inhibition of membrane localization of AKT, required for activation. Additional drugs with antitumor activity are Wortmannin (Rahn et al., 1994) and LY294002 (Hennessy et al., 2005).

Another drug targeting the mammalian rapamycin (mTOR) pathway is rapamycin (Rapamune, Wyeth Ayerst). Rapamycin specifically inhibits the mTOR and works downstream of AKT (Hay and Sonenberg, 2004). Clinical trials of mTOR inhibitors were carried out in breast cancer patients and other solid tumors (Hidalgo and Rowinsky, 2000; Chan et al., 2005; Nagata et al., 2004).

7.4.2   DATABASES INVOLVING MOLECULAR INTERACTION

Computational model preparation involves multiple steps. The first step comprises cellular processes' annotation using suitable annotation tools to build an appropriate computational network (Banik et al., 2021; Behera et al., 2020; Ghasemi et al., 2020; Kumar et al., 2020; L. Zhang et al., 2021; Lee et al., 2021; Pitsillou et al., 2020; Qi et al., 2021). These pathway databases consist of an ample amount of the current knowledge on cancer-related reactions (Duan et al., 2021; Elmezayen et al., 2021; Feng et al., 2021; Khan and Bano, 2021; L. Qi et al., 2021; Yu et al., 2021; Yuan et al., 2021; Zhang et al., 2021). Furthermore, these databases characteristically define the flow of signaling in the normal condition and thus can be employed to approximate signaling networks related to cancer.

7.4.2.1   BioCyc

The BioCyc database (Karp et al., 2005) assembles chemical data collections for compounds associated with metabolic pathways (Karp et al., 2005; http://biocyc.org). The BioCyc Open Chemical Database (BOCD) collects the data of chemical compounds from the BioCyc databases. These compounds serve as substrates for enzyme-catalyzed reactions or can act as enzyme activators or cofactors. The majority of compounds consist of their chemical structure in the database.

7.4.2.2 KEGG

KEGG database, available at http://www.genome.jp/kegg/, works for biological systems integrating chemical, genomic, and systemic information (Kanehisa et al., 2006). The KEGG PATHWAY comprises 26 maps for human ailments (Kanehisa et al., 2006). Other maps of the disease pathway are classified into four subcategories: 6 maps for neurodegenerative disorders, 3 mas for infectious diseases and metabolic disorders, and 14 maps for cancers.

7.4.2.3 Reactome

The Reactome scheme is a curated resource of essential procedures and reactions in human biology (Vastrik et al., 2007; http://reactome.org). Other multiple databases cross-reference this database's database, e.g., sequence databases available at NCBI, UniProt, Ensembl, the UCSC Genome Browser, KEGG, PubMed, and GO. It is a free online tool and its software is open source. The Reactome database consists of several cellular pathways (Joshi-Tope et al., 2005) with different signaling pathways associated with cancer.

7.4.2.4 ConsensusPathDB

The ConsensusPathDB database was established for integrating the interaction data more comprehensively. It was developed by Kamburov et al. (2009). This database has the molecular interaction networks' integration of humans derived from various free pathway resources (http://cpdb.molgen.mpg.de). The integrated data includes diverse types of functional communications that connect diverse types of cellular entities. Presently, the database covers human functional interactions like gene regulations and physical and biochemical reactions found from the integration of 12 publicly available databases.

7.4.2.5 TRANSPATH^R

This database is based on regulatory network data, primarily in human, rat, and mouse cells. TRANSPATH^R emphasizes signal transduction pathways that target the molecules involved in gene regulation, such as transcription factors. However, it also focuses on metabolic enzymes and systemic proteins (Krull et al., 2006). In addition, collected intracellular regulatory data are annotated and deposited at different levels of abstraction, suitable for molecular interactions (Wingender et al., 2007).

7.4.3 ANNOTATION TOOLS

Complex and multifaceted molecular processes of cancer from the primary state can be searched from pathway models. To correlate the network information map data with computer-readable systems, curated model data could be considered the foremost step (de Bernard, 2009; Bouwmeester et al., 2004; Kerrien et al., 2007). There are different tools available that provide this service. The Reactome Curator tool, a software application, has been applied in Java. It is developed for data annotations linked to biological pathways. This tool allows the annotation of pathways based on the current Reactome database data. A pathway can be designed from a set of data (de Bernard, 2009).

7.4.4 MODELING TOOLS

Modeling and simulation methods are used to comprehensively study multifaceted biological systems associated with cancer beginning and progression (Austin et al., 2017; Asadi et al., 2019; Ghita et al., 2020; Ozturk Mizrak et al., 2020; Palmaria et al., 2020; Phillips and Schoen, 2020; Yap et al., 2020; Safavi et al., 2021). Computational modeling needs the pathway schema conversion to computer models that can transfer information on model components' concentrations and reaction kinetics of the components involved. This process involves the appropriate computer object designing, reactions execution, and kinetic and model analysis (Klipp et al., 2005; Batooei et al., 2021; Das et al., 2021; Iranpour et al., 2021; Jayasekera et al., 2021; Newman and Osmundson, 2021; Swaminath et al., 2021; Tambas et al., 2021; Yusupov et al., 2021).

A modeling tool permits the kinetic study of the system reactions under investigation. For example, model components variations in time upon agitations such as mutations or effects of any drug or external stimuli like growth factors.

Such a computational model can be used to achieve precise *in silico* case studies after establishment. Perturbations are integrated strategically in the model after doing modifications in the model parameter. For example, the rate constant value of a reaction exhibits the ability to activate or deactivate a model component via phosphorylation. This phenomenon can show significant changes. Similarly, a higher degradation rate constant value could lead to a continuous signal.

Consequently, the model structure should be constructed based on different parameter values. Validation of these resulting changes should be performed with experimental results. Computational tools supporting the model population, its simulation study, and analysis create the systems' biology base (Wierling et al., 2007).

Different modeling tools are available that are used for molecular network analysis. For example, E-Cell (Tomita et al., 1999), GEPASI (Mendes, 1993, 1997), Virtual Cell (Loew and Schaff, 2001; Slepchenko et al., 2003), Cell Designer (Funahashi et al., 2003), JDesigner (Novikov and Barillot, 2008), and the integrated Systems Biology Workbench tools (Hucka et al., 2001). Advancement and analysis of small models can be done using these tools, even though cancer-related models should comprise different components, reactions, and kinetic parameters in high numbers. Therefore, such a model entails computerization of the reaction network annotations, its cohort, numerical integration of differential equations, model simulation, and then the visualization of the results (Klamt et al., 2006).

7.4.5 Computational Models for Processes Associated with Cancer

7.4.5.1 BioModels Database

The BioModels Database (Le Nov`ere et al., 2006) permits scientists to interchange their computational models and offers a free, central, open-access annotated computational models' repository in the SBML format and other designed formats. These are connected to appropriate data resources and databases of compounds and process pathways.

7.4.5.2 Detailed Kinetic Models: Cancer

Behavior changes in human cells characterize protein structures and networks (Bellomo and Delitala, 2008; Steffes et al., 2017; Calzone et al., 2018; Andris, Seidel and Hubbuch, 2019; O'Brien et al., 2019; Csercsik and Kovács, 2020; Malik et al., 2020; Marín-Hernández et al., 2020; Matsukuma et al., 2020; Mathon et al., 2021; Mohammadzadeh-Asl et al., 2020; Preziosi et al., 2021). It occurs in both, i.e., in cell to cell and the same cell with time. Kinetic models can be used to define such variability. A set of first-order nonlinear differential equations describes the kinetic models of biochemical structures. These biochemical systems contain a broad range of unknown parameters, essential dynamics, and indeterminate connectivity (Bertolet et al., 2021; El-Wakil et al., 2021; Kashani et al., 2021; Khanna et al., 2021; Kopp et al., 2021; Li et al., 2021; Suthers et al., 2021; van Rosmalen et al., 2021).

In 2003, Brown and Sethna (Brown and Sethna, 2003) have prepared a statistical scheme for investigating the model's behavior so that maximum predictive information can be extracted out. The authors present a combined procedure for constructing, assessing, and applying models with numerous unknown parameters. Experimental measurements taken from living cells and the theoretical models help understand the dynamics and inconsistency of protein structures. For example, in a study by Geva-Zatorsky et al. (2006), fluorescently labeled *p53* and MDM2 dynamics were measured. Furthermore, they have worked on a corresponding model considering the negative feedback loop among the tumor suppressor gene *p53* and the oncogene MDM2.

Another pathway responsible for multiple human cancer progression is epidermal growth factor (EGF) signaling (Jones et al., 2006). The model constructed by Birtwistle et al. (2007) offers a quantitative explanation of the critical downstream protein stimulation in the EGF pathway and the activation of the extracellular-signal-regulated kinase (ERK) and AKT. However, EGF pathway activation with different ligands results in erratic signaling behavior based on the model and experimental proof analysis. Therefore, after applying the ERK cascade inhibitor, i.e., U0126, Heregulin-induced ERK activity will be inclined at a lower rate than EGF-induced ERK activity. These variations in EGF-induced ERK activity appeared due to phosphatidylinositol 3-kinase (PI3K) regulation.

The computation model of Schoeberl et al. (2002) was based on components involved in the signal pathway of epidermal growth factor (EGF) receptor. The model offers perception about the signal-response relationships that developed due to the EGF binding with its receptor at the surface of the cell and protein activation in the signaling pathway. In this model, the receptor activation's preliminary velocity was an essential parameter for the signal. The predictions of this model were validated further with an experimental study.

In another study, Sasagawa et al. (2005) have established a model consisting of ERK signaling networks by confining *in silico* dynamics of PC12 cells. Through this model, authors have predicted that temporary ERK activation depends on rapid upsurges of epidermal and nerve growth factors and independent on their final concentrations and further, it was validated. However, constant ERK activation was found to be dependent on the final neural growth factor concentrations.

The Wnt and the ERK pathways are known to be involved in the cancer pathogenesis of different types. Kim et al. (2007) presented that due to a positive feedback loop in the crosstalk of signaling pathways between the Wnt and the ERK, any change in protein due to gene mutation leads to pathway variations beyond the actual action site of the pathway. Therefore, crosstalk amid signaling pathways can lead to changes in the system's characteristics at a larger scale. Based on these experimental reports, authors have established that this positive feedback loop can produce bistability in both the signaling pathways, i.e., Wnt and the ERK. Precisely, due to the higher production of beta-catenin and decrease in the MAP kinase phosphatase(s) velocity, sustained activation of both the pathways was caused. Due to this, Wnt and ERK pathways are upheld at a high level without a tenacious extracellular signal.

TGF beta signaling pathway plays a central role in growth, apoptosis, and cell division processes. Therefore, investigation of TGF-β signaling dynamics is required to understand the events related to cancer. Zi and Klipp (2007) have proposed a mathematical model for the SMAD-dependent TGF-β signaling pathway. It was based on the constraint modeling method and constructed by fitting the experimental data and integrating qualitative constraints of the empirical study. This type of modeling method is also helpful for quantitative modeling of other signaling cascades.

Another computational model was developed by Yamada et al. (2003) related to the *JAK/STAT* signaling network. In this model, the role of the cytokine signaling-1 (*SOCS1*) suppressor was investigated, which is a negative regulator of the *JAK-STAT* pathway (Janus kinase-signal transducer and activator of transcription). Different values of the parameters were used for simulation in this model. Additionally, various initial concentrations and parameter values were compared and also the steady-state concentration of activated transcription factors (STAT1) was analyzed.

Dysregulation of the apoptotic programmed cell death pathway is another characteristic of cancer. Caspases are crucial for apoptosis activation. Caspase proteins are from the cysteine proteases family, which has essential roles in apoptosis. The inhibitors of apoptosis (IAP) can constrain caspases. Legewie et al. (2006) have developed a model based on the caspase-3 and caspase-9 inhibition by IAPs. This results in the generation of implicit positive feedback leading to bistability and irreversibility in caspase activation.

7.5 *IN SILICO* REPURPOSING OF CANCER DRUGS

In anticancer pharmaceutics, repurposing the obsolete FDA-approved drugs for novel therapeutic use became a fascinating approach (Hsieh et al., 2019; Lo and Torres, 2019; Melge et al., 2019; Costa et al., 2020; Dinić et al., 2020; GNS et al., 2020; Irham et al., 2020; De et al., 2021; Mahdian et al., 2021). This process is also called drug repurposing. The primary benefit of this process is associated with the cost of drug and time issues. Drug repurposing significantly reduces drug development's associated risk and expense and curtails the time gap from drug finding to its availability for patients because of having suitable and appropriate pharmacokinetics and clinical data (Shaughnessy, 2011; Papapetropoulos and Szabo, 2018; Fayed et al., 2021; Issa et al., 2021; Karaman Mayack and Sippl, 2021; Li et al., 2021; Mottini et al., 2021; Sankhe et al., 2021; Sohraby and Aryapour, 2021; Zhang et al., 2021). Generally, experimental-driven drug repurposing is just a matter of uncertainty; it can't be driven hypothetically. Instead, it can be obtained from screening drug experiments or by the target similarities identification in different diseases (Wilkinson and Pritchard, 2015). In the case of cancer, examples of current drug repurposing include (i) Disulfiram, which was initially used for the treatment of alcoholism and discovered as therapeutics for cancer treatment (Iljin et al., 2009; Huang et al., 2016; Skrott et al., 2017); (ii) Valproic Acid, an antiepileptic repurposed to the anticancer drug in many clinical trials (Chateauvieux et al., 2010); (iii) Nelfinavir, originally used to treat HIV infection and now it is under clinical trials for the treatment of breast, lung, and melanoma cancers (Shim and Liu, 2014).

On the contrary, the in-silico drug repurposing method is a hypothesis-driven strategy that identifies the drug against cancer using big data (Vanhaelen et al., 2017; Ansari et al., 2020; Montes-Grajales et al., 2020; Nath et al., 2020; Gonzalez-Fierro and Dueñas-González, 2021; Juárez-López and Schcolnik-Cabrera, 2021). This has the unique advantage of the transformation of cancer morphology and targets data to druggable target prediction. *In-silico* drug repurposing method tries to discover the FDA-approved drug with inhibitory actions against cancer. A precise computational channel and algorithm are obligatory in this method for data integration.

In this channel, one crucial step is to collect and interrogate the available omics properly. These data are related to disease mechanisms and cancer biology and the mode of action of drugs. The computational analysis permits integrating multiple layers of omics data, including genomics, metabolomics, transcriptomics, and proteomics.

7.5.1 *In Silico* Drug Repurposing and Personalized Medicine in Cancer

Apart from providing novel therapeutics suggestions, *in silico* drug repurposing can also help to improve the status of personalized therapies in cancer. Providing personalized and targeted treatment to cancer patients with minimal toxicity is the main challenge of the current oncology system (Chae et al., 2017). Targeted therapies involve the selective use of drugs against the specific molecular targets of the tumor cells or inside the tumor microenvironment. Although targeting the specific mutated protein of the tumor for the treatment is the most approved and successful approach for cancer treatment, some limitations also exist associated with the effectiveness of targeted therapies in clinical methods (Kimmelman and Tannock, 2018). These could arise due to suitable experimental preclinical models' deficiency, which involves the parameters such as tumor heterogeneity, physiological parameters of patients for analysis of drug metabolism and its bioavailability (Calvo et al., 2016), and composition of the tumor microenvironment (Binnewies et al., 2018; Gillies et al., 2018), understanding the tumor immune microenvironment (TIME) for effective therapy. In complex metastatic procedures, the tumor lesions' specific biological and molecular characteristics are not well known. Hence, most of the therapeutic strategies are based on the information of primary tumors. In this complex situation, for improvement of cancer-targeted therapies, computational approaches like *in silico* drug repurposing may deliver instant advantages. To overcome these problems, the computational approach can get benefited from the tumor biology information (Waldron, 2016), results of clinical studies, and biomarkers tests with pharmacokinetics modeling of drugs and predictions for drugs efficacies alone or in combinations (Huang et al., 2014; Celebi et al., 2019; Menden et al., 2019). Indeed, information regarding the side effects of drugs can also help *in silico* drug repurposing (Kuhn et al., 2010; Von Eichborn et al., 2010; Yang and Agarwal, 2011; Pemovska et al., 2013; Ye et al., 2014; Shaked et al., 2016). Based on data mining and integration, modification in the drug repurposing is also possible in selected patients and further better respondents can be predicted without showing any resistance to drugs (Camacho et al., 2018; Gottlieb et al., 2011).

Furthermore, computational methods can also carry out the repurposing of old drugs, which are not very good respondents in certain patients. It is based on the mode of action and the better knowledge of tumor molecular profiling (Mottini et al., 2019). Thus, by using computer-aided approaches, non-targeted (old) therapies can be turned out into personalized treatments based on the selection of better responders.

7.5.2 Computational Tools for *In Silico* Drug Repurposing Related to Cancer

Different computational tools are available for anticancer drugs repurposing based on the methodologies and data type described in the literature. Classical data types consist of chemical structures of drugs, their physicochemical characters, and molecular targets. Modern computational studies have also included omics data types like transcriptional effects induced by drugs or metabolic simulations. The methodologies involve both conventional statistical methods and modern machine learning methods.

These computational drug repurposing tools can be considered to try drug repurposing predictions directly or to support the procedure. For instance, tools based on network association of drug disease can directly propose new and innovative clinical applications for phenotypes of similar diseases. Additionally, other computational tools can facilitate the drug repurposing process by including the biological insights of modes of action of drugs and the discovery of unidentified targets of existing drugs (Mottini et al., 2021).

7.5.2.1 Resources Based on the Nature of Drugs

To identify repurposing candidates, the general approach assumes that molecules with the same structures exhibit similar biological activities due to having common molecular targets (Keiser et al., 2009).

The PREDICT (Gottlieb et al., 2011) tool is used to identify novel drug-disease relationships by combining the similarity among drug structures, their side effects, and targets of the drugs with disease-disease similarity evaluated over human phenotype similarity (HPO). In drug invention and repurposing, this tool has presented a higher sensitivity and specificity. Unfortunately, though, the intricacy of the model interpretability makes it problematic.

A more straightforward model method is available, known as SDTNBI (Wu et al., 2017). It is spread as a standalone program. SDTNBI identifies the novel targets using the chemical substructures from the ChEMBL database (Gaulton et al., 2017) and the CDK Fingerprints database was used for molecular targets (Yap, 2011). It can also be used for non-approved molecules like the PREDICT. One limitation of this method is that molecules which share a small number of substructures with identified drugs in the DTI network cannot be precisely assessed. This tool helps predict AKR1C3, CA9, or CA12 inhibition by the nonsteroidal anti-inflammatory drugs (NSAIDs) as a probable plan for cancer chemoprevention.

ChemMapper is a tool that explicitly involves the intrinsic chemical features (Gong et al., 2013). It is available as an openly accessible online tool. ChemMapper predicts the mode of action and polypharmacology reactions of small molecules based on their 3D resemblances. The shape of the molecules and their pharmacophore properties define the similarity. This tool allowed the identification of Atox1-binding small molecules to inhibit cancer cell proliferation (Wang et al., 2015).

The SIDER database was found to be a good tool for drug repurposing. It contains 1,430 drug structures, 5,880 drug adverse reactions (ADRs), and 140,064 drug-ADR pairs (Kuhn et al., 2010).

The drug-ADR relations are attained by text-mining approaches obtained from literature and clinical experimental studies (Liang et al., 2017).

Another common resource is the STITCH database. It is a tool of search and information integrations used to check interactions between chemicals and proteins. This tool utilized the chemicals of PubChem database, while the relations between chemicals and chemical–protein obtained from some different resources having high-throughput methods and protein sequence matching. The STITCH database has been employed in various systematic channels. Other tools involved are mainly information-rich databases such as DrugBank, Pharos, etc. DrugBank comprehends the data of chemical, pharmacological, and pharmaceutical studies for approximately 13,363 drugs (Wishart et al., 2018). Pharos is a web interface used to explore the Target Central Resource Database (TCRD) (Nguyen et al., 2017).

7.5.2.2 Resources Based on Effects of Drug Treatment

Drug treatment effects can be examined through gene expression analysis. Due to this reason, Connectivity Map (CMap) project was used in the past for profiling of drug-induced whole-genome expression (Lamb et al., 2006). Further, it is evidenced by its latest issue in the LINCS (Library of Integrated Network-Based Cellular Signatures) (Vidović et al., 2014). CMap database includes approximately 7000 profiles obtained from the treatment of 5 cancer cell lines, while LINCS made the database of larger size up to 1,000,000 in about 40 cell lines. This was done due to L1000 technology, measuring transcription levels for 978 genes directly and around 1,000 genes via computational analysis. Due to that, alertness is required during the interpretation of L1000 data and also, a quality control assay channel of L100 is recommended for the solution of biological queries (Cheng and Li, 2016). CMap is an analytical tool that facilitates an online signature-matching approach, while the LINCS project comprises an entire landscape of online mechanisms. For example, the CMap was used to identify Digoxin as a possible therapy for medulloblastomas (groups 3 and 4) (Huang et al., 2018). Expression profiles of the LINCS project have been used to identify potential candidates for breast cancer and prostate cancer (Chen et al., 2016).

Another prevalent tool is MANTRA (Iorio et al., 2010). It is a network-based analytic tool that permits the exploration of drug locality based on the similarity among induced transcriptional reactions. The latest version also enables the user-provided profiles for the network growth in a collaborative manner (Carrella et al., 2014). MANTRA was used to identify anthelmintic drugs and play an essential role in inhibiting oncogenic *PI3K/AKT/P70S6K*-dependent signaling pathways (Carrella et al., 2016). Furthermore, MANTRA also supported identifying growth inhibitors in pancreatic tumors depending on the K-RAS oncogene activation (Mottini et al., 2019). The MANTRA model is created on symmetric Gene Set Enrichment Analysis (GSEA) of the highest and lowest expressed genes amid two drugs.

Other tools based on gene expression data for gene repurposing are DSEA (Napolitano et al., 2016) and gene2drug (Napolitano et al., 2018). DSEA aims at the importance of common pathways that are influenced by multiple drugs while gene2drug aims to identify drugs targeting a set of pathways. DSEA tool was used to identify cancer-related pathways correctly during the analysis of a group of anticancer drugs.

It is worth stating that the inconsistency of drug treatment effects is also an essential part of research because computational drug repositioning practices can take benefit from it. Regarding that, an available helpful tool is the PharmGKB database. It collects the effects of genetic variations of humans on the response of drugs (Whirl-Carrillo et al., 2012).

In the field of toxicogenomics, DrugMatrix offers the toxicogenomic description of more than 600 compounds in different tissues with an analytical tool (https://ntp.niehs.nih.gov/results/drugmatrix). In addition, TG-GATE provides toxicogenomics data obtained from in vivo and in vitro subjection to 170 compounds at multiple dosages and gene expression profiles (Igarashi et al., 2015).

7.5.2.3 Disease-Data-Based Resources

Many approaches utilize the data related to diseases for screening of drugs or drugs repurposing. For instance, in the study of Sirota et al. (2011), disease signatures were created using the gene expression data and drug-disease associations were identified after comparing them with drug-induced gene expression data. Therefore, they anticipated cimetidine (a drug) as a previously unidentified suggestion for lung adenocarcinoma. In this case, the whole-genome transcriptomic signature was used to identify drugs inducing its reverse. Thus, the transcriptomic signal of the whole genome was used to identify drugs causing the reverse effects.

Similarly, Dudley et al. (2011) have forwarded the repurposing of topiramate, an anticonvulsant used for inflammatory bowel disease; recently, a similar study was performed by Kim et al. (2019) for gastric cancer.

However, in several studies, disease-related data are used in combination with layers of heterogeneous data. For example, Yang et al. (2014) have proposed an algorithm based on fundamental inference-probabilistic matrix factorization (CI-PMF) for prediction and classification of drug-disease relations, used further for drug-repurposing estimations. The algorithm works on integrating heterogeneous databases interactions of different types, including circulatory diseases, diabetes mellitus, and tumors.

Another example of a disease data model was proposed by Li and Lu (2013). They have constructed a network by exploiting the relations between diseases, drugs, and the genes involved in the pathway. The process models the connection among the drugs, their targets, conditions, and the pathways. Hence, results obtained from this tool are mostly interpretable.

Zhang et al. (2014) have proposed another repurposing method based on chemical structures, targets, side effects of drugs, phenotypes induced, disease nature, and involved genes. The computational model generates the hypotheses for drug repurposing as a nonlinear optimization problem that used resemblance networks of multiple drugs, disease similarity networks, and drug-disease relations to discover novel potential links. The data used for the analysis was obtained from Pathway Commons (Cerami et al., 2010). It is an online web tool that assembles data from multiple public resources, including biochemical reactions, complex integrations, transport, and catalysis proceedings and other interactions such as proteins, DNA, RNA, and complexes.

In some cases, disease ontology is used to get the data of the disease. For instance, Chen et al. (2012) have developed an algorithm for predicting the first step of Anatomical Therapeutic Chemical (ATC) codes for 3,883 drugs. These predictions were based on the structural similarity of drugs and the interactions between the compounds obtained from the STITCH database. Encoding of drug- and disease-related data can be carried out using ATC classification. In this type of model or computational task, clinical drug applications are predicted by using disease ontology.

Another example is the study done by Napolitano et al. (2013), where a black-box approach was introduced, combining the transcriptional responses of drugs (Cmap), molecular structures, and identified targets. As a result, three independent layers of similarity were integrated into a single core to predict drug ATC codes using an SVM amplifier. From this study, possible repurposing of anthelmintic drugs as antineoplastics was recommended.

Another black-box method using ATC codes is NetPredATC (Wang et al., 2013). It was also based on an SVM classifier exploiting structural similarity with the resemblance of target protein sequences. Thus, ATC codes work as perfect and straightforward labeling for classification-modeling functions.

7.6 TARGET LINKED DATA MODELING RELATED TO ONCOGENIC PATHWAYS

Omics studies are offering fundamental insights into the tumorigenesis and oncogenic pathways driven by MYC. Genomic and transcriptomic profiling was accomplished in Burkitt Lymphoma, which resulted in the elucidation of a complex interplay of mutational, transcriptional, and post-transcriptional events enhancing the MYC activity in cancer cells. Moreover, gene expression profiling of human primary breast tumors showed that the elevated MYC expression was due to poor prognosis and sensitized cells to cyclin-dependent kinase inhibition (Horiuchi et al., 2012) and regulated the gene expression levels of 13 different "poor-outcome" cancer signatures (Wolfer et al., 2010). Additionally, a transcriptomic data network in neuroblastoma cells exhibited the role of c-MYC as a regulator of an embryonic stem cell (ESC)–like cancer-activated signature associated with different tumor classes (Yang et al., 2017). T-cell lymphoma, genomic and transcriptomic characterization of T-cells isolated from murine MYC-driven lymphomas, has revealed a critical threshold MYC expression level that induces cell proliferation and tumor growth (Shachaf et al., 2008). The protein interaction profiling in multiple model systems has concluded that the MYC interactors needed six highly conserved regions to initiate tumorigenesis (Kalkat et al., 2018). Protein-mRNA correlation characterization of breast tumors unveiled a relationship among a subset of proteins and an MYC-dependent signature, thus categorizing tumors into two subgroups with different survival outcomes (Tang et al., 2018). Through proteomic profiling, c-MYC has appeared to be an essential regulatory protein in Panin-3 lesions and it interacts directly with the deregulated proteins intricate in cancer growth (Pan et al., 2009); comparative proteomic characterization of MYC overexpression in MYC-deficient fibroblast cells have shed light onto the role of MYC in cell growth and cytoskeletal function control (Shiio et al., 2002). Additionally, the gene expression profile and proteomic assay established coordination among regulation of transcription and translation after modulation of MYC activity by enhancing oncogenic transformation, migration, and invasiveness in inducible-MYC U2OS cells (Elkon et al., 2015). Moreover, the transcriptional and proteomic profiles in breast carcinoma cell lines identified an MYC/G9a H3K9-methyltransferase complex, which significantly promoted MYC-mediated transcriptional repression and tumor initiation (Tu et al., 2018).

A combined analysis of metabolomics and transcriptomic profiling of carcinoma colorectal demonstrated that MYC up-regulation enhanced mitochondrial metabolism and resulted in higher metabolic rates of glucose, glutamine, and amino acids (Tarrado-Castellarnau et al., 2017). Metabolomics profiling of lung cancer cells confirmed an aberrant lipid metabolism dependent on MYC up-regulation responsible for lung cancer cell survival and proliferation (Hall et al., 2016). In contrast, activation of MYC in breast cancer resulted in the accumulation of oncometabolite responsible for cellular proliferation and apoptosis inhibition (Terunuma et al., 2014). Remarkably, a multi-omics analysis of colorectal cancer patients showed that MYC regulated the global metabolic reprogramming and exaggerated the pyrimidine synthesis (Satoh et al., 2017). Metabolomics tactics on MYC-driven triple-negative breast cancer (TNBC) models established a lipid metabolism-associated gene signature. They recognized a crucial role of fatty acid oxidation (FAO) sustaining cancer stem cells' sub-population (Camarda et al., 2016). MYC-overexpression also enhanced glutamine uptake in cancer cells by elevating the expression of glutamine synthase, an essential enzyme for the de novo glutamine synthesis (Bott et al., 2015; Dejure and Eilers, 2017), exhibiting the auxiliary role of MYC in cancer through the nutrient uptake control and other anaplerotic reactions of cellular metabolism.

The functional molecular networks linked with oncogenic β-catenin signaling in colorectal cancer have been recognized by transcriptomic and proteomic profiling (Chen et al., 2010; Ewing et al., 2018), as well as networks related to the regulation of metabolism and energy homeostasis in breast cancer (Vergara et al., 2017). Although transcriptomic analysis has been performed in colorectal cancer cells, the silencing of various WNT components, they recognized β-catenin-dependent and -independent signaling mechanisms specific to tumor development (Voloshanenko et al., 2018) and identified β-catenin particular targets. Mass spectrometry studies identified new potential β-catenin interacting proteins alongside coatomer complex subunits linked with retrograde transport at the Golgi (Semaan et al., 2019). By complementing genomic and proteomic tactics in various cancer cell lines expressing a constitutively active, oncogenic β-catenin, it was identified that the regulators of β-catenin and defined functional signaling networks are necessary for β-catenin-dependent survival of cancer cells (Rosenbluh et al., 2012). Genomic and transcriptomic profiling in cancer stem cells (CSCs) showed a crucial role of β-catenin in the epithelial-to-mesenchymal transition (EMT) by conjugating several effectors part in the EMT process. The β-catenin-dependent transcription profile was linked with a gene signature identified from patients with high-grade and metastatic lung cancer, thus showing a predictor for poor prognosis (Chang et al., 2015). In recent times, transcriptome studies demonstrated that the β-catenin-dependent signaling enhanced tumor immune evasion and a lower immune response in melanoma (Spranger et al., 2015; Nsengimana et al., 2018; Wang et al., 2018). These studies identified a unique functional framework for the tumor-promoting activity of oncogenic β-catenin protein.

Murine mice models have been used to a greater extent to investigate tumor suppressor mechanisms associated with *p53*. They are regarded as good preclinical models to identify critical targets and drug responses (Pfefferle et al., 2016). Genomic profiling was studied by using high-throughput sequencing of primary mouse embryonic fibroblasts concerning DNA damage and detected a close linking among *p53* and autophagy to initiate apoptosis and tumor suppression (Broz et al., 2013); further, the transcription profiling of irradiated mouse models overexpressing *p53* has allowed recognizing that the tumor suppressor controls mRNA expression of various genes linked with DNA damage response (Tanikawa et al., 2017). The transcriptomic and metabolomics profiling using the CAFs and lung cancer cells signified that the transcriptional program braced by *p53* was transformed considerably in CAFs about the normal fibroblasts, suggesting that tumor progression can involve a non-mutational conversion of stromal *p53*, from tumor suppressive to supportive tumor function (Arandkar et al., 2018). Additionally, a gene expression profiling of melanoma tumor samples showed that *p53* family gene deregulation was linked to the poor survival of the patients (Badal et al., 2017). Transcriptome data in colorectal cells validated that *p53* activity was controlled by lncRNAs regulating the *p53* transcriptional network (Sánchez et al., 2014), while gene signature profiling studied accomplished in murine mesenchymal stem cells showed that mutant *p53*-dependent embryonic stem cell gene signature is linked to the poor prognosis of patients and elevated tumorigenesis in embryonic stem cells (Koifman et al., 2018). Transcriptomic and proteomic profiling established the fact that molecular pathways for activation of lung adenocarcinoma cells were linked to *p53* mutational status (Taguchi et al., 2014), and they also play a significant role of miRNAs mediated control in the expression of *p53* target genes in colorectal cancer cells (Hünten et al., 2015). Notably, the proteomic profiling in lung cancer cell lines expressing *p53* confirmed the *p53*-mediated development of exosomes that affected the microenvironment and enhanced invasiveness and migration (Novo et al., 2018). Moreover, transcriptome, proteome, and interactome studies in triple-negative breast cancers have recognized the proteasome machinery as a target of mutant *p53* missense proteins (Walerych et al., 2016). Multi-omics studies accomplished in colorectal, breast carcinoma cells, and osteosarcoma cells demarcated that *p53* can control a definite transcriptional gene network despite cell-type diversity (Andrysik et al., 2017).

Metabolites studies in mouse embryonic fibroblasts (MEFs) expressing wild-type or mutant *p53* have allowed for the identification of an elevated induction of *p53*-dependent genes in retort to the stress, highlighting the significance of *p53* in the regulation of stress related to metabolism (Lowman et al., 2019). The *p53*-induced protein TIGER links *p53* activity with cell metabolism, inhibiting glycolysis and lowering mitochondria-induced ROS levels. In addition, TIGAR expression protected cells from the accumulation of genomic damage (Bensaad et al., 2006). Alongside the *p53*-specific metabolism control, a metabolomic profile recognized niclosamide as a probable drug that persuaded mitochondrial uncoupling in *p53* mutant colorectal cancer cells and rendered cells susceptible to apoptosis (Kumar et al., 2018).

7.7 TECHNOLOGIES FOR CANCER IMMUNITY INVESTIGATION

The utilization of NGS for genomic, transcriptomic, and epigenomic profiling of tumors is constructing the primary data source enabling the identification of hallmark characteristics. The employment of these tools on bulk tissue and the computational tools that will allow the NGS data to be extracted have been previously studied (Thorsson et al., 2018). In addition, the microbiome studies (Knight et al., 2018) have progressed faster and now deliver the methods to study the microbiota composition and function from 16S ribosomal RNA sequencing, metagenomics, and metatranscriptomics data. The recent critical advancements include the development of new tools for single-cell analysis and multiplexed spatial cellular phenotyping. These tools are of precise significance for cancer immunity as they permit the detailed analysis of cancer immunity, involving the depiction of the cellular composition of cancerous and normal tissue, quantification of the immune contexture, and coupling of α chains and β chains of individual T cell receptors (TCRs) and B cell receptors (BCRs). The data types, the transitional studies, and the immunogenomic studies are shown. A fair comparison of the data is necessary to understand the practical steps that resulted in the generated data to analyze the origins of the critical characteristic and possible biases of the resultant data.

7.7.1 SINGLE-CELL OMICS STUDY OF ISOLATED CELLS

The bulk RNA sequencing (RNA-seq) data permit the restructuring of an average transcriptome of various mixed cell populations and newly developed scRNA-seq technologies can be employed to re-create the transcriptomes of individual cells, generating new possibilities for the analysis of heterogeneity, plasticity, and functional diversity of the immune system (Finotello and Eduati, 2018; Giladi and Amit, 2018). The tools used to apprehend single cells can be allocated to either plate- or microfluidics-based methods (Kolodziejczyk et al., 2015). Plate-based methods such as Smart-seq2 (Picelli et al., 2013) categorize cells into separate wells via fluorescence-activated cell sorting (FACS). They create full-length transcripts from single cells and possess better sensitivity than microfluidics-based techniques, but lower throughput of the sequenced cells due to the complexity of the single-cell isolation methods (Ziegenhain et al., 2017). On the other hand, microfluidics-based systems such as the 10X Chromium (Zheng et al., 2017) produce nanoliter-sized droplets consisting of single cells and barcoded bead reagents necessary for the downstream processes.

These systems are cost efficient and allow the profiling of higher quantities of cells compared to plate-based systems. An alternate process for the study of single cells is CyTOF (Bendall and Nolan, 2012), the analyses cells as per their cell-surface-expressed proteins. In this method, the metal-isotope-conjugated antibodies are employed for cellular staining and are then subjected to a quadrupole time-of-flight (TOF) mass spectrometer characterization. The key benefit compared to the traditional fluorescence-based flow cytometry is the removal of spectral overlaps and hence the number of analyzed markers can be higher.

7.7.2 COMPUTATIONAL TOOLS FOR THE PREDICTION OF NEOANTIGENS

In silico identification of presumed neoantigens from mutated genes comprises three main computational steps: first, recognition of somatic mutations by utilizing whole-genome sequencing (WGS) or whole-exome sequencing (WES) data from coupled tumor and normal tissue and restructuring of mutated peptides; second, the genotyping of the patient's HLA genes from tumor RNA-seq or WES data; and, third, estimating of peptides conjugating to the patient's HLA molecules. Mutated peptides ascending from somatic mutations can be identified by comparing tumor and normal-tissue NGS data from the same patient. NGS data for neoantigen identification are created preferentially from WES, which delivers the deepest mutation coverage by confining the assay only to protein-coding regions of the genome. The computational studies comprise the data pre-processing and quality control, recognition of somatic mutations by implementing tools for variant detection, and identification of the affected proteins and functional effect using public sources of genomic, transcriptomic, and proteomic sequences (Ding et al., 2014; Xu, 2018).

The state-of-the-art techniques for HLA typing from NGS data are established and extensively used, such as OptiType (Szolek et al., 2014) and Polysolver (Shukla et al., 2015), that exhibited better accuracy in four-digit HLA typing, alongside seq2HLA (Boegel et al., 2013), to compute both HLA types and allele-specific expression.

Recently developed methods including Kourami (Lee and Kingsford, 2018), HLA*LA (Dilthey et al., 2019), arcasHLA (Orenbuch et al., 2020), xHLA (Xie et al., 2017), HLA-HD (Kawaguchi et al., 2017), and HLAProfiler (Buchkovich et al., 2017), are capable of performing class II HLA typing. Unfortunately, though, the unbiased benchmarking of these techniques is not available and can be beneficial for analyzing their accurateness in class II typing, due to the minimal validation functional.

The techniques for identifying peptides capable of binding with HLA molecules use machine-learning algorithms trained on large *in vitro* peptide–HLA binding data sets. NetMHC (Nielsen et al., 2003) and its pan-allele version NetMHCpan (Jurtz et al., 2017) is constructed on artificial neural networks and presently are the most extensively utilized tools due to their high performance. Both tools are capable of identifying the binding affinity as the half-maximal inhibitory concentration (IC50) articulated in nanomolar units and the rank of predicted affinity in comparison with a set of random natural peptides. Strong binders are typically designated because of a binding affinity or rank lower than 500 nM or 0.5%, respectively. Recent developments in deep learning have nurtured new machine-learning tools constructed on deep convolutional neural networks, such as HLA-CNN (Han and Kim, 2017) and DeepSeqPan (Liu et al., 2019). Simultaneously, a pan-allele tool called PSSMHCpan has been developed to identify the binding domains for analyzing the peptide binding affinity for the presently under-represented HLA alleles (Liu et al., 2017). The recently developed pVAC tools suite for the estimation and ranking of potential neoantigens (Hundal et al., 2020) involves a modified version of the pVACseq (Hundal et al., 2016) pipeline capable of computing binding-affinity with different state-of-the-art machine-learning algorithms, alongside quantification of features related to antigen pre-processing and identification.

Notably, only ~1–5% of the class I binders was identified *in silico* by deploying different computational tools and are experimentally validated (Lee et al., 2018). The likely cause for the discrepancy among predicted and experimentally validated neoantigens is the low sensitivity of mass spectrometry (MS)–based protocols to identify the binding peptides directly. Despite this limitation, the MS measurements of eluted HLA-binding peptides can be utilized to investigate the human immunopeptidome now, and the set of peptides presented on HLA molecules, and to permit restructuring of antigen profiles demonstrated *in vivo* that could not be apprehended from previous *in vitro* affinity studies (Gfeller et al., 2018; O'Donnell et al., 2018). Novel methods such as MHCflurry 1.2.0 (Boehm et al., 2019), ForestMHC (Bassani-Sternberg et al., 2017), MixMHCpred (Gfeller et al., 2018), and EDGE (Bulik-Sullivan et al., 2019), as well as the current version of

NetMHC (Andreatta and Nielsen, 2016), have also been trained on MS data from HLA-eluted peptides. The growing amount of HLA–ligand MS data available in databases like IEDB (Vita et al., 2015), PRIDE (Vizcaíno et al., 2016), or SysteMHC Atlas (Shao et al., 2018) is capable of providing rich training data sets for the next-generation predictors. However, MS analysis with two main limitations: first, the obligation for a large quantity of starting material (~1 × 108 cells) and, second, the dependability on protein sequence databases for data characterization, thus limiting the prediction of peptides to the annotated human proteome. The latter problem can be resolved with computational methods for refreshing the reference databases integrating predicted non-canonical neoantigens, such as those developed from non-exonic regions, insertions or deletions (indels), gene fusions, alternative splicing, or post-translational modifications.

However, when augmenting peptide databases with non-canonical peptides, attention must be taken to avoid false positives. The potential significance of non-canonical neoantigens was re-ported in a recent study on patients with head and neck cancer treated with different immune checkpoint inhibitors, representing that gene fusion is a basis of immunogenic neoantigens that can arbitrate retorts to immunotherapy in patients with low mutational load and low pretreatment immune infiltration (Yang et al., 2019).

Despite the developments in identifying class I HLA neoantigens, the recent tools for identi-fying class II HLA neoantigens documented by CD4+ T-cells have inadequate precision and are progressing slowly due to the lack of suitable training data. The scenario of class II interpreters has advanced little, with NetMHCII and NetMHCIIpan73 still being the top-performing tools and only one unique tool proposed: MixMHC2pred75. MixMHC2pred was trained on MS-based, class II immunopeptidomics data and confirmed higher accuracy than NetMHCIIpan (Racle et al., 2019).

Inclusively, the recent advancements in deep-learning algorithms and MS-based im-munopeptidomics have shaped fruitful ground for expanding next-generation predictors of HLA presenter algorithms. Furthermore, the optimum tools of neoantigen presentation cannot identify their immunogenicity. Great collaborative efforts, including the tumor neoantigen selection alli-ance (TESLA), are tapping together large data sets for the authentication of these *in silico* methods and can help identify the best predictors for mutated peptides that bind *in vivo*. In spite of the necessity for additional optimization of tools for forecasting neoantigens, they have already es-tablished their initial clinical value. In studies on personalized vaccinations for melanoma, the forecasted neoantigens prompted effective immune responses. One study utilized synthetic pep-tides to inoculate six melanoma patients. Four out of six patients represented no disease recurrence, whereas patients suffering from metastatic disease attained ample tumor reversion with supple-mentary anti-PD1 therapy. Another study utilized the RNA-based vaccines resulting from the computationally identified neoantigens documented by CD4+ and CD8+ T-cells in 13 patients with melanoma (Ott et al., 2017).

Neoantigen-based vaccination lowered the metastatic events and resulted in a specific response among two of the five patients with metastases and complete response in a third patient treated with the vaccine in amalgamation with anti-PD1 immunotherapy.

7.7.3 Characterization of Tumor-Infiltrating Immune Cells

Bulk-tumor transcriptomics data from microarrays or RNA-seq can be utilized to enumerate di-verse cell types of the TME, alongside identifying gene signatures capable of predicting a response to immunotherapy with different checkpoint blockers. Provided the complex and multifactorial origin of the anticancer immune response and the numerous modes of immune escape (Camidge et al., 2019), it will be inspiring to authenticate the projected predictive signatures (Jiang et al., 2018) prospectively. Enumerating different cell types of the TME can be accomplished by utilizing either the methods constructed on marker genes or deconvolution-based tools and algorithms. While the techniques based on marker genes including xCell, TIminer, and MCP-counter (Finotello et al., 2019), they evaluate only semi-quantitative profusion scores; deconvolution methods can assess the segments of specific cell types in a heterocellular tissue quantitatively by considering the bulk transcriptome as the "convolution" or summation of cell-specific signatures. Briefly, the deconvolution tools express the problem as an arrangement of equations that designate the expression of each gene in a bulk sample as the weighted sum of the expression profiles of the admixed cell types. The mathematical delinquent (so-called inverse problem) of concluding the unknown cell-type segments is resolved by utilizing a signature matrix relating the expression fingerprints of sorted cell types.

CIBERSORT is a predominant deconvolution algorithm created around a signature matrix recitation of the expression profiles of 22 immune cell phenotypes developed from microarray data of sorted or enriched immune cell types (Newman et al., 2015); it utilizes support vector regression to predict the solution. Presently, CIBERSORT has been unified into CIBERSORTx, a deconvolution tool that further empowers the construction of custom signature matrices from single-cell or flow-sorted bulk transcriptomic data with the elimination of possible batch effects. Another tool, TIMER, is a multistep computational method for the computation of 6 immune cell types in 32 dissimilar cancer types (Li et al., 2016). TIMER combines the input samples to be deconvoluted with immune cell reference profiles, performs normalization to remove batch effects, derives a cancer-specific signature matrix, and computes cell abundance scores using linear least-squares deconvolution. Despite being assessed via deconvolution, TIMER scores cannot be construed as cell fractions or related across different immune cell types and data sets (Li et al., 2016). A newly reported technique, EPIC90, was evaluated using bulk and scRNA-seq data from circulating and tumor-infiltrating immune cells, CAFs and epithelial cells. EPIC offers two distinct signature matrices for evaluating blood or tumor data and, compared to previous tools, computes cell fractions mentioning the total cells in the sample, thus allowing intra- and inter-sample assessments. Also, quanTIseq is a recent algorithm for deconvolution of blood and tumor data, based on a unique signature matrix prepared from a collection of RNA-seq data sets for ten circulating immune cell types, including regulatory T-cells and classically (M1) and alternatively (M2) activated macrophages (Finotello et al., 2019). quanTIseq is precisely custom-made for RNA-seq data and outfits an entire investigative pipeline, from pre-processing raw RNA-seq data to the deconvolution of cell fractions. quanTIseq also evaluates immune cell fractions compared to and among the samples (Finotello et al., 2019). Characterization of samples from two allies of patients with melanoma treated with kinase inhibitors or immune checkpoint blockers confirmed that deconvolution algorithms could be utilized to analyze targeted agents' immunological properties and divulge immune cell composition in response to immune checkpoint blockers (Finotello et al., 2019).

Benchmarking cell-type computation algorithms are problematic due to the variances in the cell types and assessed scores/fractions. A current comparative benchmarking displayed high accuracy in the computation of CD8+ T-cells across diverse approaches but inadequate performance for heterogeneous cell types such as dendritic cells (Sturm et al., 2019). The Tumor Deconvolution Challenge, prearranged by the Dialogue on Reverse Engineering Assessment and Methods (DREAM) initiative, can predict the top performers and deliver strategies for the assortment of the best algorithm based on the cell type.

Prominently, alongside the simple computation of cell types, unique tools such as CIBERSORTx and linseed can restructure cell- and sample-specific transcriptional profiles and, thus, possess the capability to estimate the functional state of cell subpopulations in the TME.

To summarize, the assortment of the tool relies on the queries to be addressed and the type of data expected to be obtained. For example, EPIC and quanTIseq are the ideal tools to achieve cell fractions that can be evaluated both within and among samples. At the same time, MCP-counter and xCell offer higher signature specificity and lower background noise, respectively (Sturm et al., 2019). Novel tools can also be designated based on the cell type (such as CIBERSORT, xCell, and quanTIseq for M1/M2 macrophages; xCell, EPIC, and TIMER for epithelial cells).

- **3-D cellular phenotyping:** The computation of the immune contexture necessitates images from tissue slides to acquire cellular phenotypes and their spatial distribution. Discrete cells are first distinguished by thresholding and segmenting the raw images. Then their specific phenotypes are recognized and characterized based on the detection signals from the particular markers in the equivalent cellular compartment (including nucleus, cytoplasm, or cytoplasmic membrane) utilized in staining protocols. Also, commercial software packages including inForm (Perkin Elmer), Halo (Indica Labs), or StrataQuest (TissueGnostics GmbH), a rising number of open-source and free software tools, such as ImageJ, CellProfiler, and Ilastik, are (Finotello et al., 2019) available for this purpose.

  Furthermore, by uniting and encompassing their core functionalities via plug-ins, macros, or scripting, custom investigation pipelines have been developed and improved to fit the diverse multiplex imaging methods (Tsujikawa et al., 2017). As a substitute to the utilization of wholly established imaging software packages, image analysis pipelines are often executed by utilizing the image-processing procedures and libraries from MATLAB (imaging toolbox) or Python (scikit-image and opencv) (Keren et al., 2018). This is the circumstance for unique multiplex imaging tools such as IMC, MIBI-TOF, MERFISH, CODEX, seqFISH, Spatial Transcriptomics, or Slide-seq that need particular pre-processing, image restoration, and post-processing tools.

- **Imaging of single-cell data:** Imagining complex single-cell data is very stimulating, and various tools have been developed for representation and exploration. Though numerous tumor–immune cell studies are specific to tumor-infiltrating immune cells, it is necessary to note that the cellular landscape of the TME is highly heterogeneous and incorporates numerous other cell types, including CAFs and endothelial cells. Linear dimensionality reduction methods, including principal component analysis (PCA), are often capable of capturing the complex structure of single-cell data, and hence methods for nonlinear transformation in two dimensions, including t-SNE (Van der Maaten and Hinton, 2008) and its derivatives (viSNE, Amir et al., 2013), Fourier-interpolated t-SNE (Linderman et al., 2017), and hierarchical SNE (van Unen et al., 2017) have been reported. Generally, these tools are employed to graphically represent functionally associated cell groups as clusters with similar gene expression profiles in 2D plots. Though these tools are commonly utilized, they require vigilant construal as the results depend on traits balancing global and local aspects.

  Other problems include clustering performance, which can be enhanced by kernel-based similarity learning, and accuracy, which can be augmented by explicitly modeling the dropout traits.

- **Recent tools and future trends:** In recent years, various tools and techniques for evaluating cancer immunity have mellowed. Some of these tools are used for HLA typing and the tools for identifying class I HLA binding affinity from NGS data. In other areas, small advancement has been attained for different reasons. Precise forecasts of class II HLA binding affinity are challenging for both biological and technical reasons. First, the size of the binding peptides is variable (between 13 and 25 amino acids) and the peptide-flanking regions on either side of the binding core play a crucial role in peptide–HLA binding. In addition, there is a lack of both positive and negative training data sets. Hence, rather than improving algorithms to enhance their performance by a few percent (such as evaluation of class I HLA binding affinity), efforts should be focused on producing training data and evolving algorithms for applications that are evolving slowly. An additional extremely thought-provoking area is the computation of immunogenicity of neoantigens.

  Understanding the TCR identification rules for peptide– HLA complexes (pHLA) would enormously help design cancer vaccines and permit T-cell engineering for solid cancers.

Recent studies established that TCR sequences can be allocated to an antigen specificity by sequence investigation alone (Dash et al., 2017; Glanville et al., 2017). In all the studies, only a number of epitopes from common viruses were utilized; the observations endorse that the development of a generalized algorithm of TCR–pHLA identification is possible, that will be a significant phase in the direction of designing TCR sequences with neoantigen specificity and, hence, rationally engineering T-cell immunity against tumors.

Evolving NGS techniques empower simultaneous evaluation of different molecular entities such as scRNA-seq in combination with cell-surface protein expression, as in the two associated approaches of cellular indexing of transcriptomes and epitopes (CITE-seq) and RNA expression and protein sequencing (REAP-seq). Similarly, other techniques associate the spatial and molecular data like spatial transcriptomics and simultaneous detection of proteins and transcripts using IMC (Finotello et al., 2019). These and other upcoming methods will necessitate innovative computational algorithms and tools for consolidating heterogeneous data in bulk-tissue and single-cell situations. Explicitly, integrating data across diverse modalities linked with single-cell data sets such as transcriptomic, epigenomic, proteomic, and spatially determined single-cell data will be essential to acquire deep biological understanding beyond the listing of cell clusters (Stuart and Satija, 2019). In addition, transporting data from one data set to another will be enormously helpful for investigative evaluations and biological interpretation. Such original approaches using data integration and allocation learning have been recently developed (Stein-O'Brien et al., 2021). For example, an innovative tactic has been utilized for transferring scRNA-seq annotations onto chromatin accessibility data (produced using a single-cell assay for transposase accessible chromatin [scATAC-seq]), thereby disclosing finer dissimilarities within the cell types, which was not conceivable by employing solely scATAC-seq data.

REFERENCES

Akhoon, B.A., Tiwari, H., and Nargotra, A., 2019. In Silico Drug Design Methods for Drug Repurposing. *In: In Silico Drug Design* (pp. 47–84). Elsevier.

Alabi, R.O., Mäkitie, A.A., Pirinen, M., Elmusrati, M., Leivo, I., and Almangush, A., 2021. Comparison of nomogram with machine learning techniques for prediction of overall survival in patients with tongue cancer. *International Journal of Medical Informatics*, *145*. doi: 10.1016/j.ijmedinf.2020.

Alabi, R.O., Youssef, O., Pirinen, M., Elmusrati, M., Mäkitie, A.A., Leivo, I., and Almangush, A., 2021. Machine learning in oral squamous cell carcinoma: Current status, clinical concerns and prospects for future – a systematic review. *Artificial Intelligence in Medicine*. doi: 10.1016/j.artmed.2021. 102060.

Alizadeh Savareh, B., Asadzadeh Aghdaie, H., Behmanesh, A., Bashiri, A., Sadeghi, A., Zali, M., and Shams, R., 2020. A machine learning approach identified a diagnostic model for pancreatic cancer through using circulating microRNA signatures. *Pancreatology*, *20*(6), pp. 1195–1204.

Amin, M.B., Greene, F.L., Edge, S.B., Compton, C.C., Gershenwald, J.E., Brookland, R.K., Meyer, L., Gress, D.M., Byrd, D.R., and Winchester, D.P., 2017. The eighth edition AJCC cancer staging manual: continuing to build a bridge from a population-based to a more "personalized" approach to cancer staging. *CA: A Cancer Journal for Clinicians*, *67*(2), pp. 93–99.

Amir, E.A.D., Davis, K.L., Tadmor, M.D., Simonds, E.F., Levine, J.H., Bendall, S.C., Shenfeld, D.K., Krishnaswamy, S., Nolan, G.P., and Pe'er, D., 2013. viSNE enables visualization of high dimensional single-cell data and reveals phenotypic heterogeneity of leukemia. *Nature Biotechnology*, *31*(6), pp. 545–552.

Andjelkovic-Cirkovic, B., Cvetkovic, A., Ninkovic, S., Filipovic, N., 2015. An intelligent system for estimation of survival rate, relapse and metastasis – a role in individualization of breast cancer therapy. *In: IEEE 15th International Conference on Bioinformatics and Bioengineering*, BIBE 2015, Belgrade, November 2–4, 2015, pp. 171–176.

Andjelkovic Cirkovic, B.R., 2020. Machine Learning Approach for Breast Cancer Prognosis Prediction. *In: Computational Modeling in Bioengineering and Bioinformatics*(pp. 41–68). Elsevier.

Andreatta, M., and Nielsen, M., 2016. Gapped sequence alignment using artificial neural networks: application to the MHC class I system. *Bioinformatics*, *32*(4), pp. 511–517.

Andris, S., Seidel, J., and Hubbuch, J., 2019. Kinetic reaction modeling for antibody-drug conjugate process development. *Journal of Biotechnology*, *306*, pp. 71–80.

Andrysik, Z., Galbraith, M.D., Guarnieri, A.L., Zaccara, S., Sullivan, K.D., Pandey, A., MacBeth, M., Inga, A., and Espinosa, J.M., 2017. Identification of a core TP53 transcriptional program with highly distributed tumor suppressive activity. *Genome Research*, *27*(10), pp. 1645–1657.

Ansari, M.A., Jamal, Q.M.S., Rehman, S., Almatroudi, A., Alzohairy, M.A., Alomary, M.N., Tripathi, T., Alharbi, A.H., Adil, S.F., Khan, M., and Shaheer Malik, M., 2020. TAT-peptide conjugated repurposing drug against SARS-CoV-2 main protease (3CLpro): Potential therapeutic intervention to combat COVID-19. *Arabian Journal of Chemistry*, *13*(11), pp. 8069–8079.

Arandkar, S., Furth, N., Elisha, Y., Nataraj, N.B., van der Kuip, H., Yarden, Y., Aulitzky, W., Ulitsky, I., Geiger, B., and Oren, M., 2018. Altered p53 functionality in cancer-associated fibroblasts contributes to their cancer-supporting features. *Proceedings of the National Academy of Sciences*, *115*(25), pp. 6410–6415.

Asadi, M., Beik, J., Hashemian, R., Laurent, S., Farashahi, A., Mobini, M., Ghaznavi, H., and Shakeri-Zadeh, A., 2019. MRI-based numerical modeling strategy for simulation and treatment planning of nanoparticle-assisted photothermal therapy. *Physica Medica*, *66*, pp. 124–132.

Austin, A.M., Douglass, M.J.J., Nguyen, G.T., and Penfold, S.N., 2017. A radiobiological Markov simulation tool for aiding decision making in proton therapy referral. *Physica Medica*, *44*, pp. 72–82.

Badal, B., Solovyov, A., Di Cecilia, S., Chan, J.M., Chang, L.W., Iqbal, R., Aydin, I.T., Rajan, G.S., Chen, C., Abbate, F., and Arora, K.S., 2017. Transcriptional dissection of melanoma identifies a high-risk subtype underlying TP53 family genes and epigenome deregulation. *JCI insight*, *2*(9). doi: 10.1172/jci.insight.92102.

Balasubramanian, K., 2021. Computational and Artificial Intelligence Techniques for Drug Discovery and Administration. In: *Reference Module in Biomedical Sciences*. Elsevier.

Banik, A., Ghosh, K., Patil, U.K., and Gayen, S., 2021. Identification of molecular fingerprints of natural products for the inhibition of breast cancer resistance protein (BCRP). *Phytomedicine*, *85*. doi: 10.1016/j.phymed.2021.153523.

Bannigan, P., Aldeghi, M., Bao, Z., Häse, F., Aspuru-Guzik, A., and Allen, C., 2021. Machine learning directed drug formulation development. *Advanced Drug Delivery Reviews*. doi: 10.1016/j.addr.2021.05.016.

Bassani-Sternberg, M., Chong, C., Guillaume, P., Solleder, M., Pak, H., Gannon, P.O., Kandalaft, L.E., Coukos, G., and Gfeller, D., 2017. Deciphering HLA-I motifs across HLA peptidomes improves neoantigen predictions and identifies allostery regulating HLA specificity. *PLoS Computational Biology*, *13*(8), p. e1005725.

Batooei, S., Moslehi, A., and Pirayesh Islamian, J., 2021. Assessment of metallic nanoparticles as radioenhancers in gastric cancer therapy by Geant4 simulation and local effect model. *Nuclear Instruments and Methods in Physics Research, Section B: Beam Interactions with Materials and Atoms*, *488*, pp. 5–11.

Behera, A., Ashraf, R., Srivastava, A.K., and Kumar, S., 2020. Bioinformatics analysis and verification of molecular targets in ovarian cancer stem-like cells. *Heliyon*, *6*(9).

Bellomo, N., and Delitala, M., 2008. From the mathematical kinetic, and stochastic game theory to modelling mutations, onset, progression and immune competition of cancer cells. *Physics of Life Reviews*. doi: 10.1016/j.plrev.2008.07.001.

Bendall, S.C., and Nolan, G.P., 2012. From single cells to deep phenotypes in cancer. *Nature Biotechnology*, *30*(7), p. 639.

Bensaad, K., Tsuruta, A., Selak, M.A., Vidal, M.N.C., Nakano, K., Bartrons, R., Gottlieb, E., and Vousden, K.H., 2006. TIGAR, a p53-inducible regulator of glycolysis and apoptosis. *Cell*, *126*(1), pp. 107–120.

Bermejo, P., Gamez, J.A., Puerta, J.M., 2009. Incremental Wrapper-based subset Selection with replacement: an advantageous alternative to sequential forward selection. In: *IEEE Symposium on Computational Intelligence and Data Mining, CIDM* 2009.

Bertolet, A., Cortés-Giraldo, M.A., and Carabe-Fernandez, A., 2021. Implementation of the microdosimetric kinetic model using analytical microdosimetry in a treatment planning system for proton therapy. *Physica Medica*, *81*, pp. 69–76.

Bianco, G., Goodsell, D.S., and Forli, S., 2020. Selective and Effective: Current Progress in Computational Structure-Based Drug Discovery of Targeted Covalent Inhibitors. *Trends in Pharmacological Sciences*. doi: 10.1016/j.tips.2020.10.005.

Binnewies, M., Roberts, E.W., Kersten, K., Chan, V., Fearon, D.F., Merad, M., Coussens, L.M., Gabrilovich, D.I., Ostrand-Rosenberg, S., Hedrick, C.C., and Vonderheide, R.H., 2018. Understanding the tumor immune microenvironment (TIME) for effective therapy. *Nature Medicine*, 24(5), pp. 541–550.

Birtwistle, M.R., Hatakeyama, M., Yumoto, N., Ogunnaike, B.A., Hoek, J.B., and Kholodenko, B.N., 2007. Ligand-dependent responses of the ErbB signaling network: experimental and modeling analyses. *Molecular Systems Biology*, 3(1), p. 144.

Boegel, S., Lower, M., Schäfer, M., Bukur, T., De Graaf, J., Boisguérin, V., Türeci, O., Diken, M., Castle, J.C., and Sahin, U., 2013. HLA typing from RNA-Seq sequence reads. *Genome Medicine*, 4(12), pp. 1–12.

Boehm, K.M., Bhinder, B., Raja, V.J., Dephoure, N., and Elemento, O., 2019. Predicting peptide presentation by major histocompatibility complex class I: an improved machine learning approach to the immunopeptidome. *BMC Bioinformatics*, 20(1), pp. 1–11.

Bott, A.J., Peng, I.C., Fan, Y., Faubert, B., Zhao, L., Li, J., Neidler, S., Sun, Y., Jaber, N., Krokowski, D., and Lu, W., 2015. Oncogenic Myc induces expression of glutamine synthetase through promoter demethylation. *Cell Metabolism*, 22(6), pp. 1068–1077.

Bouwmeester, T., Bauch, A., Ruffner, H., Angrand, P.O., Bergamini, G., Croughton, K., Cruciat, C., Eberhard, D., Gagneur, J., Ghidelli, S., and Hopf, C., 2004. A physical and functional map of the human TNF-α/NF-κB signal transduction pathway. *Nature Cell Biology*, 6(2), pp. 97–105.

Brewster, A.M., Hortobagyi, G.N., Broglio, K.R., Kau, S.W., Santa-Maria, C.A., Arun, B., Buzdar, A.U., Booser, D.J., Valero, V., Bondy, M., and Esteva, F.J., 2008. Residual risk of breast cancer recurrence 5 years after adjuvant therapy. *Journal of the National Cancer Institute*, 100(16), pp. 1179–1183.

Brieu, N., Gavriel, C.G., Harrison, D.J., Schmidt, G., and Caie, P.D., 2018. Augmenting TNM staging with machine learning-based immune profiling for improved prognosis prediction in muscle-invasive bladder cancer patients. *Annals of Oncology*, 29, pp. viii28–viii29.

Brown, K.S., and Sethna, J.P., 2003. Statistical mechanical approaches to models with many poorly known parameters. *Physical Review E*, 68(2), p. 021904.

Broz, D.K., Mello, S.S., Bieging, K.T., Jiang, D., Dusek, R.L., Brady, C.A., Sidow, A., and Attardi, L.D., 2013. Global genomic profiling reveals an extensive p53-regulated autophagy program contributing to key p53 responses. *Genes & Development*, 27(9), pp. 1016–1031.

Buchkovich, M.L., Brown, C.C., Robasky, K., Chai, S., Westfall, S., Vincent, B.G., Weimer, E.T., and Powers, J.G., 2017. HLAProfiler utilizes k-mer profiles to improve HLA calling accuracy for rare and common alleles in RNA-seq data. *Genome Medicine*, 9(1), pp. 1–15.

Bulik-Sullivan, B., Busby, J., Palmer, C.D., Davis, M.J., Murphy, T., Clark, A., Busby, M., Duke, F., Yang, A., Young, L., and Ojo, N.C., 2019. Deep learning using tumor HLA peptide mass spectrometry datasets improves neoantigen identification. *Nature Biotechnology*, 37(1), pp. 55–63.

Burke, H.B., Goodman, P.H., Rosen, D.B., Henson, D.E., Weinstein, J.N., Harrell Jr, F.E., Marks, J.R., Winchester, D.P., and Bostwick, D.G., 1997. Artificial neural networks improve the accuracy of cancer survival prediction. *Cancer*, 79(4), pp. 857–862.

Burki, T.K., 2016. Predicting lung cancer prognosis using machine learning. *The Lancet. Oncology*. doi: 10.1016/S1470-2045(16)30436-3.

Calvo, E., Walko, C., Dees, E.C., and Valenzuela, B., 2016. Pharmacogenomics, pharmacokinetics, and pharmacodynamics in the era of targeted therapies. *American Society of Clinical Oncology Educational Book*, 36, pp. e175–e184.

Calzone, L., Barillot, E., and Zinovyev, A., 2018. Logical versus kinetic modeling of biological networks: Applications in cancer research. *Current Opinion in Chemical Engineering*. doi: 10.1016/j.coche.2018.02.005.

Camacho, D.M., Collins, K.M., Powers, R.K., Costello, J.C., and Collins, J.J., 2018. Next-generation machine learning for biological networks. *Cell*, 173(7), pp. 1581–1592.

Camarda, R., Zhou, A.Y., Kohnz, R.A., Balakrishnan, S., Mahieu, C., Anderton, B., Eyob, H., Kajimura, S., Tward, A., Krings, G., and Nomura, D.K., 2016. Inhibition of fatty acid oxidation as a therapy for MYC-overexpressing triple-negative breast cancer. *Nature medicine*, 22(4), pp. 427–432.

Camidge, D.R., Doebele, R.C., and Kerr, K.M., 2019. Comparing and contrasting predictive biomarkers for immunotherapy and targeted therapy of NSCLC. *Nature Reviews Clinical Oncology*, 16(6), pp. 341–355.

Carrella, D., Manni, I., Tumaini, B., Dattilo, R., Papaccio, F., Mutarelli, M., Sirci, F., Amoreo, C.A., Mottolese, M., Iezzi, M., and Ciolli, L., 2016. Computational drugs repositioning identifies inhibitors of oncogenic PI3K/AKT/P70S6K-dependent pathways among FDA-approved compounds. *Oncotarget*, *7*(37), p. 58743.

Carrella, D., Napolitano, F., Rispoli, R., Miglietta, M., Carissimo, A., Cutillo, L., Sirci, F., Gregoretti, F., and Di Bernardo, D., 2014. Mantra 2.0: an online collaborative resource for drug mode of action and repurposing by network analysis. *Bioinformatics*, *30*(12), pp. 1787–1788.

Catrina, I., O'Brien, P.J., Purcell, J., Nikolic-Hughes, I., Zalatan, J.G., Hengge, A.C., and Herschlag, D., 2007. Probing the origin of the compromised catalysis of E. coli alkaline phosphatase in its promiscuous sulfatase reaction. *Journal of the American Chemical Society*, *129*(17), pp. 5760–5765.

Celebi, R., Don't Walk, O.B., Movva, R., Alpsoy, S., and Dumontier, M., 2019. In-silico prediction of synergistic anti-cancer drug combinations using multi-omics data. *Scientific Reports*, *9*(1), pp. 1–10.

Cerami, E.G., Gross, B.E., Demir, E., Rodchenkov, I., Babur, O., Anwar, N., Schultz, N., Bader, G.D., and Sander, C., 2010. Pathway Commons, a web resource for biological pathway data. *Nucleic Acids Research*, *39*(suppl_1), pp. D685–D690.

Chae, Y.K., Pan, A.P., Davis, A.A., Patel, S.P., Carneiro, B.A., Kurzrock, R., and Giles, F.J., 2017. Path toward precision oncology: review of targeted therapy studies and tools to aid in defining "actionability" of a molecular lesion and patient management support. *Molecular Cancer Therapeutics*, *16*(12), pp. 2645–2655.

Chai, H., Zhou, X., Zhang, Z., Rao, J., Zhao, H., and Yang, Y., 2021. Integrating multi-omics data through deep learning for accurate cancer prognosis prediction. *Computers in Biology and Medicine*, *134*. doi: 10.1016/j.compbiomed.2021.104481.

Chan, S., Scheulen, M.E., Johnston, S., Mross, K., Cardoso, F., Dittrich, C., Eiermann, W., Hess, D., Morant, R., Semiglazov, V., and Borner, M., 2005. Phase II study of temsirolimus (CCI-779), a novel inhibitor of mTOR, in heavily pretreated patients with locally advanced or metastatic breast cancer. *Journal of Clinical Oncology*, *23*(23), pp. 5314–5322.

Chang, Y.W., Su, Y.J., Hsiao, M., Wei, K.C., Lin, W.H., Liang, C.J., Chen, S.C., and Lee, J.L., 2015. Diverse targets of β-catenin during the epithelial–mesenchymal transition define cancer stem cells and predict disease relapse. *Cancer research*, *75*(16), pp. 3398–3410.

Chari, S., Sridhar, K., Walenga, R., and Kleinstreuer, C., 2021. Computational analysis of a 3D mucociliary clearance model predicting nasal drug uptake. *Journal of Aerosol Science*, *155*. doi: 10.1016/j.jaerosci.2021.105757.

Chateauvieux, S., Morceau, F., Dicato, M., and Diederich, M., 2010. Molecular and therapeutic potential and toxicity of valproic acid. *Journal of Biomedicine and Biotechnology*, *2010*. doi: 10.1155/2010/479364.

Chawla, N.V., Bowyer, K.W., Hall, L.O., and Kegelmeyer, W.P., 2002. SMOTE: synthetic minority over-sampling technique. *Journal of Artificial Intelligence Research*, *16*, pp. 321–357.

Chen, H.R., Sherr, D.H., Hu, Z., and DeLisi, C., 2016. A network based approach to drug repositioning identifies plausible candidates for breast cancer and prostate cancer. *BMC Medical Genomics*, *9*(1), pp. 1–11.

Chen, K., Wang, K., Kirichian, A.M., Al Aowad, A.F., Iyer, L.K., Adelstein, S.J., and Kassis, A.I., 2006. In silico design, synthesis, and biological evaluation of radioiodinated quinazolinone derivatives for alkaline phosphatase–mediated cancer diagnosis and therapy. *Molecular Cancer Therapeutics*, *5*(12), pp. 3001–3013.

Chen, L., Zeng, W.M., Cai, Y.D., Feng, K.Y., and Chou, K.C., 2012. Predicting anatomical therapeutic chemical (ATC) classification of drugs by integrating chemical-chemical interactions and similarities. *PloS One*, *7*(4), p. e35254.

Chen, Y., Gruidl, M., Remily-Wood, E., Liu, R.Z., Eschrich, S., Lloyd, M., Nasir, A., Bui, M.M., Huang, E., Shibata, D., and Yeatman, T., 2010. Quantification of β-catenin signaling components in colon cancer cell lines, tissue sections, and microdissected tumor cells using reaction monitoring mass spectrometry. *Journal of Proteome Research*, *9*(8), pp. 4215–4227.

Cheng, L., and Li, L., 2016. Systematic quality control analysis of LINCS data. *CPT: Pharmacometrics & Systems Pharmacology*, *5*(11), pp. 588–598.

Chunyu, L., Ran, L., Junteng, Z., Miye, W., Jing, X., Lan, S., Yixuan, Z., Rui, Z., Yizhou, F., Chen, W., Hongmei, Y., and Qing, Z., 2020. Characterizing the critical features when personalizing anti-hypertensive drugs using spectrum analysis and machine learning methods. *Artificial Intelligence in Medicine*, *104*. doi: 10.1016/j.artmed.2020.101841.

Costa, B., Amorim, I., Gärtner, F., and Vale, N., 2020. Understanding Breast cancer: from conventional therapies to repurposed drugs. *European Journal of Pharmaceutical Sciences*. doi: 10.1016/j.ejps.2020.105401.

Costantino, J.P., Gail, M.H., Pee, D., Anderson, S., Redmond, C.K., Benichou, J., and Wieand, H.S., 1999. Validation studies for models projecting the risk of invasive and total breast cancer incidence. *Journal of the National Cancer Institute*, *91*(18), pp. 1541–1548.

Cruz, J.A., and Wishart, D.S., 2006. Applications of machine learning in cancer prediction and prognosis. *Cancer Informatics, 2*, p. 117693510600200030.

Csercsik, D., and Kovács, L., 2020. Reaction kinetic interpretation of mechanisms related to vascular tumor growth with respect to structural identifiability. *IFAC-PapersOnLine, 53*(2), pp. 16106–16111.

D'Souza, S., Prema, K. V., and Balaji, S., 2020. Machine learning models for drug–target interactions: current knowledge and future directions. *Drug Discovery Today.* doi: 10.1016/j.drudis.2020.03.003.

Dagogo-Jack, I., and Shaw, A.T., 2018. Tumour heterogeneity and resistance to cancer therapies. *Nature Reviews Clinical Oncology, 15*(2), p. 81.

Dang, C.V., 2012. MYC on the path to cancer. *Cell, 149*(1), pp. 22–35.

Das, P., Das, S., Das, P., Rihan, F.A.A., Uzuntarla, M., and Ghosh, D., 2021. Optimal control strategy for cancer remission using combinatorial therapy: A mathematical model-based approach. *Chaos, Solitons and Fractals, 145.* doi: 10.1016/j.chaos.2021.110789.

Dash, P., Fiore-Gartland, A.J., Hertz, T., Wang, G.C., Sharma, S., Souquette, A., Crawford, J.C., Clemens, E.B., Nguyen, T.H., Kedzierska, K., and La Gruta, N.L., 2017. Quantifiable predictive features define epitope-specific T cell receptor repertoires. *Nature, 547*(7661), pp. 89–93.

de Bono, B., 2009. The Breadth and Depth of BioMedical Molecular Networks: The Reactome Perspective. *In: Handbook of Research on Systems Biology Applications in Medicine* (pp. 714–729). IGI Global.

De, P., Chakraborty, I., Karna, B., and Mazumder, N., 2021. Brief review on repurposed drugs and vaccines for possible treatment of COVID-19. *European Journal of Pharmacology.* doi: 10.1016/j.ejphar. 2021.173977.

Dejure, F.R., and Eilers, M., 2017. MYC and tumor metabolism: chicken and egg. *The EMBO Journal, 36*(23), pp. 3409–3420.

Dilthey, A.T., Mentzer, A.J., Carapito, R., Cutland, C., Cereb, N., Madhi, S.A., Rhie, A., Koren, S., Bahram, S., McVean, G., and Phillippy, A.M., 2019. HLA* LA–HLA typing from linearly projected graph alignments. *Bioinformatics, 35*(21), pp. 4394–4396.

Ding, L., Wendl, M.C., McMichael, J.F., and Raphael, B.J., 2014. Expanding the computational toolbox for mining cancer genomes. *Nature Reviews Genetics, 15*(8), pp. 556–570.

Dinić, J., Efferth, T., García-Sosa, A.T., Grahovac, J., Padrón, J.M., Pajeva, I., Rizzolio, F., Saponara, S., Spengler, G., and Tsakovska, I., 2020. Repurposing old drugs to fight multidrug resistant cancers. *Drug Resistance Updates.* doi: 10.1016/j.drup.2020.100713.

Dong, J., Gao, H., and Ouyang, D., 2021. PharmSD: A novel AI-based computational platform for solid dispersion formulation design. *International Journal of Pharmaceutics, 604*, p. 120705.

Duan, F., Li, H., Liu, W., Zhao, J., Yang, Z., and Zhang, J., 2021. Long non-coding RNA FOXD2-AS1 serve as a potential prognostic biomarker for patients with cancer: A meta-analysis and database testing. *The American Journal of the Medical Sciences.* doi: 10.1016/j.amjms.2021.01.020.

Dudley, J.T., Sirota, M., Shenoy, M., Pai, R.K., Roedder, S., Chiang, A.P., Morgan, A.A., Sarwal, M.M., Pasricha, P.J., and Butte, A.J., 2011. Computational repositioning of the anticonvulsant topiramate for inflammatory bowel disease. *Science translational medicine, 3*(96), pp. 96ra76–96ra76.

Elbadawi, M., Gaisford, S., and Basit, A.W., 2021. Advanced machine-learning techniques in drug discovery. *Drug Discovery Today.* doi: 10.1016/j.drudis.2020.12.003.

Elkon, R., Loayza-Puch, F., Korkmaz, G., Lopes, R., Van Breugel, P.C., Bleijerveld, O.B., Altelaar, A.M., Wolf, E., Lorenzin, F., Eilers, M., and Agami, R., 2015. Myc coordinates transcription and translation to enhance transformation and suppress invasiveness. *EMBO Reports, 16*(12), pp. 1723–1736.

Elmezayen, A.D., Al-Obaidi, A., and Yelekçi, K., 2021. Discovery of novel isoform-selective histone dea-cetylases 5 and 9 inhibitors through combined ligand-based pharmacophore modeling, molecular mocking, and molecular dynamics simulations for cancer treatment. *Journal of Molecular Graphics and Modelling, 106*, p. 107937.

El-Wakil, M.H., Meheissen, M.A., and Abu-Serie, M.M., 2021. Nitrofurazone repurposing towards design and synthesis of novel apoptotic-dependent anticancer and antimicrobial agents: Biological evaluation, kinetic studies and molecular modeling. *Bioorganic Chemistry, 113*, p. 104971.

Eun, M.Y., Jeon, E.T., Seo, K.D., Lee, D., and Jung, J.M., 2021. Reperfusion therapy in acute ischemic stroke with active cancer: A meta-analysis aided by machine learning. *Journal of Stroke and Cerebrovascular Diseases, 30*(6). doi: 10.1016/j.jstrokecerebrovasdis.2021.105742.

Ewing, R.M., Song, J., Gokulrangan, G., Bai, S., Bowler, E.H., Bolton, R., Skipp, P., Wang, Y., and Wang, Z., 2018. Multiproteomic and transcriptomic analysis of oncogenic β-Catenin molecular networks. *Journal of Proteome Research, 17*(6), pp. 2216–2225.

Fahami, M.A., Roshanzamir, M., Izadi, N.H., Keyvani, V., and Alizadehsani, R., 2021. Detection of effective genes in colon cancer: A machine learning approach. *Informatics in Medicine Unlocked, 24*, p. 100605.

Farizhandi, A.A.K., Alishiri, M., and Lau, R., 2021. Machine learning approach for carrier surface design in carrier-based dry powder inhalation. *Computers & Chemical Engineering*, *151*, p. 107367.

Fayed, M.A.A., El-Behairy, M.F., Abdallah, I.A., Abdel-Bar, H.M., Elimam, H., Mostafa, A., Moatasim, Y., Abouzid, K.A.M., and Elshaier, Y.A.M.M., 2021. Structure- and ligand-based in silico studies towards the repurposing of marine bioactive compounds to target SARS-CoV-2. *Arabian Journal of Chemistry*, *14*(4). doi: 10.1016/j.arabjc.2021.103092.

Feng, R., Yu, F., Xu, J., and Hu, X., 2021. Knowledge gaps in immune response and immunotherapy involving nanomaterials: Databases and artificial intelligence for material design. *Biomaterials*. 10.1016/j.biomaterials.2020.120469.

Filipovic, N., 2020. Machine Learning Approach for Breast Cancer Prognosis Prediction. *In: Computational Modeling In Bioengineering and Bioinformatics* (pp. 41–68). Academic Press, New York.

Finotello, F., and Eduati, F., 2018. Multi-omics profiling of the tumor microenvironment: Paving the way to precision immuno-oncology. *Frontiers in Oncology*, *8*, p. 430.

Finotello, F., Rieder, D., Hackl, H., and Trajanoski, Z., 2019. Next-generation computational tools for interrogating cancer immunity. *Nature Reviews Genetics*, *20*(12), pp. 724–746.

Funahashi, A., Morohashi, M., Kitano, H., and Tanimura, N., 2003. CellDesigner: A process diagram editor for gene-regulatory and biochemical networks. *Biosilico*, *1*(5), pp. 159–162.

Gail, M.H., Brinton, L.A., Byar, D.P., Corle, D.K., Green, S.B., Schairer, C., and Mulvihill, J.J., 1989. Projecting individualized probabilities of developing breast cancer for white females who are being examined annually. *JNCI: Journal of the National Cancer Institute*, *81*(24), pp. 1879–1886.

Gaulton, A., Hersey, A., Nowotka, M., Bento, A.P., Chambers, J., Mendez, D., Mutowo, P., Atkinson, F., Bellis, L.J., Cibrián-Uhalte, E., and Davies, M., 2017. The ChEMBL database in 2017. *Nucleic Acids Research*, *45*(D1), pp. D945–D954.

Gaur, N.K., Goyal, V.D., Kulkarni, K., and Makde, R.D., 2021. Machine learning classifiers aid virtual screening for efficient design of mini-protein therapeutics. *Bioorganic and Medicinal Chemistry Letters*, 38. doi: 10.1016/j.bmcl.2021.127852.

Geva-Zatorsky, N., Rosenfeld, N., Itzkovitz, S., Milo, R., Sigal, A., Dekel, E., Yarnitzky, T., Liron, Y., Polak, P., Lahav, G., and Alon, U., 2006. Oscillations and variability in the p53 system. *Molecular Systems Biology*, *2*(1). doi: 10.1038/msb4100068.

Gfeller, D., and Bassani-Sternberg, M., 2018. Predicting antigen presentation – what could we learn from a million peptides?. *Frontiers in immunology*, *9*, p. 1716.

Gfeller, D., Guillaume, P., Michaux, J., Pak, H.S., Daniel, R.T., Racle, J., Coukos, G., and Bassani-Sternberg, M., 2018. The length distribution and multiple specificity of naturally presented HLA-I ligands. *The Journal of Immunology*, *201*(12), pp. 3705–3716.

Ghasemi, T., Khalaj-Kondori, M., Hosseinpour feizi, M.A., and Asadi, P., 2020. lncRNA-miRNA-mRNA interaction network for colorectal cancer; an in silico analysis. *Computational Biology and Chemistry*, *89*. doi: 10.1016/j.compbiolchem.2020.107370.

Ghetiu, T., Polack, F., and Bown, J., 2011. Safety-critical systems argumentation and validation in computational modeling for drug design. *Current Opinion in Biotechnology*, *22*, p. S29.

Ghita, M., Drexler, D.A., Kovács, L., Copot, D., Muresan, C.I., and Ionescu, C.M., 2020. Model-Based Management of Lung Cancer Radiation Therapy. *IFAC-PapersOnLine*, *53*(2), p. 15928–15933.

Giladi, A., and Amit, I., 2018. Single-cell genomics: A stepping stone for future immunology discoveries. *Cell*, *172*(1-2), pp. 14–21.

Gillies, R.J., Brown, J.S., Anderson, A.R., and Gatenby, R.A., 2018. Eco-evolutionary causes and consequences of temporal changes in intratumoural blood flow. *Nature Reviews Cancer*, *18*(9), pp. 576–585.

Glanville, J., Huang, H., Nau, A., Hatton, O., Wagar, L.E., Rubelt, F., Ji, X., Han, A., Krams, S.M., Pettus, C., and Haas, N., 2017. Identifying specificity groups in the T cell receptor repertoire. *Nature*, *547*(7661), pp. 94–98.

Gns, H., PrasannaMarise, V.L., Pai, R.R., Mariam Jos, S., Krishna Murthy, M., and Saraswathy, G.R., 2020. Unveiling potential anticancer drugs through in silico drug repurposing approaches. *In: Drug Repurposing in Cancer Therapy* (pp. 81–119). Elsevier.

Gong, J., Cai, C., Liu, X., Ku, X., Jiang, H., Gao, D., and Li, H., 2013. ChemMapper: a versatile web server for exploring pharmacology and chemical structure association based on molecular 3D similarity method. *Bioinformatics*, *29*(14), pp. 1827–1829.

Gonzalez-Fierro, A., and Dueñas-González, A., 2021. Drug repurposing for cancer therapy, easier said than done. *Seminars in Cancer Biology*. doi: 10.1016/j.semcancer.2019.12.012.

Gopal, V.N., Al-Turjman, F., Kumar, R., Anand, L., and Rajesh, M., 2021. Feature selection and classification in breast cancer prediction using IoT and machine learning. *Measurement: Journal of the International Measurement Confederation*, 178. doi: 10.1016/j.measurement.2021.109442.

Gottlieb, A., Stein, G.Y., Ruppin, E., and Sharan, R., 2011. PREDICT: a method for inferring novel drug indications with application to personalized medicine. *Molecular Systems Biology*, 7(1), p. 496.

Guo, C., Wang, J., Wang, Y., Qu, X., Shi, Z., Meng, Y., Qiu, J., and Hua, K., 2021. Novel artificial intelligence machine learning approaches to precisely predict survival and site-specific recurrence in cervical cancer: A multi-institutional study. *Translational Oncology*, 14(5). doi: 10.1016/j.tranon.2021.101032.

Hall, Z., Ament, Z., Wilson, C.H., Burkhart, D.L., Ashmore, T., Koulman, A., Littlewood, T., Evan, G.I., and Griffin, J.L., 2016. Myc expression drives aberrant lipid metabolism in lung cancer. *Cancer Research*, 76(16), pp. 4608–4618.

Han, Y., and Kim, D., 2017. Deep convolutional neural networks for pan-specific peptide-MHC class I binding prediction. *BMC Bioinformatics*, 18(1), pp. 1–9.

Hanahan, D., and Weinberg, R.A., 2000. The hallmarks of cancer. *Cell*, 100(1), pp. 57–70.

Hathout, R.M., 2021. Machine learning methods in drug delivery. In: *Applications of Artificial Intelligence in Process Systems Engineering* (pp. 361–380). Elsevier.

Hay, N., and Sonenberg, N., 2004. Upstream and downstream of mTOR. *Genes & development*, 18(16), pp. 1926–1945.

Hennessy, B.T., Smith, D.L., Ram, P.T., Lu, Y., and Mills, G.B., 2005. Exploiting the PI3K/AKT pathway for cancer drug discovery. *Nature reviews Drug discovery*, 4(12), pp. 988–1004.

Hidalgo, M., and Rowinsky, E.K., 2000. The rapamycin-sensitive signal transduction pathway as a target for cancer therapy. *Oncogene*, 19(56), pp. 6680–6686.

Ho, N.H., Harapanhalli, R.S., Dahman, B.A., Chen, K., Wang, K., Adelstein, S.J., and Kassis, A.I., 2002. Synthesis and biologic evaluation of a radioiodinated quinazolinone derivative for enzyme-mediated insolubilization therapy. *Bioconjugate Chemistry*, 13(2), pp. 357–364.

Horiuchi, D., Kusdra, L., Huskey, N.E., Chandriani, S., Lenburg, M.E., Gonzalez-Angulo, A.M., Creasman, K.J., Bazarov, A.V., Smyth, J.W., Davis, S.E., and Yaswen, P., 2012. MYC pathway activation in triple-negative breast cancer is synthetic lethal with CDK inhibition. *Journal of Experimental Medicine*, 209(4), pp. 679–696.

Hsieh, Y.Y., Liu, T.P., and Yang, P.M., 2019. In silico repurposing the Rac1 inhibitor NSC23766 for treating PTTG1-high expressing clear cell renal carcinoma. *Pathology Research and Practice*, 215(6). doi: 10.1016/j.prp.2019.03.002.

Huang, H., Zhang, P., Qu, X.A., Sanseau, P., and Yang, L., 2014. Systematic prediction of drug combinations based on clinical side-effects. *Scientific Reports*, 4(1), pp. 1–7.

Huang, J., Campian, J.L., Gujar, A.D., Tran, D.D., Lockhart, A.C., DeWees, T.A., Tsien, C.I., and Kim, A.H., 2016. A phase I study to repurpose disulfiram in combination with temozolomide to treat newly diagnosed glioblastoma after chemoradiotherapy. *Journal of Neuro-Oncology*, 128(2), pp. 259–266.

Huang, L., Injac, S.G., Cui, K., Braun, F., Lin, Q., Du, Y., Zhang, H., Kogiso, M., Lindsay, H., Zhao, S., and Baxter, P., 2018. Systems biology–based drug repositioning identifies digoxin as a potential therapy for groups 3 and 4 medulloblastoma. *Science Translational Medicine*, 10(464). doi: 10.1126/scitranslmed.aat0150.

Hucka, M., Finney, A., Sauro, H.M., Bolouri, H., Doyle, J., and Kitano, H., 2001. The ERATO Systems Biology Workbench: Enabling Interaction and Exchange Between Software Tools for Computational Biology. In: *Biocomputing 2002* (pp. 450–461). World Scientific Publishing Co Pte Ltd.

Hundal, J., Carreno, B.M., Petti, A.A., Linette, G.P., Griffith, O.L., Mardis, E.R., and Griffith, M., 2016. pVAC-Seq: A genome-guided in silico approach to identifying tumor neoantigens. *Genome Medicine*, 8(1), pp. 1–11.

Hundal, J., Kiwala, S., McMichael, J., Miller, C.A., Xia, H., Wollam, A.T., Liu, C.J., Zhao, S., Feng, Y.Y., Graubert, A.P., and Wollam, A.Z., 2020. pVACtools: A computational toolkit to identify and visualize cancer neoantigens. *Cancer Immunology Research*, 8(3), pp. 409–420.

Hünten, S., Kaller, M., Drepper, F., Oeljeklaus, S., Bonfert, T., Erhard, F., Dueck, A., Eichner, N., Friedel, C.C., Meister, G., and Zimmer, R., 2015. p53-regulated networks of protein, mRNA, miRNA, and lncRNA expression revealed by integrated pulsed stable isotope labeling with amino acids in cell culture (pSILAC) and next generation sequencing (NGS) analyses. *Molecular & Cellular Proteomics*, 14(10), pp. 2609–2629.

Igarashi, Y., Nakatsu, N., Yamashita, T., Ono, A., Ohno, Y., Urushidani, T., and Yamada, H., 2015. Open TG-GATEs: a large-scale toxicogenomics database. *Nucleic Acids Research*, 43(D1), pp. D921–D927.

Iljin, K., Ketola, K., Vainio, P., Halonen, P., Kohonen, P., Fey, V., Grafström, R.C., Perälä, M., and Kallioniemi, O., 2009. High-throughput cell-based screening of 4910 known drugs and drug-like small molecules identifies disulfiram as an inhibitor of prostate cancer cell growth. *Clinical Cancer Research*, *15*(19), pp. 6070–6078.

Iorio, F., Bosotti, R., Scacheri, E., Belcastro, V., Mithbaokar, P., Ferriero, R., Murino, L., Tagliaferri, R., Brunetti-Pierri, N., Isacchi, A., and di Bernardo, D., 2010. Discovery of drug mode of action and drug repositioning from transcriptional responses. *Proceedings of the National Academy of Sciences*, *107*(33), pp. 14621–14626.

Iranpour, S., Bahrami, A.R., Sh. Saljooghi, A., and Matin, M.M., 2021. Application of smart nanoparticles as a potential platform for effective colorectal cancer therapy. *Coordination Chemistry Reviews*. doi: 10.1016/j.ccr.2021.213949.

Irham, L.M., Wong, H.S.C., Chou, W.H., Adikusuma, W., Mugiyanto, E., Huang, W.C., and Chang, W.C., 2020. Integration of genetic variants and gene network for drug repurposing in colorectal cancer. *Pharmacological Research*, *161*. doi: 10.1016/j.phrs.2020.105203.

Issa, N.T., Stathias, V., Schürer, S., and Dakshanamurthy, S., 2021. Machine and deep learning approaches for cancer drug repurposing. *Seminars in Cancer Biology*. doi: 10.1016/j.semcancer.2019.12.011.

Jacobi, C.E., de Bock, G.H., Siegerink, B., and van Asperen, C.J., 2009. Differences and similarities in breast cancer risk assessment models in clinical practice: which model to choose?. *Breast Cancer Research and Treatment*, *115*(2), pp. 381–390.

Jayasekera, J., Mandelblatt, J., and Schechter, C., 2021. PCN23 The Development and Validation of a Simulation Model-based Clinical Decision Tool to Support Oncology Practice. *Value in Health*, *24*, p. S22.

Jiang, P., Gu, S., Pan, D., Fu, J., Sahu, A., Hu, X., Li, Z., Traugh, N., Bu, X., Li, B., and Liu, J., 2018. Signatures of T cell dysfunction and exclusion predict cancer immunotherapy response. *Nature Medicine*, *24*(10), pp. 1550–1558.

Jones, R.B., Gordus, A., Krall, J.A., and MacBeath, G., 2006. A quantitative protein interaction network for the ErbB receptors using protein microarrays. *Nature*, *439*(7073), pp. 168–174.

Joshi-Tope, G., Gillespie, M., Vastrik, I., D'Eustachio, P., Schmidt, E., de Bono, B., Jassal, B., Gopinath, G.R., Wu, G.R., Matthews, L., and Lewis, S., 2005. Reactome: a knowledgebase of biological pathways. *Nucleic Acids Research*, *33*(suppl_1), pp. D428–D432.

Juárez-López, D., and Schcolnik-Cabrera, A., 2021. Drug Repurposing: Considerations to Surpass While Re-directing Old Compounds for New Treatments. *Archives of Medical Research*. doi: 10.1016/j.arcmed.2020.10.021.

Jung, S., Ahn, E., Koh, S.B., Lee, S.-H., and Hwang, G.-S., 2021. Purine metabolite-based machine learning models for risk prediction, prognosis, and diagnosis of coronary artery disease. *Biomedicine & Pharmacotherapy*, *139*, p. 111621.

Jurtz, V., Paul, S., Andreatta, M., Marcatili, P., Peters, B., and Nielsen, M., 2017. NetMHCpan-4.0: improved peptide–MHC class I interaction predictions integrating eluted ligand and peptide binding affinity data. *The Journal of Immunology*, *199*(9), pp. 3360–3368.

Kalinli, A., Sarikoc, F., Akgun, H., and Ozturk, F., 2013. Performance comparison of machine learning methods for prognosis of hormone receptor status in breast cancer tissue samples. *Computer Methods and Programs in Biomedicine*, *110*(3), pp. 298–307.

Kalkat, M., Resetca, D., Lourenco, C., Chan, P.K., Wei, Y., Shiah, Y.J., Vitkin, N., Tong, Y., Sunnerhagen, M., Done, S.J., and Boutros, P.C., 2018. MYC protein interactome profiling reveals functionally distinct regions that cooperate to drive tumorigenesis. *Molecular Cell*, *72*(5), pp. 836–848.

Kamburov, A., Wierling, C., Lehrach, H., and Herwig, R., 2009. ConsensusPathDB – a database for integrating human functional interaction networks. *Nucleic Acids Research*, *37*(suppl_1), pp. D623–D628.

Kamerzell, T.J., and Middaugh, C.R., 2021. Prediction machines: Applied machine learning for therapeutic protein design and development. *Journal of Pharmaceutical Sciences*. doi: 10.1016/j.xphs.2020.11.034.

Kanehisa, M., Goto, S., Hattori, M., Aoki-Kinoshita, K.F., Itoh, M., Kawashima, S., Katayama, T., Araki, M., and Hirakawa, M., 2006. From genomics to chemical genomics: new developments in KEGG. *Nucleic Acids Research*, *34*(suppl_1), pp. D354–D357.

Karaman Mayack, B., and Sippl, W., 2021. Current In Silico Drug Repurposing Strategies. *In: Systems Medicine* (pp. 257–268). Elsevier.

Karp, P.D., Ouzounis, C.A., Moore-Kochlacs, C., Goldovsky, L., Kaipa, P., Ahrén, D., Tsoka, S., Darzentas, N., Kunin, V., and López-Bigas, N., 2005. Expansion of the BioCyc collection of pathway/genome databases to 160 genomes. *Nucleic Acids Research*, *33*(19), pp. 6083–6089.

Kashani, H.M., Madrakian, T., and Afkhami, A., 2021. Development of modified polymer dot as stimuli-sensitive and 67Ga radio-carrier, for investigation of in vitro drug delivery, in vivo imaging and drug release kinetic. *Journal of Pharmaceutical and Biomedical Analysis, 114217.* doi: 10.1016/j.jpba.2021.114217.

Kassis, A.I., Korideck, H., Wang, K., Pospisil, P., and Adelstein, S.J., 2008. Novel prodrugs for targeting diagnostic and therapeutic radionuclides to solid tumors. *Molecules, 13*(2), pp. 391–404.

Kaushik, M., Chandra Joshi, R., Kushwah, A.S., Gupta, M.K., Banerjee, M., Burget, R., and Dutta, M.K., 2021. Cytokine gene variants and socio-demographic characteristics as predictors of cervical cancer: A machine learning approach. *Computers in Biology and Medicine, 134*, p. 104559.

Kawaguchi, S., Higasa, K., Shimizu, M., Yamada, R., and Matsuda, F., 2017. HLA-HD: An accurate HLA typing algorithm for next-generation sequencing data. *Human Mutation, 38*(7), pp. 788–797.

Keiser, M.J., Setola, V., Irwin, J.J., Laggner, C., Abbas, A.I., Hufeisen, S.J., Jensen, N.H., Kuijer, M.B., Matos, R.C., Tran, T.B., and Whaley, R., 2009. Predicting new molecular targets for known drugs. *Nature, 462*(7270), pp. 175–181.

Keren, L., Bosse, M., Marquez, D., Angoshtari, R., Jain, S., Varma, S., Yang, S.R., Kurian, A., Van Valen, D., West, R., and Bendall, S.C., 2018. A structured tumor-immune microenvironment in triple negative breast cancer revealed by multiplexed ion beam imaging. *Cell, 174*(6), pp. 1373–1387.

Kerrien, S., Alam-Faruque, Y., Aranda, B., Bancarz, I., Bridge, A., Derow, C., Dimmer, E., Feuermann, M., Friedrichsen, A., Huntley, R., and Kohler, C., 2007. IntAct – open source resource for molecular interaction data. *Nucleic Acids Research, 35*(suppl_1), pp. D561–D565.

Khan, A.A., and Bano, Y., 2021. Salmonella enterica subsp. enterica host-pathogen interactions and their implications in gallbladder cancer. *Microbial Pathogenesis,* 157, pp. 105011.

Khanna, K., Mandal, S., Blanchard, A.T., Tewari, M., Johnson-Buck, A., and Walter, N.G., 2021. Rapid kinetic fingerprinting of single nucleic acid molecules by a FRET-based dynamic nanosensor. *Biosensors and Bioelectronics,* 113433. doi: 10.1016/j.bios.2021.113433.

Kim, D., Rath, O., Kolch, W., and Cho, K.H., 2007. A hidden oncogenic positive feedback loop caused by crosstalk between Wnt and ERK pathways. *Oncogene, 26*(31), pp. 4571–4579.

Kim, I.W., Jang, H., Kim, J.H., Kim, M.G., Kim, S., and Oh, J.M., 2019. Computational drug repositioning for gastric cancer using reversal gene expression profiles. *Scientific Reports, 9*(1), pp. 1–10.

Kim, W., Kim, K.S., Lee, J.E., Noh, D.Y., Kim, S.W., Jung, Y.S., Park, M.Y., and Park, R.W., 2012. Development of novel breast cancer recurrence prediction model using support vector machine. *Journal of Breast Cancer, 15*(2), pp. 230–238.

Kimmelman, J., and Tannock, I., 2018. The paradox of precision medicine. *Nature Reviews Clinical Oncology, 15*(6), pp. 341–342.

Klamt, S., Saez-Rodriguez, J., Lindquist, J.A., Simeoni, L., and Gilles, E.D., 2006. A methodology for the structural and functional analysis of signaling and regulatory networks. *BMC Bioinformatics, 7*(1), pp. 1–26.

Klipp, E., Herwig, R., Kowald, A., Wierling, C., and Lehrach, H., 2005. *Systems Biology in Practice: Concepts, Implementation and Application.* John Wiley & Sons.

Knight, R., Vrbanac, A., Taylor, B.C., Aksenov, A., Callewaert, C., Debelius, J., Gonzalez, A., Kosciolek, T., McCall, L.I., McDonald, D., and Melnik, A.V., 2018. Best practices for analysing microbiomes. *Nature Reviews Microbiology, 16*(7), pp. 410–422.

Koifman, G., Shetzer, Y., Eizenberger, S., Solomon, H., Rotkopf, R., Molchadsky, A., Lonetto, G., Goldfinger, N., and Rotter, V., 2018. A mutant p53-dependent embryonic stem cell gene signature is associated with augmented tumorigenesis of stem cells. *Cancer Research, 78*(20), pp. 5833–5847.

Kolodziejczyk, A.A., Kim, J.K., Svensson, V., Marioni, J.C., and Teichmann, S.A., 2015. The technology and biology of single-cell RNA sequencing. *Molecular Cell, 58*(4), pp. 610–620.

Kononenko, I., 1993. Inductive and Bayesian learning in medical diagnosis. *Applied Artificial Intelligence an International Journal, 7*(4), pp. 317–337.

Kopp, B., Mein, S., Tessonnier, T., Besuglow, J., Harrabi, S., Heim, E., Abdollahi, A., Haberer, T., Debus, J., and Mairani, A., 2021. Rapid effective dose calculation for raster-scanning 4He ion therapy with the modified microdosimetric kinetic model (mMKM). *Physica Medica, 81*, pp. 273–284.

Kourou, K., Exarchos, T.P., Exarchos, K.P., Karamouzis, M.V., and Fotiadis, D.I., 2015. Machine learning applications in cancer prognosis and prediction. *Computational and Structural Biotechnology Journal, 13*, pp. 8–17.

Kowalewski, J., and Ray, A., 2020. Predicting novel drugs for SARS-CoV-2 using machine learning from a >10 million chemical space. *Heliyon, 6*(8). doi: 10.1016/j.heliyon.2020.e04639.

Krull, M., Pistor, S., Voss, N., Kel, A., Reuter, I., Kronenberg, D., Michael, H., Schwarzer, K., Potapov, A., Choi, C., and Kel-Margoulis, O., 2006. TRANSPATH®: An information resource for storing and visualizing signaling pathways and their pathological aberrations. *Nucleic Acids Research*, *34*(suppl_1), pp. D546–D551.

Kuhn, M., Campillos, M., Letunic, I., Jensen, L.J., and Bork, P., 2010. A side effect resource to capture phenotypic effects of drugs. *Molecular Systems Biology*, *6*(1), p. 343.

Kumar, R., Coronel, L., Somalanka, B., Raju, A., Aning, O.A., An, O., Ho, Y.S., Chen, S., Mak, S.Y., Hor, P.Y., and Yang, H., 2018. Mitochondrial uncoupling reveals a novel therapeutic opportunity for p53-defective cancers. *Nature Communications*, *9*(1), pp. 1–13.

Kumar, R., Lathwal, A., Kumar, V., Patiyal, S., Raghav, P.K., and Raghava, G.P.S., 2020. CancerEnD: A database of cancer associated enhancers. *Genomics*, *112*(5), pp. 3696–3702.

Lamb, J., Crawford, E.D., Peck, D., Modell, J.W., Blat, I.C., Wrobel, M.J., Lerner, J., Brunet, J.P., Subramanian, A., Ross, K.N., and Reich, M., 2006. The Connectivity Map: Using gene-expression signatures to connect small molecules, genes, and disease. *science*, *313*(5795), pp. 1929–1935.

Le Novere, N., Bornstein, B., Broicher, A., Courtot, M., Donizelli, M., Dharuri, H., Li, L., Sauro, H., Schilstra, M., Shapiro, B., and Snoep, J.L., 2006. BioModels Database: A free, centralized database of curated, published, quantitative kinetic models of biochemical and cellular systems. *Nucleic Acids Research*, *34*(suppl_1), pp. D689–D691.

Lee, C., Light, A., Alaa, A., Thurtle, D., van der Schaar, M., and Gnanapragasam, V.J., 2021. Application of a novel machine learning framework for predicting non-metastatic prostate cancer-specific mortality in men using the Surveillance, Epidemiology, and End Results (SEER) database. *The Lancet Digital Health*, *3*(3), pp. e158–e165.

Lee, C.H., Yelensky, R., Jooss, K., and Chan, T.A., 2018. Update on tumor neoantigens and their utility: why it is good to be different. *Trends in Immunology*, *39*(7), pp. 536–548.

Lee, H., and Kingsford, C., 2018. Kourami: graph-guided assembly for novel human leukocyte antigen allele discovery. *Genome Biology*, *19*(1), pp. 1–16.

Legewie, S., Blüthgen, N., and Herzel, H., 2006. Mathematical modeling identifies inhibitors of apoptosis as mediators of positive feedback and bistability. *PLoS Computational Biology*, *2*(9), p. e120.

Li, B., Severson, E., Pignon, J.C., Zhao, H., Li, T., Novak, J., Jiang, P., Shen, H., Aster, J.C., Rodig, S., and Signoretti, S., 2016. Comprehensive analyses of tumor immunity: Implications for cancer immunotherapy. *Genome Biology*, *17*(1), pp. 1–16.

Li, H., Yuan, H., Middleton, A., Li, J., Nicol, B., Carmichael, P., Guo, J., Peng, S., and Zhang, Q., 2021. Next generation risk assessment (NGRA): Bridging in vitro points-of-departure to human safety assessment using physiologically-based kinetic (PBK) modelling – A case study of doxorubicin with dose metrics considerations. *Toxicology in Vitro*, *74*. doi: 10.1016/j.tiv.2021.105171.

Li, J., and Lu, Z., 2013. Pathway-based drug repositioning using causal inference. *BMC Bioinformatics*, *14*(16), pp. 1–10.

Li, X., Yu, J., Zhang, Z., Ren, J., Peluffo, A.E., Zhang, W., Zhao, Y., Wu, J., Yan, K., Cohen, D., and Wang, W., 2021. Network bioinformatics analysis provides insight into drug repurposing for COVID-19. *Medicine in Drug Discovery*, *10*. doi: 10.1016/j.medidd.2021.100090.

Liang, X., Zhang, P., Yan, L., Fu, Y., Peng, F., Qu, L., Shao, M., Chen, Y., and Chen, Z., 2017. LRSSL: predict and interpret drug–disease associations based on data integration using sparse subspace learning. *Bioinformatics*, *33*(8), pp. 1187–1196.

Linderman, G.C., Rachh, M., Hoskins, J.G., Steinerberger, S., and Kluger, Y., 2017. Efficient algorithms for t-distributed stochastic neighborhood embedding. *arXiv preprint arXiv:1712.09005*.

Liu, G., Li, D., Li, Z., Qiu, S., Li, W., Chao, C.C., Yang, N., Li, H., Cheng, Z., Song, X., and Cheng, L., 2017. PSSMHCpan: A novel PSSM-based software for predicting class I peptide-HLA binding affinity. *Giga Science*, *6*(5), p. gix017.

Liu, Z., Cui, Y., Xiong, Z., Nasiri, A., Zhang, A., and Hu, J., 2019. DeepSeqPan, a novel deep convolutional neural network model for pan-specific class I HLA-peptide binding affinity prediction. *Scientific Reports*, *9*(1), pp. 1–10.

Lo, Y.-C., and Torres, J.Z., 2019. In Silico Repurposing of Cell Cycle Modulators for Cancer Treatment. *In*: *In Silico Drug Design* (pp. 255–279). Elsevier.

Loew, L.M., and Schaff, J.C., 2001. The Virtual Cell: A software environment for computational cell biology. *Trends in Biotechnology*, *19*(10), pp. 401–406.

López, C., Kleinheinz, K., Aukema, S.M., Rohde, M., Bernhart, S.H., Hübschmann, D., Wagener, R., Toprak, U.H., Raimondi, F., Kreuz, M., and Waszak, S.M., 2019. Genomic and transcriptomic changes complement each other in the pathogenesis of sporadic Burkitt lymphoma. *Nature Communications*, *10*(1), pp. 1–19.

Lowman, X.H., Hanse, E.A., Yang, Y., Gabra, M.B.I., Tran, T.Q., Li, H., and Kong, M., 2019. p53 promotes cancer cell adaptation to glutamine deprivation by upregulating Slc7a3 to increase arginine uptake. *Cell Reports*, *26*(11), pp. 3051–3060.

Mahdian, S., Zarrabi, M., Panahi, Y., and Dabbagh, S., 2021. Repurposing FDA-approved drugs to fight COVID-19 using in silico methods: Targeting SARS-CoV-2 RdRp enzyme and host cell receptors (ACE2, CD147) through virtual screening and molecular dynamic simulations. *Informatics in Medicine Unlocked*, *23*. doi: 10.1016/j.imu.2021.100541.

Malik, S., Khalid, S., Ali, H., Khan, M., Mehwish, F., Javed, A., Akbar, F., Hanif, R., and Suleman, M., 2020. In-silico modeling and analysis of the therapeutic potential of miRNA-7 on EGFR associated signaling network involved in breast cancer. *Gene Reports*, *21*. 10.1016/j.genrep.2020.100938.

Marín-Hernández, A., Gallardo-Pérez, J.C., Reyes-García, M.A., Sosa-Garrocho, M., Macías-Silva, M., Rodríguez-Enríquez, S., Moreno-Sánchez, R., and Saavedra, E., 2020. Kinetic modeling of glucose central metabolism in hepatocytes and hepatoma cells. *Biochimica et Biophysica Acta – General Subjects*, *1864*(11). doi: 10.1016/j.bbagen.2020.129687.

Masuda, T., Nakaura, T., Funama, Y., Sugino, K., Sato, T., Yoshiura, T., Baba, Y., and Awai, K., 2021. Machine learning to identify lymph node metastasis from thyroid cancer in patients undergoing contrast-enhanced CT studies. *Radiography*. doi: 10.1016/j.radi.2021.03.001.

Mathon, B., Coquery, M., Liu, Z., Penru, Y., Guillon, A., Esperanza, M., Miège, C., and Choubert, J.M., 2021. Ozonation of 47 organic micropollutants in secondary treated municipal effluents: Direct and indirect kinetic reaction rates and modelling. *Chemosphere*, *262*. doi: 10.1016/j.chemosphere.2020.127969.

Matsukuma, M., Furukawa, M., Yamamoto, S., Nakamura, K., Tanabe, M., Okada, M., Iida, E., and Ito, K., 2020. The kinetic analysis of breast cancer: An investigation of the optimal temporal resolution for dynamic contrast-enhanced MR imaging. *Clinical Imaging*, *61*, pp. 4–10.

McCubrey, J.A., Steelman, L.S., Abrams, S.L., Bertrand, F.E., Ludwig, D.E., Bäsecke, J., Libra, M., Stivala, F., Milella, M., Tafuri, A., and Lunghi, P., 2008. Targeting survival cascades induced by activation of Ras/Raf/MEK/ERK, PI3K/PTEN/Akt/mTOR and Jak/STAT pathways for effective leukemia therapy. *Leukemia*, *22*(4), pp. 708–722.

Mealiffe, M.E., Stokowski, R.P., Rhees, B.K., Prentice, R.L., Pettinger, M., and Hinds, D.A., 2010. Assessment of clinical validity of a breast cancer risk model combining genetic and clinical information. *Journal of the National Cancer Institute*, *102*(21), pp. 1618–1627.

Mehta, C.H., Narayan, R., and Nayak, U.Y., 2019. Computational modeling for formulation design. *Drug Discovery Today*. doi: 10.1016/j.drudis.2018.11.018.

Melge, A.R., Manzoor, K., Nair, S. V., and Mohan, C.G., 2019. In Silico Modeling of FDA-Approved Drugs for Discovery of Anti-Cancer Agents: A Drug-Repurposing Approach. *In: In Silico Drug Design* (pp. 577–608). Elsevier.

Menden, M.P., Wang, D., Mason, M.J., Szalai, B., Bulusu, K.C., Guan, Y., Yu, T., Kang, J., Jeon, M., Wolfinger, R., and Nguyen, T., 2019. Community assessment to advance computational prediction of cancer drug combinations in a pharmacogenomic screen. *Nature communications*, *10*(1), pp. 1–17.

Mendes, P., 1993. GEPASI: a software package for modelling the dynamics, steady states and control of biochemical and other systems. *Bioinformatics*, *9*(5), pp. 563–571.

Mohammadzadeh-Asl, S., Aghanejad, A., Yekta, R., de la Guardia, M., Ezzati Nazhad Dolatabadi, J., and Keshtkar, A., 2020. Kinetic and thermodynamic insights into interaction of erlotinib with epidermal growth factor receptor: Surface plasmon resonance and molecular docking approaches. *International Journal of Biological Macromolecules*, *163*, pp. 954–958.

Mohanty, S., Rashid, M.H.A., Mohanty, C., and Swayamsiddha, S., 2021. Modern computational intelligence based drug repurposing for diabetes epidemic. *Diabetes & Metabolic Syndrome: Clinical Research & Reviews*. doi: 10.1016/j.dsx.2021.06.017.

Montes-Grajales, D., Puerta-Guardo, H., Espinosa, D.A., Harris, E., Caicedo-Torres, W., Olivero-Verbel, J., and Martínez-Romero, E., 2020. In silico drug repurposing for the identification of potential candidate molecules against arboviruses infection. *Antiviral Research*, *173*. doi: 10.1016/j.antiviral.2019.104668.

Moradi Kashkooli, F., Soltani, M., and Momeni, M.M., 2021. Computational modeling of drug delivery to solid tumors: A pilot study based on a real image. *Journal of Drug Delivery Science and Technology*, *62*. doi: 10.1016/j.jddst.2021.102347.

Morimoto-Tomita, M., Uchimura, K., Werb, Z., Hemmerich, S., and Rosen, S.D., 2002. Cloning and characterization of two extracellular heparin-degrading endosulfatases in mice and humans. *Journal of Biological Chemistry*, *277*(51), pp. 49175–49185.

Mottini, C., Napolitano, F., Li, Z., Gao, X., and Cardone, L., 2021, January. Computer-aided drug repurposing for cancer therapy: Approaches and opportunities to challenge anticancer targets. *In: Seminars in Cancer Biology* (Vol. 68, pp. 59–74). Academic Press. 10.1016/j.semcancer.2019.09.023.

Mottini, C., Napolitano, F., Li, Z., Gao, X., and Cardone, L., 2021. Computer-aided drug repurposing for cancer therapy: Approaches and opportunities to challenge anticancer targets. *Seminars in Cancer Biology*. doi: 10.1016/j.semcancer.2019.09.023.

Mottini, C., Tomihara, H., Carrella, D., Lamolinara, A., Iezzi, M., Huang, J.K., Amoreo, C.A., Buglioni, S., Manni, I., Robinson, F.S., and Minelli, R., 2019. Predictive signatures inform the effective repurposing of decitabine to treat KRAS–dependent pancreatic ductal adenocarcinoma. *Cancer Research*, *79*(21), pp. 5612–5625.

Nagata, Y., Lan, K.H., Zhou, X., Tan, M., Esteva, F.J., Sahin, A.A., Klos, K.S., Li, P., Monia, B.P., Nguyen, N.T., and Hortobagyi, G.N., 2004. PTEN activation contributes to tumor inhibition by trastuzumab, and loss of PTEN predicts trastuzumab resistance in patients. *Cancer Cell*, *6*(2), pp. 117–127.

Napolitano, F., Carrella, D., Mandriani, B., Pisonero-Vaquero, S., Sirci, F., Medina, D.L., Brunetti-Pierri, N., and Di Bernardo, D., 2018. Gene2drug: a computational tool for pathway-based rational drug repositioning. *Bioinformatics*, *34*(9), pp. 1498–1505.

Napolitano, F., Sirci, F., Carrella, D., and di Bernardo, D., 2016. Drug-set enrichment analysis: a novel tool to investigate drug mode of action. *Bioinformatics*, *32*(2), pp. 235–241.

Napolitano, F., Zhao, Y., Moreira, V.M., Tagliaferri, R., Kere, J., D'Amato, M., and Greco, D., 2013. Drug repositioning: a machine-learning approach through data integration. *Journal of Cheminformatics*, *5*(1), pp. 1–9.

Nath, J., Paul, R., Ghosh, S.K., Paul, J., Singha, B., and Debnath, N., 2020. Drug repurposing and relabeling for cancer therapy: Emerging benzimidazole antihelminthics with potent anticancer effects. *Life Sciences*, *258*. doi: 10.1016/j.lfs.2020.118189.

Newman, A.M., Liu, C.L., Green, M.R., Gentles, A.J., Feng, W., Xu, Y., Hoang, C.D., Diehn, M., and Alizadeh, A.A., 2015. Robust enumeration of cell subsets from tissue expression profiles. *Nature Methods*, *12*(5), pp. 453–457.

Newman, D.J., Hettich, S.C.L.B., Blake, C.L., and Merz, C.J., 1998. *UCI Repository of Machine Learning Databases*.

Newman, N.B., and Osmundson, E.C., 2021. Practical demonstration of time bias with administration of adjuvant therapy in lung cancer. *Lung Cancer*. doi: 10.1016/j.lungcan.2021.04.019.

Nguyen, D.T., Mathias, S., Bologa, C., Brunak, S., Fernandez, N., Gaulton, A., Hersey, A., Holmes, J., Jensen, L.J., Karlsson, A., and Liu, G., 2017. Pharos: collating protein information to shed light on the druggable genome. *Nucleic acids research*, *45*(D1), pp. D995–D1002.

Nielsen, M., Lundegaard, C., Worning, P., Lauemøller, S.L., Lamberth, K., Buus, S., Brunak, S., and Lund, O., 2003. Reliable prediction of T-cell epitopes using neural networks with novel sequence representations. *Protein Science*, *12*(5), pp. 1007–1017.

Novikov, E., and Barillot, E., 2008. Regulatory network reconstruction using an integral additive model with flexible kernel functions. *BMC Systems Biology*, *2*(1), pp. 1–14.

Novo, D., Heath, N., Mitchell, L., Caligiuri, G., MacFarlane, A., Reijmer, D., Charlton, L., Knight, J., Calka, M., McGhee, E., and Dornier, E., 2018. Mutant p53s generate pro-invasive niches by influencing exosome podocalyxin levels. *Nature Communications*, *9*(1), pp. 1–17.

Nsengimana, J., Laye, J., Filia, A., O'Shea, S., Muralidhar, S., Poźniak, J., Droop, A., Chan, M., Walker, C., Parkinson, L., and Gascoyne, J., 2018. β-Catenin–mediated immune evasion pathway frequently operates in primary cutaneous melanomas. *The Journal of Clinical Investigation*, *128*(5), pp. 2048–2063.

O'Brien, C., Allman, A., Daoutidis, P., and Hu, W.S., 2019. Kinetic model optimization and its application to mitigating the Warburg effect through multiple enzyme alterations. *Metabolic Engineering*, 56, 154–164.

O'Brien, P.J., and Herschlag, D., 1999. Catalytic promiscuity and the evolution of new enzymatic activities. *Chemistry & Biology*, *6*(4), pp. R91–R105.

O'Donnell, T.J., Rubinsteyn, A., Bonsack, M., Riemer, A.B., Laserson, U., and Hammerbacher, J., 2018. MHCflurry: Open-source class I MHC binding affinity prediction. *Cell Systems*, *7*(1), pp. 129–132.

Orenbuch, R., Filip, I., Comito, D., Shaman, J., Pe'er, I., and Rabadan, R., 2020. arcasHLA: High-resolution HLA typing from RNAseq. *Bioinformatics*, *36*(1), pp. 33–40.

Ott, P.A., Hu, Z., Keskin, D.B., Shukla, S.A., Sun, J., Bozym, D.J., Zhang, W., Luoma, A., Giobbie-Hurder, A., Peter, L., and Chen, C., 2017. An immunogenic personal neoantigen vaccine for patients with melanoma. *Nature*, *547*(7662), pp. 217–221.

Ozturk Mizrak, O., Mizrak, C., Kashkynbayev, A., and Kuang, Y., 2020. Can fractional differentiation improve stability results and data fitting ability of a prostate cancer model under intermittent androgen suppression therapy? *Chaos, Solitons and Fractals*, *131*. doi: 10.1016/j.chaos.2019.109529.

Palmaria, C., Bolderston, A., Cauti, S., and Fawcett, S., 2020. Learning From Cancer Survivors as Standardized Patients: Radiation Therapy Students' Perspective. *Journal of Medical Imaging and Radiation Sciences*, *51*(4), S78–S83.

Pan, S., Chen, R., Reimel, B.A., Crispin, D.A., Mirzaei, H., Cooke, K., Coleman, J.F., Lane, Z., Bronner, M.P., Goodlett, D.R., and McIntosh, M.W., 2009. Quantitative proteomics investigation of pancreatic intraepithelial neoplasia. *Electrophoresis*, *30*(7), pp. 1132–1144.

Panahi, R., Ebrahimie, E., Niazi, A., and Afsharifar, A., 2021. Integration of meta-analysis and supervised machine learning for pattern recognition in breast cancer using epigenetic data. *Informatics in Medicine Unlocked*, *24*, p. 100629.

Papapetropoulos, A., and Szabo, C., 2018. Inventing new therapies without reinventing the wheel: The power of drug repurposing. *British Journal of Pharmacology*, *175*(2), p. 165.

Paredes-Ramos, M., Sabín-López, A., Peña-García, J., Pérez-Sánchez, H., López-Vilariño, J.M., and Sastre de Vicente, M.E., 2020. Computational aided acetaminophen – phthalic acid molecularly imprinted polymer design for analytical determination of known and new developed recreational drugs. *Journal of Molecular Graphics and Modelling*, *100*. doi: 10.1016/j.jmgm.2020.107627.

Park, K., Ali, A., Kim, D., An, Y., Kim, M., and Shin, H., 2013. Robust predictive model for evaluating breast cancer survivability. *Engineering Applications of Artificial Intelligence*, *26*(9), pp. 2194–2205.

Pemovska, T., Kontro, M., Yadav, B., Edgren, H., Eldfors, S., Szwajda, A., Almusa, H., Bespalov, M.M., Ellonen, P., Elonen, E., and Gjertsen, B.T., 2013. Individualized systems medicine strategy to tailor treatments for patients with chemorefractory acute myeloid leukemia. *Cancer Discovery*, *3*(12), pp. 1416–1429.

Pfefferle, A.D., Agrawal, Y.N., Koboldt, D.C., Kanchi, K.L., Herschkowitz, J.I., Mardis, E.R., Rosen, J.M., and Perou, C.M., 2016. Genomic profiling of murine mammary tumors identifies potential personalized drug targets for p53-deficient mammary cancers. *Disease Models & Mechanisms*, *9*(7), pp. 749–757.

Phillips, C.J., and Schoen, R.E., 2020. Screening for colorectal cancer in the age of simulation models: A historical lens. *Gastroenterology*, *159*(4), pp. 1201–1204.

Picelli, S., Björklund, Å.K., Faridani, O.R., Sagasser, S., Winberg, G., and Sandberg, R., 2013. Smart-seq2 for sensitive full-length transcriptome profiling in single cells. *Nature Methods*, *10*(11), pp. 1096–1098.

Pitsillou, E., Liang, J., Hung, A., and Karagiannis, T.C., 2020. Chromatin modification by olive phenolics: In silico molecular docking studies utilising the phenolic groups categorised in the OliveNet™ database against lysine specific demethylase enzymes: Chromatin modification by olive phenolics. *Journal of Molecular Graphics and Modelling*, *97*. doi: 10.1016/j.jmgm.2020.107575.

Pospisil, P., and Kassis, A.I., 2014. Computational and biological evaluation of radioiodinated quinazolinone prodrug for targeting pancreatic cancer. *In: Molecular Diagnostics and Treatment of Pancreatic Cancer* (pp. 385–403). Academic Press.

Pospisil, P., Wang, K., Al Aowad, A.F., Iyer, L.K., Adelstein, S.J., and Kassis, A.I., 2007. Computational modeling and experimental evaluation of a novel prodrug for targeting the extracellular space of prostate tumors. *Cancer Research*, *67*(5), pp. 2197–2205.

Poustforoosh, A., Hashemipour, H., Tüzün, B., Pardakhty, A., Mehrabani, M., and Nematollahi, M.H., 2021. Evaluation of potential anti-RNA-dependent RNA polymerase (RdRP) drugs against the newly emerged model of COVID-19 RdRP using computational methods. *Biophysical Chemistry*, *272*. doi: 10.1016/j.bpc.2021.106564.

Praiss, A.M., Huang, Y., St. Clair, C.M., Tergas, A.I., Melamed, A., Khoury-Collado, F., Hou, J.Y., Hu, J., Hur, C., Hershman, D.L., and Wright, J.D., 2020. Using machine learning to create prognostic systems for endometrial cancer. *Gynecologic Oncology*, *159*(3), pp. 744–750.

Preziosi, L., Toscani, G., and Zanella, M., 2021. Control of tumor growth distributions through kinetic methods. *Journal of Theoretical Biology*, *514*. doi: 10.1016/j.jtbi.2021.110579.

Qi, L., Zhang, Y., Song, F., Han, Y., and Ding, Y., 2021. A newly identified small molecular compound acts as a protein kinase inhibitor to suppress metastasis of colorectal cancer. *Bioorganic Chemistry*, *107*. doi: 10.1016/j.bioorg.2021.104625.

Qi, W.J., Sheng, W.S., Peng, C., Xiaodong, M., and Yao, T.Z., 2021. Investigating into anti-cancer potential of lycopene: Molecular targets. *Biomedicine and Pharmacotherapy*. doi: 10.1016/j.biopha.2021.111546.

Racle, J., Michaux, J., Rockinger, G.A., Arnaud, M., Bobisse, S., Chong, C., Guillaume, P., Coukos, G., Harari, A., Jandus, C., and Bassani-Sternberg, M., 2019. Deep motif deconvolution of HLA-II peptidomes for robust class II epitope predictions. *bioRxiv*, p. 539338. doi: 10.1038/s41587-019-0289-6.

Rahn, T., Ridderstråle, M., Tornqvist, H., Manganiello, V., Fredrikson, G., Belfrage, P., and Degerman, E., 1994. Essential role of phosphatidylinositol 3-kinase in insulin-induced activation and phosphorylation of the cGMP-inhibited cAMP phosphodiesterase in rat adipocytes studies using the selective inhibitor wortmannin. *FEBS Letters, 350*(2-3), pp. 314–318.

Rajput, A., Thakur, A., Mukhopadhyay, A., Kamboj, S., Rastogi, A., Gautam, S., Jassal, H., and Kumar, M., 2021. Prediction of repurposed drugs for Coronaviruses using artificial intelligence and machine learning. *Computational and Structural Biotechnology Journal, 19*, pp. 3133–3148.

Réda, C., Kaufmann, E., and Delahaye-Duriez, A., 2020. Machine learning applications in drug development. *Computational and Structural Biotechnology Journal.* doi: 10.1016/j.csbj.2019.12.006.

Rosenbluh, J., Mercer, J., Shrestha, Y., Oliver, R., Tamayo, P., Doench, J.G., Tirosh, I., Piccioni, F., Hartenian, E., Horn, H., and Fagbami, L., 2016. Genetic and proteomic interrogation of lower confidence candidate genes reveals signaling networks in β-catenin-active cancers. *Cell Systems, 3*(3), pp. 302–316.

Rosenbluh, J., Nijhawan, D., Cox, A.G., Li, X., Neal, J.T., Schafer, E.J., Zack, T.I., Wang, X., Tsherniak, A., Schinzel, A.C., & Shao, D.D. 2012. β-Catenin-driven cancers require a YAP1 transcriptional complex for survival and tumorigenesis. *Cell, 151*(7), pp. 1457–1473.

Safavi, A., Ghodousi, E.S., Ghavamizadeh, M., Sabaghan, M., Azadbakht, O., veisi, A., Babaei, H., Nazeri, Z., Darabi, M.K., and Zarezade, V., 2021. Computational investigation of novel farnesyltransferase inhibitors using 3D-QSAR pharmacophore modeling, virtual screening, molecular docking and molecular dynamics simulation studies: A new insight into cancer treatment. *Journal of Molecular Structure, 1241*, p. 130667.

Sánchez, Y., Segura, V., Marín-Béjar, O., Athie, A., Marchese, F.P., González, J., Bujanda, L., Guo, S., Matheu, A., and Huarte, M., 2014. Genome-wide analysis of the human p53 transcriptional network unveils a lncRNA tumour suppressor signature. *Nature Communications, 5*(1), pp. 1–13.

Sankhe, R., Rathi, E., Manandhar, S., Kumar, A., Pai, S.R.K., Kini, S.G., and Kishore, A., 2021. Repurposing of existing FDA approved drugs for Neprilysin inhibition: An in-silico study. *Journal of Molecular Structure, 1224*. doi: 10.1016/j.molstruc.2020.129073.

Sasagawa, S., Ozaki, Y.I., Fujita, K., and Kuroda, S., 2005. Prediction and validation of the distinct dynamics of transient and sustained ERK activation. *Nature Cell Biology, 7*(4), pp. 365–373.

Sasahara, K., Shibata, M., Sasabe, H., Suzuki, T., Takeuchi, K., Umehara, K., and Kashiyama, E., 2021a. Feature importance of machine learning prediction models shows structurally active part and important physicochemical features in drug design. *Drug Metabolism and Pharmacokinetics, 39*, p. 100401.

Sasahara, K., Shibata, M., Sasabe, H., Suzuki, T., Takeuchi, K., Umehara, K., and Kashiyama, E., 2021b. Predicting drug metabolism and pharmacokinetics features of in-house compounds by a hybrid machine-learning model. *Drug Metabolism and Pharmacokinetics, 39*, p. 100395.

Satoh, K., Yachida, S., Sugimoto, M., Oshima, M., Nakagawa, T., Akamoto, S., Tabata, S., Saitoh, K., Kato, K., Sato, S., and Igarashi, K., 2017. Global metabolic reprogramming of colorectal cancer occurs at adenoma stage and is induced by MYC. *Proceedings of the National Academy of Sciences, 114*(37), pp. E7697–E7706.

Schoeberl, B., Eichler-Jonsson, C., Gilles, E.D., and Müller, G., 2002. Computational modeling of the dynamics of the MAP kinase cascade activated by surface and internalized EGF receptors. *Nature Biotechnology, 20*(4), pp. 370–375.

Semaan, C., Henderson, B.R., and Molloy, M.P., 2019. Proteomic screen with the proto-oncogene beta-catenin identifies interaction with Golgi coatomer complex I. *Biochemistry and Biophysics Reports, 19*, p. 100662.

Shachaf, C.M., Gentles, A.J., Elchuri, S., Sahoo, D., Soen, Y., Sharpe, O., Perez, O.D., Chang, M., Mitchel, D., Robinson, W.H., and Dill, D., 2008. Genomic and proteomic analysis reveals a threshold level of MYC required for tumor maintenance. *Cancer Research, 68*(13), pp. 5132–5142.

Shaikh, F.J., and Rao, D.S., 2021. Predication of cancer disease using machine learning approach. *Materials Today: Proceedings.* 10.1016/j.matpr.2021.03.625

Shaked, I., Oberhardt, M.A., Atias, N., Sharan, R., and Ruppin, E., 2016. Metabolic network prediction of drug side effects. *Cell Systems, 2*(3), pp. 209–213.

Shamsi, M., Mohammadi, A., Manshadi, M.K.D., and Sanati-Nezhad, A., 2019. Mathematical and computational modeling of nano-engineered drug delivery systems. *Journal of Controlled Release.* doi: 10.1016/j.jconrel.2019.06.014.

Shao, W., Pedrioli, P.G., Wolski, W., Scurtescu, C., Schmid, E., Vizcaíno, J.A., Courcelles, M., Schuster, H., Kowalewski, D., Marino, F., and Arlehamn, C.S.L., 2018. The SysteMHC atlas project. *Nucleic Acids Research, 46*(D1), pp. D1237–D1247.

Shao, W., Wang, T., Sun, L., Dong, T., Han, Z., Huang, Z., Zhang, J., Zhang, D., and Huang, K., 2020. Multi-task multi-modal learning for joint diagnosis and prognosis of human cancers. *Medical Image Analysis, 65*. doi: 10.1016/j.media.2020.101795.

Shaughnessy, A.F., 2011. Old drugs, new tricks. *BMJ*, *342*. doi: 10.1136/bmj.d741.

Shiio, Y., Donohoe, S., Eugene, C.Y., Goodlett, D.R., Aebersold, R., and Eisenman, R.N., 2002. Quantitative proteomic analysis of Myc oncoprotein function. *The EMBO Journal*, *21*(19), pp. 5088–5096.

Shim, J.S., and Liu, J.O., 2014. Recent advances in drug repositioning for the discovery of new anticancer drugs. *International Journal of Biological Sciences*, *10*(7), p. 654.

Shukla, S.A., Rooney, M.S., Rajasagi, M., Tiao, G., Dixon, P.M., Lawrence, M.S., Stevens, J., Lane, W.J., Dellagatta, J.L., Steelman, S., and Sougnez, C., 2015. Comprehensive analysis of cancer-associated somatic mutations in class I HLA genes. *Nature biotechnology*, *33*(11), pp. 1152–1158.

Sirota, M., Dudley, J.T., Kim, J., Chiang, A.P., Morgan, A.A., Sweet-Cordero, A., Sage, J., and Butte, A.J., 2011. Discovery and preclinical validation of drug indications using compendia of public gene expression data. *Science Translational Medicine*, *3*(96), pp. 96ra77–96ra77.

Skrott, Z., Mistrik, M., Andersen, K.K., Friis, S., Majera, D., Gursky, J., Ozdian, T., Bartkova, J., Turi, Z., Moudry, P., and Kraus, M., 2017. Alcohol-abuse drug disulfiram targets cancer via p97 segregase adaptor NPL4. *Nature*, *552*(7684), pp. 194–199.

Slepchenko, B.M., Schaff, J.C., Macara, I., and Loew, L.M., 2003. Quantitative cell biology with the Virtual Cell. *Trends in Cell Biology*, *13*(11), pp. 570–576.

Sohraby, F., and Aryapour, H., 2021. Rational drug repurposing for cancer by inclusion of the unbiased molecular dynamics simulation in the structure-based virtual screening approach: Challenges and breakthroughs. *Seminars in Cancer Biology*. doi: 10.1016/j.semcancer.2020.04.007.

Spranger, S., Bao, R., and Gajewski, T.F., 2015. Melanoma-intrinsic β-catenin signalling prevents anti-tumour immunity. *Nature*, *523*(7559), pp. 231–235.

Steffes, V.M., Murali, M.M., Park, Y., Fletcher, B.J., Ewert, K.K., and Safinya, C.R., 2017. Distinct solubility and cytotoxicity regimes of paclitaxel-loaded cationic liposomes at low and high drug content revealed by kinetic phase behavior and cancer cell viability studies. *Biomaterials*, *145*, pp. 242–255.

Stein-O'Brien, G.L., Clark, B., Sherman, T., Zibetti, C., Hu, Q., Sealfon, R., Liu, S., Qian, J., Colantuoni, C., Blackshaw, S., and Goff, L.A., 2021. Erratum: Decomposing cell identity for transfer learning across cellular measurements, platforms, tissues, and species. *Cell Systems*, *12*(2), p. 203.

Štrumbelj, E., Bosnić, Z., Kononenko, I., Zakotnik, B., and Kuhar, C.G., 2010. Explanation and reliability of prediction models: the case of breast cancer recurrence. *Knowledge and Information Systems*, *24*(2), pp. 305–324.

Stuart, T., and Satija, R., 2019. Integrative single-cell analysis. *Nature Reviews Genetics*, *20*(5), pp. 257–272.

Sturm, G., Finotello, F., Petitprez, F., Zhang, J.D., Baumbach, J., Fridman, W.H., List, M., and Aneichyk, T., 2019. Comprehensive evaluation of transcriptome-based cell-type quantification methods for immuno-oncology. *Bioinformatics*, *35*(14), pp. i436–i445.

Suthers, P.F., Foster, C.J., Sarkar, D., Wang, L., and Maranas, C.D., 2021. Recent advances in constraint and machine learning-based metabolic modeling by leveraging stoichiometric balances, thermodynamic feasibility and kinetic law formalisms. *Metabolic Engineering*, *63*, pp. 13–33.

Swaminath, A., Cheung, P., Glicksman, R.M., Donovan, E.K., Niglas, M., Vesprini, D., Kapoor, A., Erler, D., and Chu, W., 2021. Patient-reported quality of life following stereotactic body radiation therapy for primary kidney cancer – results from a prospective cohort study. *Clinical Oncology*, *33*(7), pp. 468–475.

Szolek, A., Schubert, B., Mohr, C., Sturm, M., Feldhahn, M., and Kohlbacher, O., 2014. OptiType: Precision HLA typing from next-generation sequencing data. *Bioinformatics*, *30*(23), pp. 3310–3316.

Taguchi, A., Delgado, O., Çeliktaş, M., Katayama, H., Wang, H., Gazdar, A.F., and Hanash, S.M., 2014. Proteomic signatures associated with p53 mutational status in lung adenocarcinoma. *Proteomics*, *14*(23-24), pp. 2750–2759.

Tambas, M., van der Laan, H.P., Rutgers, W., van den Hoek, J.G.M., Oldehinkel, E., Meijer, T.W.H., van der Schaaf, A., Scandurra, D., Free, J., Both, S., Steenbakkers, R.J.H.M., and Langendijk, J.A., 2021. Development of advanced preselection tools to reduce redundant plan comparisons in model-based selection of head and neck cancer patients for proton therapy. *Radiotherapy and Oncology*, *160*, pp. 61–68.

Tang, W., Zhou, M., Dorsey, T.H., Prieto, D.A., Wang, X.W., Ruppin, E., Veenstra, T.D., and Ambs, S., 2018. Integrated proteotranscriptomics of breast cancer reveals globally increased protein-mRNA concordance associated with subtypes and survival. *Genome Medicine*, *10*(1), pp. 1–14.

Tanikawa, C., Zhang, Y.Z., Yamamoto, R., Tsuda, Y., Tanaka, M., Funauchi, Y., Mori, J., Imoto, S., Yamaguchi, R., Nakamura, Y., and Miyano, S., 2017. The transcriptional landscape of p53 signalling pathway. *EBioMedicine*, *20*, pp. 109–119.

Tarrado-Castellarnau, M., de Atauri, P., Tarragó-Celada, J., Perarnau, J., Yuneva, M., Thomson, T.M., and Cascante, M., 2017. De novo MYC addiction as an adaptive response of cancer cells to CDK 4/6 inhibition. *Molecular Systems Biology*, *13*(10), p. 940.

Terunuma, A., Putluri, N., Mishra, P., Mathé, E.A., Dorsey, T.H., Yi, M., Wallace, T.A., Issaq, H.J., Zhou, M., Killian, J.K., and Stevenson, H.S., 2014. MYC-driven accumulation of 2-hydroxyglutarate is associated with breast cancer prognosis. *The Journal of clinical investigation*, *124*(1), pp. 398–412.

Thorsson, V., Gibbs, D.L., Brown, S.D., Wolf, D., Bortone, D.S., Yang, T.H.O., Porta-Pardo, E., Gao, G.F., Plaisier, C.L., Eddy, J.A., and Ziv, E., 2018. The immune landscape of cancer. *Immunity*, *48*(4), pp. 812–830.

Tomita, M., Hashimoto, K., Takahashi, K., Shimizu, T.S., Matsuzaki, Y., Miyoshi, F., Saito, K., Tanida, S., Yugi, K., Venter, J.C., and Hutchison 3rd, C.A., 1999. E-CELL: Software environment for whole-cell simulation. *Bioinformatics (Oxford, England)*, *15*(1), pp. 72–84.

Toss, A., and Cristofanilli, M., 2015. Molecular characterization and targeted therapeutic approaches in breast cancer. *Breast Cancer Research*, *17*(1), p. 60.

Tsujikawa, T., Kumar, S., Borkar, R.N., Azimi, V., Thibault, G., Chang, Y.H., Balter, A., Kawashima, R., Choe, G., Sauer, D., and El Rassi, E., 2017. Quantitative multiplex immunohistochemistry reveals myeloid-inflamed tumor-immune complexity associated with poor prognosis. *Cell Reports*, *19*(1), pp. 203–217.

Tu, W.B., Shiah, Y.J., Lourenco, C., Mullen, P.J., Dingar, D., Redel, C., Tamachi, A., Ba-Alawi, W., Aman, A., Al-Awar, R., and Cescon, D.W., 2018. MYC interacts with the G9a histone methyltransferase to drive transcriptional repression and tumorigenesis. *Cancer Cell*, *34*(4), pp. 579–595.

Van Der Maaten, L., 2014. Accelerating t-SNE using tree-based algorithms. *The Journal of Machine Learning Research*, *15*(1), pp. 3221–3245.

Van der Maaten, L., and Hinton, G., 2008. Visualizing data using t-SNE. *Journal of Machine Learning Research*, *9*(11), pp. 2579–2605.

van Rosmalen, R.P., Smith, R.W., Martins dos Santos, V.A.P., Fleck, C., and Suarez-Diez, M., 2021. Model reduction of genome-scale metabolic models as a basis for targeted kinetic models. *Metabolic Engineering*, *64*, pp. 74–84.

Van Ummersen, L., Binger, K., Volkman, J., Marnocha, R., Tutsch, K., Kolesar, J., Arzoomanian, R., Alberti, D., and Wilding, G., 2004. A phase I trial of perifosine (NSC 639966) on a loading dose/maintenance dose schedule in patients with advanced cancer. *Clinical Cancer Research*, *10*(22), pp. 7450–7456.

van Unen, V., Höllt, T., Pezzotti, N., Li, N., Reinders, M.J., Eisemann, E., Koning, F., Vilanova, A., and Lelieveldt, B.P., 2017. Visual analysis of mass cytometry data by hierarchical stochastic neighbour embedding reveals rare cell types. *Nature Communications*, *8*(1), pp. 1–10.

Vanhaelen, Q., Mamoshina, P., Aliper, A.M., Artemov, A., Lezhnina, K., Ozerov, I., Labat, I., and Zhavoronkov, A., 2017. Design of efficient computational workflows for in silico drug repurposing. *Drug Discovery Today*. doi: 10.1016/j.drudis.2016.09.019.

Vastrik, I., D'Eustachio, P., Schmidt, E., Joshi-Tope, G., Gopinath, G., Croft, D., de Bono, B., Gillespie, M., Jassal, B., Lewis, S., and Matthews, L., 2007. Reactome: A knowledge base of biologic pathways and processes. *Genome Biology*, *8*(3), pp. 1–13.

Vergara, D., Stanca, E., Guerra, F., Priore, P., Gaballo, A., Franck, J., Simeone, P., Trerotola, M., De Domenico, S., Fournier, I., and Bucci, C., 2017. β-Catenin knockdown affects mitochondrial biogenesis and lipid metabolism in breast cancer cells. *Frontiers in Physiology*, *8*, p. 544.

Vidović, D., Koleti, A., and Schürer, S.C., 2014. Large-scale integration of small molecule-induced genome-wide transcriptional responses, Kinome-wide binding affinities and cell-growth inhibition profiles reveal global trends characterizing systems-level drug action. *Frontiers in Genetics*, *5*, p. 342.

Vita, R., Overton, J.A., Greenbaum, J.A., Ponomarenko, J., Clark, J.D., Cantrell, J.R., Wheeler, D.K., Gabbard, J.L., Hix, D., Sette, A., and Peters, B., 2015. The immune epitope database (IEDB) 3.0. *Nucleic Acids Research*, *43*(D1), pp. D405–D412.

Vivanco, I., and Sawyers, C.L., 2002. The phosphatidylinositol 3-kinase–AKT pathway in human cancer. *Nature Reviews Cancer*, *2*(7), pp. 489–501.

Vizcaíno, J.A., Csordas, A., Del-Toro, N., Dianes, J.A., Griss, J., Lavidas, I., Mayer, G., Perez-Riverol, Y., Reisinger, F., Ternent, T., and Xu, Q.W., 2016. 2016 update of the PRIDE database and its related tools. *Nucleic Acids Research*, *44*(D1), pp. D447–D456.

Voloshanenko, O., Schwartz, U., Kranz, D., Rauscher, B., Linnebacher, M., Augustin, I., and Boutros, M., 2018. β-catenin-independent regulation of Wnt target genes by RoR2 and ATF2/ATF4 in colon cancer cells. *Scientific Reports*, *8*(1), pp. 1–14.

Von Eichborn, J., Murgueitio, M.S., Dunkel, M., Koerner, S., Bourne, P.E., and Preissner, R., 2010. PROMISCUOUS: a database for network-based drug-repositioning. *Nucleic Acids Research*, *39*(suppl_1), pp. D1060–D1066.

Wadanambi, P.M., and Mannapperuma, U., 2021. Computational study to discover potent phytochemical inhibitors against drug target, squalene synthase from Leishmania donovani. *Heliyon*, 7(6), p. e07178.

Waldron, D., 2016. A multi-layer omics approach to cancer. *Nature Reviews Genetics*, 17(8), pp.437-437.

Walerych, D., Lisek, K., Sommaggio, R., Piazza, S., Ciani, Y., Dalla, E., Rajkowska, K., Gaweda-Walerych, K., Ingallina, E., Tonelli, C., and Morelli, M.J., 2016. Proteasome machinery is instrumental in a common gain-of-function program of the p53 missense mutants in cancer. *Nature cell biology*, 18(8), pp. 897–909.

Wang, B., Tian, T., Kalland, K.H., Ke, X., and Qu, Y., 2018. Targeting Wnt/β-catenin signaling for cancer immunotherapy. *Trends in Pharmacological Sciences*, 39(7), pp. 648–658.

Wang, G., Zhang, G., Choi, K.S., Lam, K.M., and Lu, J., 2020. Output based transfer learning with least squares support vector machine and its application in bladder cancer prognosis. *Neurocomputing*, 387, pp. 279–292.

Wang, J., Luo, C., Shan, C., You, Q., Lu, J., Elf, S., Zhou, Y., Wen, Y., Vinkenborg, J.L., Fan, J., and Kang, H., 2015. Inhibition of human copper trafficking by a small molecule significantly attenuates cancer cell proliferation. *Nature Chemistry*, 7(12), p. 968.

Wang, K., Kirichian, A.M., Al Aowad, A.F., Adelstein, S.J., and Kassis, A.I., 2007. Evaluation of chemical, physical, and biologic properties of tumor-targeting radioiodinated quinazolinone derivative. *Bioconjugate Chemistry*, 18(3), pp. 754–764.

Wang, Y.C., Chen, S.L., Deng, N.Y., and Wang, Y., 2013. Network predicting drug's anatomical therapeutic chemical code. *Bioinformatics*, 29(10), pp. 1317–1324.

Whirl-Carrillo, M., McDonagh, E.M., Hebert, J.M., Gong, L., Sangkuhl, K., Thorn, C.F., Altman, R.B., and Klein, T.E., 2012. Pharmacogenomics knowledge for personalized medicine. *Clinical Pharmacology & Therapeutics*, 92(4), pp. 414–417.

Wierling, C., Herwig, R., and Lehrach, H., 2007. Resources, standards and tools for systems biology. *Briefings in Functional Genomics and Proteomics*, 6(3), pp. 240–251.

Wilkinson, G.F., and Pritchard, K., 2015. In vitro screening for drug repositioning. *Journal of Biomolecular Screening*, 20(2), pp. 167–179.

Wingender, E., Crass, T., Hogan, J.D., Kel, A.E., Kel-Margoulis, O.V., and Potapov, A.P., 2007. Integrative content-driven concepts for bioinformatics "beyond the cell". *Journal of Biosciences*, 32(1), pp. 169–180.

Wishart, D.S., Feunang, Y.D., Guo, A.C., Lo, E.J., Marcu, A., Grant, J.R., Sajed, T., Johnson, D., Li, C., Sayeeda, Z., and Assempour, N., 2018. DrugBank 5.0: a major update to the DrugBank database for 2018. *Nucleic Acids Research*, 46(D1), pp. D1074–D1082.

Woillard, J.B., Labriffe, M., Prémaud, A., and Marquet, P., 2021. Estimation of drug exposure by machine learning based on simulations from published pharmacokinetic models: The example of tacrolimus. *Pharmacological Research*, 167. doi: 10.1016/j.phrs.2021.105578.

Wolfer, A., Wittner, B.S., Irimia, D., Flavin, R.J., Lupien, M., Gunawardane, R.N., Meyer, C.A., Lightcap, E.S., Tamayo, P., Mesirov, J.P., and Liu, X.S., 2010. MYC regulation of a "poor-prognosis" metastatic cancer cell state. *Proceedings of the National Academy of Sciences*, 107(8), pp. 3698–3703.

Wu, Z., Cheng, F., Li, J., Li, W., Liu, G., and Tang, Y., 2017. SDTNBI: an integrated network and chemoinformatics tool for systematic prediction of drug–target interactions and drug repositioning. *Briefings in Bioinformatics*, 18(2), pp. 333–347.

Xie, C., Yeo, Z.X., Wong, M., Piper, J., Long, T., Kirkness, E.F., Biggs, W.H., Bloom, K., Spellman, S., Vierra-Green, C., and Brady, C., 2017. Fast and accurate HLA typing from short-read next-generation sequence data with xHLA. *Proceedings of the National Academy of Sciences*, 114(30), pp. 8059–8064.

Xu, C., 2018. A review of somatic single nucleotide variant calling algorithms for next-generation sequencing data. *Computational and Structural Biotechnology Journal*, 16, pp. 15–24.

Yamada, S., Shiono, S., Joo, A., and Yoshimura, A., 2003. Control mechanism of JAK/STAT signal transduction pathway. *FEBS letters*, 534(1-3), pp. 190–196.

Yang, J., Li, Z., Fan, X., and Cheng, Y., 2014. Drug–disease association and drug-repositioning predictions in complex diseases using causal inference–probabilistic matrix factorization. *Journal of Chemical Information and Modeling*, 54(9), pp. 2562–2569.

Yang, L., and Agarwal, P., 2011. Systematic drug repositioning based on clinical side-effects. *PloS One*, 6(12), p. e28025.

Yang, W., Lee, K.W., Srivastava, R.M., Kuo, F., Krishna, C., Chowell, D., Makarov, V., Hoen, D., Dalin, M.G., Wexler, L., and Ghossein, R., 2019. Immunogenic neoantigens derived from gene fusions stimulate T cell responses. *Nature Medicine*, 25(5), pp. 767–775.

Yang, X.H., Tang, F., Shin, J., and Cunningham, J.M., 2017. A c-Myc-regulated stem cell-like signature in high-risk neuroblastoma: A systematic discovery (Target neuroblastoma ESC-like signature). *Scientific Reports*, 7(1), pp. 1–15.

Yap, C.W., 2011. PaDEL-descriptor: An open source software to calculate molecular descriptors and fingerprints. *Journal of Computational Chemistry, 32*(7), pp. 1466–1474.

Yap, J., Resta-López, J., Kacperek, A., Schnuerer, R., Jolly, S., Boogert, S., and Welsch, C., 2020. Beam characterisation studies of the 62 MeV proton therapy beamline at the Clatterbridge Cancer Centre. *Physica Medica*, 77, 108–120.

Ye, H., Liu, Q., and Wei, J., 2014. Construction of drug network based on side effects and its application for drug repositioning. *PloS One, 9*(2), p. e87864.

Yeung, M.W., Benjamins, J.-W., van der Harst, P., and Juarez-Orozco, L.E., 2021. Machine Learning in Cardiovascular Genomics, Proteomics, and Drug Discovery. *In: Machine Learning in Cardiovascular Medicine* (pp. 325–352). Elsevier.

Yu, S., Gao, W., Zeng, P., Chen, C., Zhang, Z., Liu, Z., and Liu, J., 2021. Exploring the effect of Gupi Xiaoji Prescription on hepatitis B virus-related liver cancer through network pharmacology and in vitro experiments. *Biomedicine & Pharmacotherapy*, 139, pp. 111612.

Yuan, C., Wang, M.H., Wang, F., Chen, P.Y., Ke, X.G., Yu, B., Yang, Y.F., You, P.T., and Wu, H.Z., 2021. Network pharmacology and molecular docking reveal the mechanism of Scopoletin against non-small cell lung cancer. *Life Sciences, 270.* doi: 10.1016/j.lfs.2021.119105.

Yusupov, M., Privat-Maldonado, A., Cordeiro, R.M., Verswyvel, H., Shaw, P., Razzokov, J., Smits, E., and Bogaerts, A., 2021. Oxidative damage to hyaluronan–CD44 interactions as an underlying mechanism of action of oxidative stress-inducing cancer therapy. *Redox Biology, 43.* doi: 10.1016/j.redox.2021.101968.

Zhang, G.P., 2000. Neural networks for classification: a survey. *IEEE Transactions on Systems, Man, and Cybernetics, Part C (Applications and Reviews), 30*(4), pp. 451–462.

Zhang, J., Yan, S., Li, R., Wang, G., Kang, S., Wang, Y., Hou, W., Wang, C., and Tian, W., 2021. CRMarker: A manually curated comprehensive resource of cancer RNA markers. *International Journal of Biological Macromolecules, 174*, pp. 263–269.

Zhang, L., Yang, K., Wang, M., Zeng, L., Sun, E., Zhang, F., Cao, Z., Zhang, X., Zhang, H., and Guo, Z., 2021. Exploring the mechanism of Cremastra Appendiculata (SUANPANQI) against breast cancer by network pharmacology and molecular docking. *Computational Biology and Chemistry.* doi: 10.1016/j.compbiolchem.2020.107396.

Zhang, P., Wang, F., and Hu, J., 2014. Towards drug repositioning: a unified computational framework for integrating multiple aspects of drug similarity and disease similarity. In *AMIA Annual Symposium Proceedings* (Vol. 2014, p. 1258). American Medical Informatics Association.

Zhang, Z., Liu, J., Liu, Y., Shi, D., He, Y., and Zhao, P., 2021. Virtual screening of the multi-gene regulatory molecular mechanism of Si-Wu-tang against non-triple-negative breast cancer based on network pharmacology combined with experimental validation. *Journal of Ethnopharmacology*, 269. doi: 10.1016/j.jep.2020.113696.

Zhao, L., Ciallella, H.L., Aleksunes, L.M., and Zhu, H., 2020. Advancing computer-aided drug discovery (CADD) by big data and data-driven machine learning modeling. *Drug Discovery Today.* doi: 10.1016/j.drudis.2020.07.005.

Zheng, G.X., Terry, J.M., Belgrader, P., Ryvkin, P., Bent, Z.W., Wilson, R., Ziraldo, S.B., Wheeler, T.D., McDermott, G.P., Zhu, J., and Gregory, M.T., 2017. Massively parallel digital transcriptional profiling of single cells. *Nature Communications, 8*(1), pp. 1–12.

Zhu, H., Zhu, L., Sun, Z., and Khan, A., 2021. Machine learning based simulation of an anti-cancer drug (busulfan) solubility in supercritical carbon dioxide: ANFIS model and experimental validation. *Journal of Molecular Liquids, 338*, p. 116731.

Zi, Z., and Klipp, E., 2007. Constraint-based modeling and kinetic analysis of the Smad dependent TGF-β signaling pathway. *PloS One, 2*(9), p. e936.

Ziegenhain, C., Vieth, B., Parekh, S., Reinius, B., Guillaumet-Adkins, A., Smets, M., Leonhardt, H., Heyn, H., Hellmann, I., and Enard, W., 2017. Comparative analysis of single-cell RNA sequencing methods. *Molecular Cell, 65*(4), pp. 631–643.

Zoudani, E.L., and Soltani, M., 2020. A new computational method of modeling and evaluation of dissolving microneedle for drug delivery applications: Extension to theoretical modeling of a novel design of microneedle (array in array) for efficient drug delivery. *European Journal of Pharmaceutical Sciences, 150.* doi: 10.1016/j.ejps.2020.105339.

Index

For Product Safety Concerns and Information please contact our EU
representative GPSR@taylorandfrancis.com
Taylor & Francis Verlag GmbH, Kaufingerstraße 24, 80331 München, Germany

www.ingramcontent.com/pod-product-compliance
Ingram Content Group UK Ltd.
Pitfield, Milton Keynes, MK11 3LW, UK
UKHW050926180425
457613UK00003B/39